INTRODUCTION TO TENSORS, SPINORS, AND RELATIVISTIC WAVE-EQUATIONS

(Relation Structure)

BY

E. M. CORSON

Adjunct Professor in Theoretical Physics, Graduate School of Arts and Science, New York University. Senior Scientist in Mathematical Physics, Institute of Mathematical Sciences, New York University. Research Associate in Psychiatry, New York State Psychiatric Institute. Formerly Fulbright Exchange Professor, Department of Mathematical Physics, University of Edinburgh.

CHELSEA PUBLISHING COMPANY
NEW YORK, N. Y.

SECOND (UNALTERED) EDITION

THE PRESENT WORK IS A SECOND (TEXTUALLY UNALTERED) EDITION OF A WORK BY EDWARD MICHAEL CORSON, ORIGINALLY PUBLISHED AT GLASGOW IN 1953. SEVERAL ESSENTIALLY BLANK (BUT NUMBERED) PAGES HAVE BEEN OMITTED IN THIS EDITION. THESE OMITTED PAGES SERVED TO SEPARATE THE VARIOUS PARTS OF THE BOOK. APART FROM THIS, THE TEXT IS ENTIRELY UNCHANGED. THIS EDITION IS PUBLISHED AT NEW YORK, N.Y. IT IS PRINTED ON SPECIAL 'LONG-LIFE' ACID-FREE (PERMANENT) PAPER.

INTERNATIONAL STANDARD BOOK NUMBER 0-8284-0315-5

LIBRARY OF CONGRESS CATALOG NUMBER 80-85523

PRINTED IN THE UNITED STATES OF AMERICA

PREFACE

"*It was the best of times, it was the worst of times; it was the age of wisdom, it was the age of foolishness; it was the epoch of belief, it was the epoch of incredulity; it was the season of Light, it was the season of Darkness; it was the spring of hope, it was the winter of despair; we had everything before us, we had nothing before us. . . .*"

DICKENS, *Tale of Two Cities.*

There exists at present a veritable jungle of literature on the theories of the so-called elementary particles and their interactions—including nuclear forces. The tremendous volume of publications on these topics attests the interest which they enjoy among theoretical physicists, but, by the same token, clearly indicates the unsettled and problematical nature of this frontier.

The very concept "*elementary particle*" is not unambiguously defined. Recent experiments in the high-energy region, the explosion-like processes, the probability that in sufficiently energetic collisions particles of any type can be created*—all indicate that, while our qualitative concepts are probably correct, the present formalism may be wholly inadequate. Apparently what is needed (*inter alia*) is neither a *one*-particle nor a *many*-particle theory, but rather an *all*-particle theory which contains all particles, or their representative wave-fields, simultaneously—and how many species does "*all*" comprise? Perhaps in an over-all theory lies the limitation on the possible species, their rest-masses, and charges. It seems fairly certain that something going far beyond the welding of special relativity to quantum mechanics will be required to explain the nature, relation and origin of elementary particles, or equivalently, the concepts of mass, charge, etc. (which Nature appears to have assigned so sparingly) as quantization processes.

Because of these facts, the theoretical physicist today faces a formidable and often discouraging task on entering this field of research. As regards the canonical theory of quantized wave-fields (q-number theory), there exists the excellent volume by Wentzel, but for full appreciation and understanding, this formalism should be preceded or supplemented by a study of tensor-spinor analysis and relativistic wave-equations (c-number theory), which are primary, and on which topics no text exists.

The very fact that field theory is in a phase of almost amœboid growth—though perhaps intrinsically in a morbid state—raises doubts as to the appropriateness of such a text at this time. The labour of preparing such a volume must be weighed in terms of its possible usefulness and the probability of transition to out-of-dateness. We take the view that theoretical physics will continue to undergo fundamental changes in varying measure; indeed, it would become a dull field if there were

* Experimentally, we know that there are transitions between nucleons and mesons; mesons, electrons and neutrinos; protons, electrons and photons, etc., so that seemingly any particle form can be attained from any other, at least by intermediate steps.

indications of closure in view. However, the basic frame of our investigations is essentially *relation-structure*: the intrinsic requirements on any field theory which find their origin in considerations of covariance, group theory, and the elementary properties of space-time. These aspects of present-day theory have shown their usefulness, quite apart from assumptions of special formalisms, and may be expected to be prerequisite to understanding future developments—if not preserved *in toto*.

Just such considerations, particularly a more careful analysis and reformulation of global covariance, have lead to the recent developments in quantum electrodynamics. Here, too, an advance has been made, a number of important experimental findings have been theoretically reproduced, *but*, little light has been cast on most of the fundamental issues noted above, and we still have *subtraction physics—be it ever so covariant*.

The study which we have termed *relation-structure* is now more a way-station than an end in itself. It has become sufficiently formalized to be, perhaps, classified as mathematical physics as distinct from theoretical physics. As such, it might be viewed as a subject to be noted and then all but forgotten, except as a tool. Yet, as evidenced by the preceding remarks, the fundamentals are by no means exhausted.

Our work here is divided into essentially two parts: Mathematical Foundation and Physical Principles. The separation is, of course, only nominal, and in developing the mathematical methods we must occasionally refer to physical concepts, e.g. spin characterization, which are not strictly defined until the work of the second part. This appears unavoidable, but not so undesirable as would be a complete jumbling of mathematical and physical methods—which would interrupt more seriously the continuity of development.

Chapter I reviews briefly relevant topics in Tensor Analysis, and elementary concepts in group theory. We do not attempt to record the long string of definitions and axioms which may be found in any textbook on groups.

Chapter II deals, in rather more detail, with Spinor Analysis, in terms of both the 2- and 4-spinor formulations, the connection with World-Tensors, and Group Theory.

As is well known, Spinor Analysis includes the usual tensor scheme. Basically, it is not more difficult—at least as regards physical applications, although the connection with (approximately) visualizable entities (e.g. real space-time, metric tensor, etc.) is sometimes less obvious or entirely absent, and the notation is more complicated. Our presentation is based on the view that the theoretical physicist is primarily interested in the intelligent application of the rules of the spinor calculus, rather than in the more abstruse geometrical theory of the pure mathematician (which is not intended to detract from the necessity and beauty of the latter development).

Chapter III develops, *from the canonical standpoint*, the essential field-theoretic concepts: the variational equations, the field-variable transformation properties, the conservation laws for mechanical and electrical entities—all of which stem from elementary requirements of invariance under wider and wider groups of transformations.

Chapters IV and V, respectively, present the Field or Tensor-Spinor and Particle or Matrix-Algebraic aspects of Relativistic Wave-Equations for particles of arbitrary spin and rest-mass. Again the separation is nominal, and the complete correspondence of the two aspects is seen in an intermingling of concepts which, eventually, are entirely blended into one formalism within the frame of a suitably wide mathematical structure. Here we find some of the most beautiful aspects of the theory treated in rather complete detail, viz. its basis in abstract algebra and group theory. The application of these latter methods permits a most elegant and concise derivation of many important results here obtained.

Here, too, we find the conditions, in terms of the several current theories, imposed on the wave-fields representing particles: the distinction between field quanta of non-zero and zero rest-mass, the difficulties of the usual Lagrangian formalism for fields corresponding to quanta of spin greater than one, the sharp separation of characteristics of integral and half-integral spin fields, and, eventually, a connection between spin and statistics, and the need for second quantization.

It is hoped that here the physicist may find the answers as to how and why a particular covariant formalism is suitable for the description of particles of given spin, and how and why the so-called field aspect of a given genus of particles is related to the so-called particle aspect—although this field-particle division is somewhat loose, or even redundant, outside the framework of the second quantization. As mentioned before, much of this material may be found in the literature, but it is hoped that the present review will sufficiently weed out and organize the subject-matter as to provide a better basis for entry into current literature and research.

. . . .

The writing of this volume has had a somewhat unusual course. It has taken us through several countries, and from the Institute for Advanced Study and the Armour Research Foundation to the University of Edinburgh and finally to the New York University—in which sanctuary the work was completed. It is the author's pleasure to express appreciation to these several Institutes for the hospitality extended; to the Armour Research Foundation and The Research Corporation for early support and encouragement of this work; and to the State Department for the Fulbright appointment which made it possible to accept Professor Born's invitation to Edinburgh.

Thanks are also due Professors M. Born, N. Kemmer, and G. Wentzel who read parts of the MS. and contributed helpful suggestions and criticism, and to Professors W. Pauli, F. J. Belinfante, and A. H. Taub for correspondence on specific questions of principle; though, of course, responsibility for any errors or shortcomings rests solely with the author.

Finally, I would mark my inexpressible gratitude to Professor A. Einstein, Professor R. Courant, Dr. S. C. Post, and Dr. H. T. Heald for their rare kindness and understanding.

E. M. C.

UNITS AND NOTATION

The postulates of the special relativity and quantum theory form the basis of all theories which we consider here, and it would, therefore, be most natural (as especially emphasized by Pauli and Heisenberg) to take $\hbar$, the Dirac action quantum, and c, the velocity of light in vacuo, as units. With such a system of "natural" units, all quantities are considered to be rationalized to the dimension of a length by multiplication with appropriate powers of $\hbar$ and c.

Undoubtedly this programme simplifies the form of most of the equations and is more logical than the conventional units in current use. Nonetheless, we shall employ the latter, since the above procedure would not in itself add to the clarity of an introductory review, nor facilitate comparison with the vast majority of current literature.

As time coordinate, we use either the light-path length $x_4 = ct$ or, when desirable (and explicitly noted), the imaginary $x_0 = ict$. In general, the metric tensor is taken to be of signature $(- - - +)$ in the preferred (= Galilean) coordinate systems, unless otherwise indicated. The lower-case Greek letters α, β, γ, . . . ranging, in general, from 1 to 4 will label 4-tensors; the lower-case Roman indices i, j, k, . . . (from the middle of the alphabet) ranging from 1 to 3 will label 3-tensors as, for example, in the $3 + 1$ dimensional representation of 4-tensors when convenient and feasible; the lower-case Roman letters p, q, r, . . . being reserved for general labelling and also sometimes as matrix suffixes. The capital Roman letters A, B, . . . will range from 1 to 2 and label 2-spinors; the corresponding lower-case letters a, b, . . . are reserved for labelling 4-spinors and hence range from 1 to 4. The Einstein summation convention is used throughout.

The ordinary scalar product will, as usual, be indicated by a dot symbol ($\cdot$); the vector product by a bold carat symbol ($\wedge$), and the matrix direct product by a bold cross ($\times$).

Bibliographical references in the text give only the significant figures of the year of publication; e.g. footnote 2, page 11, reads cf. v. der Waerden (32); Bargmann (47)*; Pauli (33).

CONTENTS

PART I. — MATHEMATICAL INTRODUCTION

Part II.—PHYSICAL PRINCIPLES

CHAPTER I

Tensor Analysis

§ 1. Characterization of the Minkowski Space and the Lorentz Group

The Minkowski space of the special relativity is a real (metric) space, R_4, of four dimensions which, for specified origin, contains a special invariant locus

$$-(x^1)^2 - (x^2)^2 - (x^3)^2 + (x^4)^2 = g_{\mu\nu}x^\mu x^\nu = 0 \tag{1.1}$$

the " light cone " with vertex at the origin. The points (events) on the future and past branches of the light cone—characterized by the conditions

$$g_{\mu\nu}x^\mu x^\nu = 0 \qquad x^4 > 0 \tag{1.2a}$$
$$g_{\mu\nu}x^\mu x^\nu = 0 \qquad x^4 < 0 \tag{1.2b}$$

can be connected with the origin by a light signal; the points which do not lie on the light cone fall into three disconnected sets: the absolute future, the absolute past, and the absolute elsewhere—characterized respectively by the conditions

$$g_{\mu\nu}x^\mu x^\nu > 0 \qquad x^4 > 0 \tag{1.2c}$$
$$g_{\mu\nu}x^\mu x^\nu > 0 \qquad x^4 < 0 \tag{1.2d}$$
$$g_{\mu\nu}x^\mu x^\nu < 0 \tag{1.2e}$$

Of these the first two, enclosed by the positive and negative branches respectively, furnish intervals termed " time-like "—corresponding to possible paths of motion with velocity less than c—in contradistinction to the last which lies outside the light cone and is termed " space-like ".

The homogeneous Lorentz group is defined to be the set of all *real* linear transformations

$$x^{\mu\prime} = L^\mu{}_\nu x^\nu \tag{1.3}$$

which leave invariant the fundamental indefinite quadratic form $g_{\mu\nu}x^\mu x^\nu$ and *do not* interchange past and future; i.e.

$$g_{\mu\nu}L^\mu{}_\alpha L^\nu{}_\beta = g_{\alpha\beta} \qquad L^4{}_4 > 0 \tag{1.4}$$[1]

The latter condition implies that the point (0, 0, 0, 1) of R_4 which is a point in the absolute future is transformed into a point in the absolute future. The determinant of equation (1.4) gives

$$| L^\mu{}_\nu | = \pm 1 \tag{1.5}$$

[1] i.e. the Lorentz transformations carry the metric tensor over into itself.

and the transformations of the (full) Lorentz group for which $| L^\mu{}_\nu | = +1$ form an invariant subgroup called the restricted or proper Lorentz group. Those for which $| L^\mu{}_\nu | = -1$ are called improper transformations and obviously do not form a group. A typical example of the latter type is

$$L^1{}_1 = -L^2{}_2 = L^3{}_3 = L^4{}_4 = 1 \quad \text{all other coefficients zero.} \tag{1.6}$$

The transformations of this type cause a space reflection in each of the 3-spaces, $x^4 =$ constant, and this interchange of right- and left-handed systems of coordinate axes typifies improper transformations; whereas the proper transformations preserve the distinction between right- and left-handed axes in the space-like sections of R_4. The totality of improper transformations is obtained by multiplying the proper ones by a single improper one, such as (1.6).

If we relax the restriction $L^4{}_4 > 0$, leaving only the condition of invariance (1.4), the group so defined is called the extended Lorentz group. By definition it allows the interchange of past and future as well as right- to left-handed coordinate axes. The extended group is generated by adjoining to the Lorentz group the transformation

$$x^{\mu\prime} = -x^\mu \tag{1.7}$$

Thus, starting from the proper Lorentz group the elements of which satisfy

$$| L^\mu{}_\nu | = +1 \qquad \text{and} \qquad L^4{}_4 > 0 \tag{1.8a}$$

the Lorentz group obtains by adding transformations for which

$$| L^\mu{}_\nu | = -1 \qquad \text{and} \qquad L^4{}_4 > 0 \tag{1.8b}$$

while the extended Lorentz group contains in addition the transformations satisfying

$$| L^\mu{}_\nu | = +1 \qquad \text{and} \qquad L^4{}_4 < 0 \tag{1.8c}$$

and

$$| L^\mu{}_\nu | = -1 \qquad \text{and} \qquad L^4{}_4 < 0 \tag{1.8d}$$

We shall also have need to consider the inhomogeneous (proper) Lorentz group which, for our purposes, consists of transformations the prototype of which, $\Lambda = (a, L)$, is the product (sequential performance) of a homogeneous proper Lorentz transformation followed by a translation by the real vector a. It is readily verified that the product of two inhomogeneous Lorentz transformations (a, L_1) and (b, L_2) is again an inhomogeneous transformation (c, L); viz.

$$(a, L_1)\,(b, L_2) = (c, L) \tag{1.9a}$$

where

$$L^\lambda{}_\nu = L_{(1)}{}^\lambda{}_\mu L_{(2)}{}^\mu{}_\nu \qquad c^\lambda = a^\lambda + L_{(1)}{}^\lambda{}_\mu b^\mu \tag{1.9b}$$

§ 2. Covariant and Contravariant Tensors

(a) *Definition and Transformation Laws*

Confining our considerations, for the present, to the Galilean case, it is well known that the contravariant components of the metric tensor are defined by

$$g^{\lambda\mu}g_{\mu\kappa} = \delta^{\lambda}{}_{\kappa} \tag{2.1}$$

where $\delta^{\lambda}{}_{\kappa}$ is the constant or numerical Kronecker tensor.[1]

By definition covariant and contravariant tensors[2] transform under constant linear transformations according to the laws

$$X'_{\alpha\beta\ldots} = a^{\mu}{}_{\alpha}a^{\nu}{}_{\beta}\ldots X_{\mu\nu\ldots} \tag{2.2a}$$

and

$$X^{\alpha\beta\ldots\prime} = A^{\alpha}{}_{\mu}A^{\beta}{}_{\nu}\ldots X^{\mu\nu\ldots} \tag{2.2b}$$

respectively, where

$$A^{\alpha}{}_{\mu}a^{\mu}{}_{\beta} = \delta^{\alpha}{}_{\beta} \tag{2.3}$$

In our further work we shall usually associate the capital Roman symbols $A, \ldots$ (reserving L for Lorentz) with contravariant components and the corresponding lower-case symbols $a, \ldots$ with covariant components when dealing with constant linear transformations.

Thus, in general, we have for a mixed absolute tensor

$$X^{\alpha\beta\ldots\prime}_{\gamma\delta\ldots} = A^{\alpha}{}_{\kappa}A^{\beta}{}_{\lambda}\ldots a^{\mu}{}_{\gamma}a^{\nu}{}_{\delta}\ldots X^{\kappa\lambda\ldots}_{\mu\nu\ldots} \tag{2.4a}$$

and the extension to general coordinate transformations is essentially automatic, viz.

$$X^{\alpha\beta\ldots\prime}_{\gamma\delta\ldots} = \frac{\partial x^{\alpha'}}{\partial x^{\kappa}}\frac{\partial x^{\beta'}}{\partial x^{\lambda}}\ldots\frac{\partial x^{\mu}}{\partial x^{\gamma'}}\frac{\partial x^{\nu}}{\partial x^{\delta'}}\ldots X^{\kappa\lambda\ldots}_{\mu\nu\ldots} \tag{2.4b}$$

Considering now a constant linear transformation with matrix $\| a^{\mu}{}_{\nu} \|$, the transformation law for the covariant metric tensor is

$$g'_{\mu\nu} = a^{\alpha}{}_{\mu}a^{\beta}{}_{\nu}g_{\alpha\beta} \tag{2.5}$$

and taking the determinant of both sides we find

$$g' = | g'_{\mu\nu} | = a^2 g \tag{2.6}$$

where $a = | a^{\mu}{}_{\nu} |$. Equation (2.6) is a simple well-known example of the transformation law of a relative scalar of weight 2, and the general definition of a relative

[1] We shall usually refer to tensors and tensor densities (cf. § 2*b*) which have the same components in all coordinate systems as numerical tensors, etc.

[2] The geometrical entities, one at each point, which obey these laws constitute strictly speaking a tensor field, but we omit the latter term both in conformity with general usage and to avoid possible confusion with the term " field " in its later physical context.

tensor of weight ω is a form or geometrical object transforming according to the law

$$X^{\alpha\beta}_{\gamma\delta}{}^{\cdots\prime}_{\cdots} = A^{\alpha}{}_{\kappa}A^{\beta}{}_{\lambda}\ldots a^{\mu}{}_{\gamma}a^{\nu}{}_{\delta}\ldots a^{\omega}X^{\kappa\lambda\cdots}_{\mu\nu\cdots} \qquad (2.7)^3$$

(b) Association (of) by Metric Tensor and by Antisymmetric ϵ-Systems

The rule $X_\alpha = g_{\alpha\beta}X^\beta$ is the familiar way of associating a covariant tensor with a contravariant one, or alternatively, covariant and contravariant components of the entity X. A less familiar mode of association uses the covariant relative tensor (Levi-Civita densities) $\epsilon_{\alpha\beta\gamma\delta}$ of weight -1. In any one coordinate system we define $\epsilon_{\alpha\beta\gamma\delta}$ to be $+1$ if α, β, γ, δ is an even permutation of say 1, 2, 3, 4; -1 if odd, and zero otherwise. Then the well-known expansion of a fourth-rank determinant $|a^{\mu}{}_{\nu}| = a$ is given by

$$a\epsilon_{\alpha\beta\gamma\delta} = a^{\kappa}{}_{\alpha}a^{\lambda}{}_{\beta}a^{\mu}{}_{\gamma}a^{\nu}{}_{\delta}\,\epsilon_{\kappa\lambda\mu\nu} \qquad (2.8)$$

which ensures that the tensor shall have the same components in any other system, and the weight of $\epsilon_{\alpha\beta\gamma\delta}$ is seen to be -1. A contravariant tensor X^α then determines a covariant tensor

$$\check{X}_{\alpha\beta\gamma} = \epsilon_{\alpha\beta\gamma\delta}X^\delta \qquad (2.9)$$

which amounts to a mere relabelling of the components as indicated by $\check{X}_{123} = -\check{X}_{213} = \ldots = \check{X}_{231} = X^4$, etc. While the components of $\check{X}_{\alpha\beta\gamma}$ are simply equal to plus or minus the corresponding components of X^δ it is nevertheless true that if X^δ is a contravariant vector then $\check{X}_{\alpha\beta\gamma}$ is a covariant relative tensor. If the weight of X^δ is ω, $\check{X}_{\alpha\beta\gamma}$ is of weight $\omega - 1$.

For the corresponding operation of raising indices we use a relative tensor $\epsilon^{\alpha\beta\gamma\delta}$ of weight $+1$ defined exactly as its covariant counterpart. Thus, the expansion of $|A^{\alpha}{}_{\beta}| = A$ can be written in the form

$$A\epsilon^{\alpha\beta\gamma\delta} = A^{\alpha}{}_{\kappa}A^{\beta}{}_{\lambda}A^{\gamma}{}_{\mu}A^{\delta}{}_{\nu}\epsilon^{\kappa\lambda\mu\nu} \qquad (2.10)$$

or

$$\epsilon^{\alpha\beta\gamma\delta} = A^{\alpha}{}_{\kappa}A^{\beta}{}_{\lambda}A^{\gamma}{}_{\mu}A^{\delta}{}_{\nu}a\epsilon^{\kappa\lambda\mu\nu}$$

since $A = a^{-1}$, and this identity proves that if $\epsilon^{\alpha\beta\gamma\delta}$ is a contravariant tensor of weight $+1$ then its components have the same values in all coordinate systems. Accordingly, a covariant index is converted into three contravariant ones by

$$\check{X}^{\beta\gamma\delta} = \epsilon^{\alpha\beta\gamma\delta}X_\alpha \qquad (2.11)$$

and if X_α is of weight ω, $\check{X}^{\beta\gamma\delta}$ is of weight $\omega + 1$.

We may also sum off two of the indices of $\epsilon^{\alpha\beta\gamma\delta}$ against two covariant indices as in

$$\check{X}^{\gamma\delta} = \tfrac{1}{2}\epsilon^{\alpha\beta\gamma\delta}X_{\alpha\beta} \qquad (2.12)$$

which states that $\check{X}^{12} = \frac{1}{2}(X_{34} - X_{43})$, $\check{X}^{13} = \frac{1}{2}(X_{42} - X_{24})$, etc. However, since the right-hand members of these equations involve only the combinations $X_{\alpha\beta} - X_{\beta\alpha}$,

[3] In this connection it is important to note that we shall not usually use the terms scalar density, vector density, etc., associated with various relative tensors of weight 1 in order to avoid confusion with the "density" tensors of field theory (cf. Chapter III).

the equations will not have a unique inverse unless the components $X_{\alpha\beta}$ are restricted in some way. The simplest restriction is that of antisymmetry, viz.

$$X_{\alpha\beta} = -X_{\beta\alpha} \tag{2.13}$$

and we shall apply (2.12) only to such antisymmetric (skew) tensors. For these the inverse to (2.12) is

$$X_{\alpha\beta} = \tfrac{1}{2}\epsilon_{\alpha\beta\gamma\delta}\check{X}^{\gamma\delta} \tag{2.14}$$

since

$$\tfrac{1}{2}\epsilon_{\alpha\beta\gamma\delta}\epsilon^{\kappa\lambda\gamma\delta} = \delta^{\kappa\lambda}_{\alpha\beta} = \delta^{\kappa}{}_{\alpha}\delta^{\lambda}{}_{\beta} - \delta^{\kappa}{}_{\beta}\delta^{\lambda}{}_{\alpha} \tag{2.15}$$

the proof of which is essentially a simple matter of counting, and is the defining equation of the first generalized Kronecker tensor $\delta^{\kappa\lambda}_{\alpha\beta}$ which typifies the generalized δ's (cf. equations 2.18). $\delta^{\kappa\lambda}_{\alpha\beta}$ has the value $+1$ or -1 according as an even or odd permutation is required to arrange the superscripts in the same order as the subscripts, if the superscripts are all distinct and the subscripts are the same set of numbers; otherwise its value is zero. The inverse to equation (2.11) is similarly obtained by observing that

$$\frac{1}{3!}\,\epsilon^{\alpha\beta\gamma\delta}\epsilon_{\alpha\beta\gamma\lambda} = \delta^{\delta}{}_{\lambda} \tag{2.16}$$

from which the inverse to $\check{X}_{\alpha\beta\gamma} = \epsilon_{\alpha\beta\gamma\delta}X^{\delta}$ is

$$X^{\lambda} = \frac{1}{3!}\,\check{X}_{\alpha\beta\gamma}\,\epsilon^{\alpha\beta\gamma\lambda} \tag{2.17}$$

For convenience of future reference we list here the properties of the ϵ-, δ-systems for the 4-space case which is of primary interest to us:

$$\begin{aligned}
\epsilon^{\mu\nu\sigma\tau}\epsilon_{\alpha\beta\gamma\delta} &= \delta^{\mu\nu\sigma\tau}_{\alpha\beta\gamma\delta} \\
\epsilon^{\mu\nu\sigma\tau}\epsilon_{\mu\alpha\beta\gamma} &= \delta^{\nu\sigma\tau}_{\alpha\beta\gamma} \\
\epsilon^{\mu\nu\sigma\tau}\epsilon_{\mu\nu\alpha\beta} &= 2\delta^{\sigma\tau}_{\alpha\beta} \\
\epsilon^{\mu\nu\sigma\tau}\epsilon_{\mu\nu\sigma\alpha} &= 3!\ \delta^{\tau}{}_{\alpha}
\end{aligned} \tag{2.18}$$

We see that all the ϵ- and δ-systems are examples of numerical relative tensors and tensors, respectively. By the same argument as before, the laws of transformation of the ϵ's for general coordinate transformations are:

$$\epsilon^{\alpha\beta\cdots} = \frac{\partial x^{\alpha'}}{\partial x^{\mu}}\frac{\partial x^{\beta'}}{\partial x^{\nu}}\cdots\left|\frac{\partial x}{\partial x'}\right|\epsilon^{\mu\nu\cdots} \tag{2.19a}$$

$$\epsilon_{\alpha\beta\cdots} = \frac{\partial x^{\mu}}{\partial x^{\alpha'}}\frac{\partial x^{\nu}}{\partial x^{\beta'}}\cdots\left|\frac{\partial x}{\partial x'}\right|^{-1}\epsilon_{\mu\nu\cdots} \tag{2.19b}$$

where $\left|\dfrac{\partial x}{\partial x'}\right|$ is the functional or Jacobian determinant.

§ 3. Special Forms—Dual Tensors

For our further work we define the following " dual tensor systems ", which differ only slightly from the cases considered in the preceding section. A second-rank antisymmetric tensor $G_{\sigma\tau}$ defines a dual[1] which is also of second rank and antisymmetric, according to

$$\check{G}^{\mu\nu} = \tfrac{1}{2} g^{-1/2} \epsilon^{\mu\nu\sigma\tau} G_{\sigma\tau} \tag{3.1a}$$

A third-rank antisymmetric tensor $U_{\nu\sigma\tau}$ defines a dual $\check{U}^{\mu}$ according to

$$\check{U}^{\mu} = \frac{1}{3!} g^{-1/2} \epsilon^{\mu\nu\sigma\tau} U_{\nu\sigma\tau} \tag{3.2a}$$

A fourth-rank antisymmetric tensor $\phi_{\mu\nu\sigma\tau}$ defines a dual $\check{\phi}$ according to

$$\check{\phi} = \frac{1}{4!} g^{-1/2} \epsilon^{\mu\nu\sigma\tau} \phi_{\mu\nu\sigma\tau} \tag{3.3a}$$

where $g^{1/2}$ is the square root of the determinant of the metric tensor $g_{\mu\nu}$, and is $+(-1)^{1/2}$ in a preferred ($=$ Galilean) coordinate system.

From the properties of the ϵ-, δ-systems it follows that the inverse relations are, respectively

$$G_{\mu\nu} = \tfrac{1}{2} g^{1/2} \epsilon_{\mu\nu\sigma\tau} \check{G}^{\sigma\tau} \tag{3.1b}$$

$$U_{\nu\sigma\tau} = g^{1/2} \epsilon_{\mu\nu\sigma\tau} \check{U}^{\mu} \tag{3.2b}$$

$$\phi_{\mu\nu\sigma\tau} = g^{1/2} \epsilon_{\mu\nu\sigma\tau} \check{\phi} \tag{3.3b}$$

The duals (3.1*a*–3.3*a*) are especially important in the theory of wave fields associated with particles of spin[2] 0 and 1, and will be referred to as pseudotensor, pseudovector, and pseudoscalar, respectively. The reason for the designation " pseudo " is the fact that the duals obey a tensor transformation law if the determinant of the transformation is positive, whereas under a (general) coordinate transformation with negative determinant they transform as

$$\check{X}^{\alpha\beta\ldots\prime} = -\frac{\partial x^{\alpha'}}{\partial x^{\mu}} \frac{\partial x^{\beta'}}{\partial x^{\nu}} \ldots \check{X}^{\mu\nu} \tag{3.4}$$

which is to say that, in general, they differ from true (absolute) tensors in their behaviour under reflection.

It may be mentioned here that it is common in current literature to refer to the field theories involving, respectively, one antisymmetric tensor of third rank and one of the fourth rank, e.g. $\chi_{\alpha\beta\gamma}$, $\phi_{\alpha\beta\gamma\delta}$, and one antisymmetric tensor of the second rank and one of the third rank, e.g. $G_{\mu\nu}$, $U_{\alpha\beta\gamma}$, as the pseudoscalar and pseudovector theories (cf. Chapter IV). Actually, as implied by the notation, these are absolute tensors and transform as such, but their intimate connection with the corresponding

[1] We shall hereafter indicate dual tensors by the carat (v) symbol. [2] cf. Chapter IV, §§ 26 30.

duals $\check{\chi}^{\mu}$, $\check{\phi}$ and $\check{G}_{\mu\nu}$, $\check{U}^{\mu}$ justifies the somewhat unprecise terminology.[3] The reference, of course, is to the fact that the 256 components of $\phi_{\alpha\beta\gamma\delta}$, for example, reduce to but one *significant value* or *Kennzahl* which is a pseudoscalar; the 64 components of $\chi_{\alpha\beta\gamma}$ reducing to four significant values, etc.

The operation of taking the dual commutes with that of raising and lowering indices by means of the metric tensor, but does not commute with that of taking the complex conjugate, e.g. for (3.1*a*)

$$\bar{\check{G}}_{\mu\nu} = -\check{\bar{G}}_{\mu\nu} \tag{3.5}$$

An important case in physical applications is the self-dual tensor, viz.

$$\check{G}^{\mu\nu} = G^{\mu\nu} \tag{3.6}$$

and we note that an arbitrary *real* antisymmetric tensor $G^{\mu\nu}$ determines one self-dual tensor $X^{\mu\nu}$ according to

$$X^{\mu\nu} = G^{\mu\nu} + \check{G}^{\mu\nu} = \tfrac{1}{2}(X^{\mu\nu} + \bar{X}^{\mu\nu}) + \tfrac{1}{2}(X^{\mu\nu} - \bar{X}^{\mu\nu}) \tag{3.7}$$

whereas there are two self-dual tensors associated with an arbitrary complex antisymmetric tensor $G_{\mu\nu}$, viz.

$$G_{\mu\nu} + \check{G}_{\mu\nu} \tag{3.8a}$$

and

$$\bar{G}_{\mu\nu} + \check{\bar{G}}_{\mu\nu} \tag{3.8b}$$[4]

§ 4. Special Considerations on the Lorentz Group

(*a*) *Tensor Character of Rotation Matrices*

In a given coordinate system the vector mapping

$$x^{\mu\prime} = A^{\mu}{}_{\nu} x^{\nu} \tag{4.1}$$[1]

with the additional condition of " length invariance ",

$$g_{\mu\nu} x^{\mu\prime} x^{\nu\prime} = g_{\mu\nu} x^{\mu} x^{\nu} \tag{4.2a}$$

or

$$g_{\mu\nu} A^{\mu}{}_{\rho} A^{\nu}{}_{\sigma} = g_{\rho\sigma} \tag{4.2b}$$

constitutes a general rotation; and clearly the $A^{\mu}{}_{\nu}$ are the components of a mixed tensor, covariant in ν, contravariant in μ.

[3] As will be seen in Chapter IV, § 26, the field equations for the particles described in terms of the former sets (spin 0 and 1) imply and are implied by the field equations in terms of the latter. The notation for our present discussion of dual systems is chosen to conform identically with the applications of Chapter IV *et seq.*

[4] Note that $A^{\mu\nu} \bar{B}_{\mu\nu} = 0$ for any two self-dual tensors.

[1] i.e. with the interpretation of (4.1) as a geometric transformation (which amounts to a permutation of the points of the space) rather than the alternate interpretation as a coordinate transformation (which merely relabels the points of the space).

On the other hand, for a constant coordinate transformation

$$x^{\mu\prime} = A^{\mu}{}_{\nu} x^{\nu} \tag{4.3a}$$

we have, as noted before (cf. equation 2.5),

$$g'_{\mu\nu} A^{\mu}{}_{\rho} A^{\nu}{}_{\sigma} = g_{\rho\sigma} \tag{4.3b}$$

and the Lorentz transformations are precisely those under which the metric tensor is numerically invariant. Hence, according to (4.3*b*) the Lorentz matrices satisfy (4.2*b*) which is the condition for a rotation matrix, so that we set the Lorentz transformations into correspondence with the rotations (of the same matrix) and may confine our attention to the latter; every statement about a rotation $A^{\mu}{}_{\nu}$ being concomitantly a statement about a Lorentz transformation represented by $A^{\mu}{}_{\nu}$. One particular convenience of this procedure arises from the previously noted fact that the rotation matrices have tensor character.

More generally, we observe that for an arbitrary coordinate transformation $x^{\mu\prime} = x^{\mu\prime}(x^1, x^2, x^3, x^4)$,

$$A^{\mu}{}_{\nu} = \frac{\partial x^{\mu\prime}}{\partial x^{\nu}} = x^{\mu\prime}{}_{,\nu} \tag{4.4}$$

(the comma indicating derivate) the index μ in $A^{\mu}{}_{\nu}$ refers to the system of the $(x^{\mu\prime})$ with metric $g'_{\mu\nu}$; the index ν referring to the system of the (x^{ν}) with metric $g_{\mu\nu}$. Noting this it is clear that in

$$x^{\mu\prime} = A^{\mu}{}_{\nu} x^{\nu} \tag{4.5a}$$

we can raise and lower indices according to

$$x_{\mu}' = A_{\mu\nu} x^{\nu} \qquad (A_{\mu\nu} = g'_{\mu\rho} A^{\rho}{}_{\nu}) \tag{4.5b}$$

$$x_{\mu}' = A_{\mu}{}^{\nu} x_{\nu} \qquad (A_{\mu}{}^{\nu} = g'_{\mu\rho} g^{\nu\sigma} A^{\rho}{}_{\sigma}) \tag{4.5c}$$

$$x^{\mu\prime} = A^{\mu\nu} x_{\nu} \qquad (A^{\mu\nu} = g^{\nu\sigma} A^{\mu}{}_{\sigma}) \tag{4.5d}$$

so that all the transformation properties of the components of the vector x^{μ} obtain as usual when the rule of raising and lowering indices is vested in $A^{\mu}{}_{\nu}$.

(*b*) *The Infinitesimal Lorentz Transformations*

Considering again the rotation (4.1) we have from (4.2*a*, *b*)

$$A^{\mu}{}_{\rho} A_{\nu}{}^{\rho} = \delta^{\mu}{}_{\nu} \tag{4.6}$$

so that $A = | A^{\mu}{}_{\rho} | \neq 0$; every rotation has an inverse and by definition the (sequential) composition of two rotations is again a rotation. Let us put

$$(A^{-1})^{\rho}{}_{\nu} = a^{\rho}{}_{\nu} = A_{\nu}{}^{\rho} \tag{4.7a}$$

or by (4.6)

$$A^{\mu}{}_{\rho} a^{\rho}{}_{\nu} = \delta^{\mu}{}_{\nu} \qquad \text{(cf. equation 2.3)} \tag{4.7b}$$

from which $A_{\mu\nu}$ is the inverse of $A_{\nu\mu}$.

A very powerful approach to the analysis of relativistic wave equations, and indeed one of prime importance, is the method of infinitesimal Lorentz transformations—which are defined, for example, by

$$x^{\mu\prime} = L^{\mu}{}_{\nu} x^{\nu} \tag{4.8}$$

where

$$L^{\mu}{}_{\nu} = \delta^{\mu}{}_{\nu} + \omega^{\mu}{}_{\nu} \tag{4.9a}$$

or

$$L_{\mu\nu} = \delta_{\mu\nu} + \omega_{\mu\nu} \tag{4.9b}$$

and differ from the identity transformation by infinitesimals $\omega_{\mu\nu}$ such that the products of two or more ω's are negligible. From (4.6) we obtain the normalization condition characterizing the infinitesimal rotations, viz.

$$\omega_{\mu\nu} = -\omega_{\nu\mu} \tag{4.10}$$

(c) Integrability Conditions for the (Proper) Lorentz Group

In any representation of the Lorentz group the infinitesimal transformation (4.9b) is represented by

$$1 + \tfrac{1}{2}\omega^{\mu\nu}\mathfrak{T}_{\mu\nu} \qquad (4.11)^2$$

where $\mathfrak{T}^{\mu\nu}$ $(= -\mathfrak{T}^{\nu\mu})$ are the representative matrices of the (six) infinitesimal transformations. These must satisfy the so-called integrability conditions on the Lorentz group, which are

$$\begin{aligned}[\mathfrak{T}^{\kappa\lambda}, \mathfrak{T}^{\mu\nu}] &= \mathfrak{T}^{\kappa\lambda}\mathfrak{T}^{\mu\nu} - \mathfrak{T}^{\mu\nu}\mathfrak{T}^{\kappa\lambda} \\ &= g^{\lambda\mu}\mathfrak{T}^{\kappa\nu} - g^{\kappa\mu}\mathfrak{T}^{\lambda\nu} + g^{\kappa\nu}\mathfrak{T}^{\lambda\mu} - g^{\lambda\nu}\mathfrak{T}^{\kappa\mu}\end{aligned} \tag{4.12}$$

and (implicitly) define the structure constants of the group. Note that the $\mathfrak{T}^{\mu\nu}$ are not matrix elements but individual matrices ordered as components of a matrix 6-vector. These are the commutation rules (C.R.) characteristic of all infinitesimal rotations, and apart from a numerical factor are familiar as the C.R. of the components of the angular momentum 6-vector $\mathscr{J}^{\mu\nu}_{\text{op}}$ in relativistic quantum theory, viz.

$$[\mathscr{J}^{\kappa\lambda}_{\text{op}}, \mathscr{J}^{\mu\nu}_{\text{op}}] = i\hbar\{g^{\lambda\mu}\mathscr{J}^{\kappa\nu}_{\text{op}} - g^{\kappa\mu}\mathscr{J}^{\lambda\nu}_{\text{op}} + g^{\kappa\nu}\mathscr{J}^{\lambda\mu}_{\text{op}} - g^{\lambda\nu}\mathscr{J}^{\kappa\mu}_{\text{op}}\} \qquad (4.13)^3$$

Our later investigations (cf. Chapter III, § 24; Chapter IV, §§ 25, 28; Chapter V, §§ 33, 34) will make more precise the spin-relation $S^{\lambda\mu}_{\text{op}} = i\hbar\mathfrak{T}^{\lambda\mu}$, and more particularly show in an elementary physical way that the infinitesimal operator $\mathscr{J}^{\mu\nu}_{\text{op}}$, for example, generates a rotation in the (x^μ, x^ν)-plane; the components $\mathscr{J}^{23}$, $\mathscr{J}^{31}$, $\mathscr{J}^{12}$ representing respectively the x^1, x^2, x^3 components of the total angular momentum.

For the 3-dimensional case the well-known equivalence of vectors and antisymmetric tensors under the spatial orthogonal (rotation) group leads to an obvious simplification of (4.13), corresponding to the generally known C.R.

$$\mathscr{J}^k\mathscr{J}^l - \mathscr{J}^l\mathscr{J}^k = i\hbar\mathscr{J}^m \qquad (k, l, m \text{ cyclically}) \tag{4.14}$$

[2] cf. v. der Waerden (32); Bargmann (47)*; Pauli (33).
[3] cf. v. der Waerden (32); Dirac (36); Bargmann-Wigner (48)*.

§ 5. Infinitesimal Tensor Transformations

The choice of field variables suitable for the description of particles of a given spin is basically determined by the connection between the spin angular-momentum-density tensor (and hence the spin operator) and the infinitesimal transformation properties of the field variables.[1] For integral spins the field variables will be seen to have tensor character or, as we may say, belong to the tensor representations of the Lorentz group, and we wish to consider here the infinitesimal transformations of a general tensor $F_{\alpha_1\alpha_2\ldots\alpha_n} = F_{(\alpha)}$—which for convenience we consider in its covariant form.

First we note that for the infinitesimal coordinate transformation

$$x^{\mu\prime} = x^\mu + \xi^\mu(x) \tag{5.1a}$$

or

$$\delta x^\mu = x^{\mu\prime} - x^\mu = \xi^\mu(x) \tag{5.1b}$$

$$\frac{\partial x^\mu}{\partial x^{\nu\prime}} = x^\mu{}_{,\nu'} = \delta^\mu{}_\nu - \frac{\partial \xi^\mu}{\partial x^{\nu\prime}} \tag{5.2}$$

$$= \delta^\mu{}_\nu - \frac{\partial \xi^\mu}{\partial x^\nu} + \text{higher-order terms.}$$

Hence, for $F_{(\alpha)}$ we find

$$F'_{\alpha_1\alpha_2\ldots\alpha_n} = \mathop{\Pi}_s F_{\mu_1\mu_2\ldots\mu_n} \frac{\partial x^{\mu_s}}{\partial x^{\alpha_s\prime}}$$

$$= F_{\alpha_1\alpha_2\ldots\alpha_n} - \sum_{r=1}^{n} F_{\alpha_1\alpha_2\ldots\alpha_{r-1}\mu_r\alpha_{r+1}\ldots\alpha_n}\xi^{\mu_r}{}_{,\alpha_r} \tag{5.3a}$$

by (5.2), and the " local " variation of $F_{(\alpha)}$ corresponding to (5.1b) is

$$\delta F_{(\alpha)} = F'_{(\alpha)}(x') - F_{(\alpha)}(x)$$

$$= -\sum_{r=1}^{n} \delta^\mu{}_{\alpha_r} F_{\alpha_1\ldots\alpha_{r-1}\alpha\alpha_{r+1}\ldots\alpha_n}\xi^\alpha{}_{,\mu} \tag{5.3b}$$

In particular, for an infinitesimal rotation $\xi^\alpha = \delta x^\alpha = \omega^{\alpha\beta}x_\beta$ (cf. 5.4b), and

$$\delta F_{(\alpha)} = -\sum_r \omega^{\alpha\beta} g_{\beta\alpha_r} F_{\alpha_1\ldots\alpha_{r-1}\alpha\,\alpha_{r+1}\ldots\alpha_n}$$

$$= \tfrac{1}{2}\omega^{\alpha\beta}\sum_r \{g_{\alpha\alpha_r}\delta^{\beta_r}_{\beta} - g_{\beta\alpha_r}\delta^{\beta_r}_{\alpha}\}\delta^{\beta_1}_{\alpha_1}\ldots\delta^{\beta_{r-1}}_{\alpha_{r-1}}\delta^{\beta_{r+1}}_{\alpha_{r+1}}\ldots F_{\beta_1\ldots\beta_n} \tag{5.3c}$$

or in contravariant form

$$\delta F^{(\alpha)} = \tfrac{1}{2}\omega^{\mu\nu}\mathfrak{T}^{(\alpha)(\beta)}_{\mu\nu}F_{(\beta)} \tag{5.4a}$$

where

$$\mathfrak{T}^{(\alpha)(\beta)}_{\mu\nu} = \sum_r \delta^{\alpha_r\beta_r}_{\mu\;\nu} g^{\alpha_1\beta_1}\ldots g^{\alpha_{r-1}\beta_{r-1}}g^{\alpha_{r+1}\beta_{r+1}}\ldots \tag{5.4b}$$

[1] cf. Chapters III, IV, V, where the general principles, equivalent and additional criteria are developed in connection with the general principles of field theory.

When the components of $F^{(\alpha)}$ are suitably ordered in a one-column matrix form we can write this as (cf. Chapter III, § 19)

$$\delta F = \tfrac{1}{2}\omega^{\mu\nu}\mathfrak{T}_{\mu\nu\ \mathrm{op}}F \tag{5.4c}$$

The above " local " variation δF which arises from the comparison of the transformed function $F'(x')$ and the original function $F(x)$ *at a given point* P is to be contrasted with the so-called " substantial " variation

$$\delta^* F = F'(x) - F(x) = \{F'(x') - F(x)\} - \{F'(x') - F'(x)\} \tag{5.5a}$$

which compares the transformed function at a point $\mathrm{P}'(x' = x)$ whose coordinates are the same as those of the argument of the original function *at the different point* P. We readily obtain

$$\delta^* F^{(\alpha)} = \tfrac{1}{2}\omega^{\mu\nu}\mathfrak{T}_{\mu\nu}^{(\alpha)(\beta)}F_{(\beta)} - \omega^{\mu\nu}F^{(\alpha)}{}_{,\mu}\,x_\nu \tag{5.5b}$$

as the change produced in the variables $F^{(\alpha)}$ *as functions of their arguments* because of the infinitesimal coordinate transformation. Because the transformed $F^{(\alpha)}$ are not compared with the original values at the same world point, but with the original values at that world point which possesses the same coordinate values prior to the transformation, the above equation contains the added *transport* term. These two cases are fundamental in field theory and will be given detailed consideration in Chapter III.

CHAPTER II

Spinor Analysis

§ 6. Definition of 2-Spinors and Spin Transformations

We now consider geometrical objects, ψ_A, termed " spinors ", which are defined over a two-dimensional complex space P_1 (the spin space) and obey the transformation law

$$\psi_A{}' = t^B{}_A \psi_B \qquad (A, B = 1, 2) \tag{6.1}$$

with the transformation matrix $\| t^B_A \|$ in general complex and non-singular. For the purposes of physical applications of the spinor theory (in the restricted relativity) we could, as will be seen,[1] restrict ourselves to transformations of spin coordinates or frames with unimodular matrices $\| s^A{}_B \|$. However, noting that in every pencil of non-singular matrices $\| \rho t^A{}_B \|$ there are just *two* unimodular ones which are

$$s^A{}_B = t^A{}_B t^{-1/2} \qquad t = | t^A{}_B | \tag{6.2}$$

the two values of $| s^A{}_B |$ arising from the ambiguity in sign of $t^{-1/2}$; we shall continue to make use of the more general transformation (6.1) which is a composite of a transformation with matrix (6.2) and the transformation

$$\psi_A{}' = \psi_A t^{1/2} \tag{6.3}$$

which multiplies the coordinates of each point in the spin space by a factor which does not change the coordinate system. The effect of restricting ourselves to unimodular transformations (6.2) is achieved by assigning a suitable weight to ψ_A; thus, if ψ_A is to be of weight $-\frac{1}{2}$, its transformation law is

$$\psi_A{}' = t^B{}_A t^{-1/2} \psi_B = s^B{}_A \psi_B \tag{6.4}$$

and we call quantities transforming in this way " covariant regular spinors of first rank ". Similarly, a geometric object ψ^A of weight $\frac{1}{2}$ transforming contragrediently to ψ_A according to

$$\psi^{A\prime} = T^A{}_B t^{1/2} \psi^B = S^A{}_B \psi^B \tag{6.5}$$

where

$$T^A{}_B t^B{}_C = S^A{}_B s^B{}_C = \delta^A{}_C \tag{6.6}$$

will be called a " contravariant regular spinor of first rank ".

[1] The abstract space which we are here considering axiomatically is referred to as " spin space " in brief, but the term " spin " must not be taken literally in the ordinary sense; the name " spinor " implies only that these entities are suitable for describing particles with spin (cf. Chapter IV). Further, we shall see (§ 8) that a two-dimensional complex spin space is associated with each point of the metric space with real coordinates x^μ.

By analogy with the ϵ-systems in tensor theory we associate covariant and contravariant components according to the rule

$$\psi_A = \epsilon_{AB}\psi^B \tag{6.7}$$

where

$$\| \epsilon_{AB} \| = \begin{Vmatrix} 0 & 1 \\ -1 & 0 \end{Vmatrix} = \| \epsilon^{AB} \| \tag{6.8}$$

For the moment these may be considered as a special kind of ϵ-system, but we shall see (below) that they are the so-called fundamental numerical metric spinors of second rank. We have now no fundamental quadratic form in this space, so that the above rule is our only way of lowering indices. For this reason we write ψ_A on the left of (6.7) instead of $\breve{\psi}_A$ as we should in strict analogy with (2.9), and refer simply to spinors rather than relative spinors. The equations inverse to (6.7) are then

$$\psi^A = \epsilon^{BA}\psi_B = \epsilon^{BA}\epsilon_{BC}\psi^C = \delta^A{}_C\psi^C \qquad (6.9)^2$$

The formula for the expansion of a second-rank determinant is analogous to (2.8) and (2.10), and therefore the transformations laws of ϵ_{AB} and ϵ^{AB} are

$$\epsilon_{AB} = t^C{}_A t^D{}_B t^{-1}\epsilon_{CD} = s^C{}_A s^D{}_B \epsilon_{CD} \tag{6.10a}$$

$$\epsilon^{AB} = T^A{}_C T^B{}_D t\epsilon^{CD} = S^A{}_C S^B{}_D \epsilon^{CD} \tag{6.10b}$$

from which it follows that relative to the general constant-spin (frame) transformations, ϵ_{AB} and ϵ^{AB} are of weight -1 and $+1$ respectively.

The higher-rank spinors are defined as quantities which transform as products of simple (first-rank) spinors, so that the metric spinors ϵ_{AB} and ϵ^{AB} are of second rank as already noted. Moreover, the weights $\frac{1}{2}$ for ψ^A and $-\frac{1}{2}$ for ψ_A are consistent with the rules

$$\psi_A = \epsilon_{AB}\psi^B \tag{6.11}$$

$$\psi^A = \epsilon^{BA}\psi_B$$

It follows that

$$\phi^A\psi_A = -\phi_A\psi^A \tag{6.12a}$$

so that every odd-rank spinor has zero absolute value, viz.

$$\phi_A\phi^A = 0 \tag{6.12b}$$

$$\psi_{ABC}\psi^{ABC} = 0 \tag{6.12c}$$

etc., and in particular

$$\phi^A\psi_A\chi_B + \phi_A\psi_B\chi^A + \phi_B\psi^A\chi_A = 0 \tag{6.12d}$$

The geometric object which has the components ψ_1, ψ_2 is closely analogous to a covariant vector but the coefficients of the transformations (6.1) being, in general, complex instead of real, allow the more general transformation law

$$\chi'_A = t^B{}_A t^w \bar{t}^d \chi_B \tag{6.13}$$

for a geometric object with components χ_A. The indices w and d are usually called

[2] Note that in (6.7) we sum on the second index of ϵ_{AB} while in (6.9) we sum on the first index of ϵ^{BA}.

the weight and anti-weight respectively, and when $w = d$ the transformation may be written as

$$\chi'_A = t^B{}_A \mid t \mid^{2w} \chi_B \tag{6.14}$$

where $\mid t \mid = (t\bar{t})^{1/2}$ is the absolute value of the determinant; in which case χ_A is said to be of absolute weight $2w$.

Consider the transformation

$$\chi^{B}_{\dot{A}}{}' = \bar{s}^C{}_A S^B{}_D \chi^D{}_{\dot{C}} \tag{6.15}$$

where the placing of the dot over the indices A, C serves as a reminder that a bar (complex conjugate) is to be placed over the corresponding $t^C{}_A$ or $s^C{}_A$ in the law of transformation,[3] e.g.

$$\chi^{A\,\ldots\,\dot{C}\,\ldots}_{E\,\ldots\,\dot{G}\,\ldots} = T^A{}_P \ldots \bar{T}^C{}_R \ldots t^T{}_E \ldots \bar{t}^U{}_G \ldots t^w \bar{t}^d \chi^{P\,\ldots\,\dot{R}\,\ldots}_{T\,\ldots\,\dot{U}\,\ldots} \tag{6.16}$$

It is always possible to choose the weights w and d so that the transformation can be written in terms of the unimodular matrices $\| s^A{}_B \|$ and $\| S^A{}_B \|$, and we usually do this.

The raising and lowering of dotted indices is accomplished with the aid of the spinors $\epsilon^{A\dot{B}}$ and $\epsilon_{A\dot{B}}$ by the rules

$$\begin{aligned} \psi_A &= \epsilon_{A\dot{B}} \psi^{\dot{B}} \\ \psi^A &= \epsilon^{\dot{B}A} \psi_{\dot{B}} \end{aligned} \tag{6.17}$$

where $\| \epsilon_{A\dot{B}} \| = \| \epsilon^{A\dot{B}} \| = \begin{Vmatrix} 0 & 1 \\ -1 & 0 \end{Vmatrix}$ and, for example, ψ_A and ψ^A are of anti-weights $-\frac{1}{2}$ and $\frac{1}{2}$ respectively. Moreover, $\epsilon_{A\dot{B}}$ and $\epsilon^{A\dot{B}}$ are of anti-weights -1 and $+1$, and have the transformation laws

$$\epsilon_{A\dot{B}} = \bar{s}^C{}_A \bar{s}^D{}_B \epsilon_{\dot{C}\dot{D}} \tag{6.18a}$$

$$\epsilon^{A\dot{B}} = \bar{S}^A{}_C \bar{S}^B{}_D \epsilon^{\dot{C}\dot{D}} \tag{6.18b}$$

It follows from the above considerations that the complex conjugate of a 2-spinor equation is obtained simply by dotting all undotted indices; a twice-dotted index reverting to plain type by the involutorial character of the operation.

§ 7. Spin-Tensors

A quantity such as $\phi_{\mu AB}$ which has both tensor and spinor indices represents a tensor-spinor[1] (in contrast to simple spinors which are scalar functions of x^μ) and transforms according to

$$\phi'_{\mu AB} = a^\nu{}_\mu \phi_{\nu AB} \tag{7.1a}$$

under coordinate transformations in R_4, and

$$\phi'_{\mu AB} = \bar{s}^C{}_A s^D{}_B \phi_{\mu\dot{C}D} \tag{7.1b}$$

[3] We shall refer to spinors with dotted indices as " conjugated " or simply as " dotted ".

[1] Often referred to as a " spin-tensor " or simply " spinor ", depending on the context.

under constant spin transformation in P_1. The corresponding contravariant transformations are

$$\phi^{\mu\dot{A}B'} = A^{\mu}{}_{\nu}\phi^{\nu\dot{A}B} \tag{7.2a}$$

and

$$\phi^{\mu\dot{A}B'} = \bar{S}^{A}{}_{C}S^{B}{}_{D}\,\phi^{\mu\dot{C}D} \tag{7.2b}$$

Under combined (constant) space- and spin-coordinate transformations these transform as

$$\phi_{\mu\dot{A}B}{}'' = a^{\nu}{}_{\mu}\phi'_{\nu\dot{A}B} = a^{\nu}{}_{\mu}\bar{s}^{C}{}_{A}s^{D}{}_{B}\phi_{\nu\dot{C}D} \tag{7.3}$$

and

$$\phi^{\mu\dot{A}B''} = A^{\mu}{}_{\nu}\phi^{\nu\dot{A}B'} = A^{\mu}{}_{\nu}\bar{S}^{A}{}_{C}S^{B}{}_{D}\,\phi^{\nu\dot{C}D} \tag{7.4}$$

§ 8. Connection between 2-Spinors and World-Tensors

(a) Introductory Remarks

The relation between 2-spinors and 4-tensors may be approached from a number of different, though ultimately equivalent, points of view. Perhaps the most complete and powerful development is via the analysis of projective geometry coupled with group-theoretical considerations.[1] We shall have good reason to consider the latter aspect in some detail both in this and later chapters, but it would carry us very far afield to dwell at any length on the purely geometric considerations. The simplest alternate is to make a somewhat heuristic " Ansatz " in a purely formal way, but we should like to try to strike some compromise between the, perhaps antithetical, requirements of simplicity and completely logical development. To this end we shall consider briefly one of the more elementary (and fundamental) geometric aspects which throws considerable light on the foregoing axiomatic development, and will also serve in our further work.

(b) Space-Time and Second-Rank Hermitian Matrices.

One of the fundamental concepts underlying the theory of spinors is the fact that there exists a (1–1) correspondence between second-rank Hermitian matrices and points of R_4. This quite remarkable correspondence is readily established. We consider a Hermitian matrix $X = \| X_{\dot{A}B} \|$, where the notation anticipates the interpretation of X as a second-rank spinor of suitable weights; so that the condition of Hermiticity is $\bar{X}_{\dot{A}B} = X_{\dot{B}A}$. With the weight and anti-weight $-\frac{1}{2}$, $X_{\dot{A}B}$ transforms under general constant spin (i.e. complex binary) transformations as

$$X_{\dot{A}B}{}' = \bar{t}^{C}{}_{A}t^{D}{}_{B}t^{-1/2}\bar{t}^{-1/2}X_{\dot{C}D} = \bar{s}^{C}{}_{A}s^{D}{}_{B}X_{\dot{C}D} \tag{8.1}$$

and we note that the determinant of this relation yields

$$|X'_{\dot{A}B}| = |X_{\dot{C}D}| \qquad (X_{\dot{1}1}X_{\dot{2}2} - X_{\dot{1}2}X_{\dot{2}1} \text{ invariant}) \tag{8.2a}$$

[1] The wide scope of spinor analysis and the prodigious number of investigators associated with various of its developments makes it all but hopeless to list even the principal contributors. We can therefore give but a few of the important references which should be useful to the reader who wishes supplementary material or is interested in the more technical developments beyond the scope of our treatment, viz. Veblen (33, 35); Schouten (49*a*, *b*); Ruse (37); Cartan (38*a*, *b*).

i.e. the determinant is an invariant under general binary transformations. Alternatively we note that

$$| X_{\dot{A}B} | = \tfrac{1}{2}\epsilon^{\dot{C}\dot{A}}\epsilon^{DB}X_{\dot{A}B}X_{\dot{C}D} = \tfrac{1}{2}X_{\dot{A}B}X^{\dot{A}B} \tag{8.2b}$$

from which the above conclusion also follows by inspection.

The essential point now consists in noting that every Hermitian matrix of second order can be written in terms of four real parameters p, q, r, s, as, for example,

$$\| X_{\dot{A}B} \| = \begin{Vmatrix} p + q & r + is \\ r - is & p - q \end{Vmatrix} \tag{8.3}$$

(omitting a possible common real multiplicative factor which we shall refer to as the normalizer); and similarly the quadratic form

$$g_{\mu\nu}x^{\mu}x^{\nu} = -(x^1)^2 - (x^2)^2 - (x^3)^2 + (x^4)^2 \tag{8.4}$$

can be expressed in the product form

$$g_{\mu\nu}x^{\mu}x^{\nu} = (p + q)(p - q) - (r + is)(r - is)$$

and in a great variety of ways according to various possible (1–1) identifications of individuals in the sets p, q, r, s and x^1, x^2, x^3, x^4. The correspondence (8.3) is (1–1); the inverse equations are

$$\begin{aligned} p &= \tfrac{1}{2}(X_{\dot{1}1} + X_{\dot{2}2}) \\ q &= \tfrac{1}{2}(X_{\dot{1}1} - X_{\dot{2}2}) \\ r &= \tfrac{1}{2}(X_{\dot{1}2} + X_{\dot{2}1}) \\ s &= \frac{1}{2i}(X_{\dot{1}2} - X_{\dot{2}1}) \end{aligned} \tag{8.5}$$

There is obviously no *a priori* reason to choose any special normalizer or any one of the possible *identifications* above, other than as dictated by convenience, and this arbitrariness finds its corollary in the literature in the minor differences of formalism of various writers.[2]

(c) *The g-Spin Tensors*

We shall use a formalism in which the identification chosen is:

$$p = x^4,\ q = x^3,\ r = x^1,\ s = x^2 \tag{8.6}$$

with which (8.3) takes the form

$$\| X_{\dot{A}B} \| = \begin{Vmatrix} x^4 + x^3 & x^1 + ix^2 \\ x^1 - ix^2 & x^4 - x^3 \end{Vmatrix} \tag{8.7a}$$

This may be written in the form

$$X_{\dot{A}B} = g_{\mu\dot{A}B}x^{\mu}$$

[2] e.g. Fierz (39); Laporte (31); Ruse (37); Taub (39*a*, 39*b*, 40); Veblen (33, 35); v. der Waerden (29, 32)—to mention but a few; all of whom use different identifications, and normalization in the sense of a factor to (8.3). Some of these differences also arise from different choices of the signature of the metric in R_4.

where the spin-tensors $g_{\mu \dot{A} B}$ are given by the matrix equations

$$\| g_{1\dot{A}B} \| = \begin{Vmatrix} 0 & 1 \\ 1 & 0 \end{Vmatrix} \tag{8.8}$$
$$\| g_{2\dot{A}B} \| = \begin{Vmatrix} 0 & i \\ -i & 0 \end{Vmatrix}$$
$$\| g_{3\dot{A}B} \| = \begin{Vmatrix} 1 & 0 \\ 0 & -1 \end{Vmatrix}$$
$$\| g_{4\dot{A}B} \| = \begin{Vmatrix} 1 & 0 \\ 0 & 1 \end{Vmatrix}$$

The inverse equations (8.5) may be written as

$$x^\mu = \tfrac{1}{2} g^{\mu \dot{A} B} X_{\dot{A}B} \tag{8.9}$$

with

$$\| g^{1\dot{A}B} \| = \begin{Vmatrix} 0 & 1 \\ 1 & 0 \end{Vmatrix} \tag{8.10}$$
$$\| g^{2\dot{A}B} \| = \begin{Vmatrix} 0 & -i \\ i & 0 \end{Vmatrix}$$
$$\| g^{3\dot{A}B} \| = \begin{Vmatrix} 1 & 0 \\ 0 & -1 \end{Vmatrix}$$
$$\| g^{4\dot{A}B} \| = \begin{Vmatrix} 1 & 0 \\ 0 & 1 \end{Vmatrix}$$

which obtain from (8.8) by raising the tensor and spinor indices.[3] These g's are readily seen to satisfy the conditions

$$g^{\mu \dot{A} B} g_{\nu \dot{A} B} = 2\delta^\mu{}_\nu \tag{8.11a}$$
$$g^{\mu \dot{A} B} g_{\mu \dot{C} D} = 2\delta^{\dot{A}}{}_{\dot{C}}\, \delta^B{}_D \tag{8.11b}$$

Multiplying the latter by $\epsilon^{\dot{C}\dot{F}} \epsilon^{DG}$ and summing, we obtain the further useful relations

$$g^{\mu \dot{A} B} g_\mu{}^{\dot{F}G} = 2\epsilon^{\dot{A}\dot{F}} \epsilon^{BG} \tag{8.11c}$$

and from either (8.8) or (8.10) we obtain the additional forms $g^{\mu A}{}_B$, viz.

$$\| g^{1A}{}_B \| = \begin{Vmatrix} 1 & 0 \\ 0 & -1 \end{Vmatrix} \tag{8.12}$$
$$\| g^{2A}{}_B \| = \begin{Vmatrix} -i & 0 \\ 0 & -i \end{Vmatrix}$$
$$\| g^{3A}{}_B \| = \begin{Vmatrix} 0 & -1 \\ -1 & 0 \end{Vmatrix}$$
$$\| g^{4A}{}_B \| = \begin{Vmatrix} 0 & -1 \\ 1 & 0 \end{Vmatrix}$$

[3] By means of the metric tensor and the (skew) metric spinors, respectively.

From (8.8) and (8.10) it is readily verified that

$$\bar{g}_{\mu \dot{A} B} = g_{\mu \dot{B} A} \qquad \text{and} \qquad \bar{g}^{\mu \dot{A} B} = g^{\mu \dot{B} A} \tag{8.13}$$

that is, each of the matrices $\| g_{\mu \dot{A} B} \|$ and $\| g^{\mu \dot{A} B} \|$ is Hermitian.[4]

Another important relation satisfied by the g-spinors is

$$g^{\mu \dot{A} B} g_{\nu \dot{A} C} + g_{\nu}{}^{\dot{A} B} g^{\mu}{}_{\dot{A} C} = 2 \delta^{\mu}{}_{\nu} \delta^{B}{}_{C} \tag{8.14}$$

from which (8.11a) obtains by contraction on the indices B, C. This relation may be easily expressed in the equivalent forms

$$g_{\mu \dot{A} B} g_{\nu}{}^{\dot{A} C} + g_{\nu \dot{A} B} g_{\mu}{}^{\dot{A} C} = 2 g_{\mu\nu} \delta^{C}{}_{B} \tag{8.15a}$$

or

$$\bar{g}_{\mu \dot{B} A} g_{\nu}{}^{\dot{A} C} + \bar{g}_{\nu \dot{B} A} g_{\mu}{}^{\dot{A} C} = 2 g_{\mu\nu} \delta^{C}{}_{B} \tag{8.15b}$$

by (8.13). Similarly, from (8.14) and (8.13) we obtain

$$\bar{g}^{\mu \dot{B}}{}_{A} g^{\nu \dot{A}}{}_{C} + \bar{g}^{\nu \dot{B}}{}_{A} g^{\mu \dot{A}}{}_{C} = -2 g^{\mu\nu} \delta^{B}{}_{C} \tag{8.15c}$$

which may be written in the matrix form

$$\bar{g}^{\mu} g^{\nu} + \bar{g}^{\nu} g^{\mu} = -2 g^{\mu\nu} \qquad (\text{cf. } 8.12) \tag{8.15d}$$
[5]

with $g^{\mu} = \| g^{\mu \dot{A}}{}_{B} \|$.

These forms are of great importance for the linearization of quadratic forms, in particular in connection with the Dirac equations of the electron-positron field (cf. Chapter IV).

§ 9. Symbolic Development of the g-Spin Tensors

The properties of the g-systems have been derived in large measure by prescribing explicitly the numbers $g_{\mu \dot{A} B}$, etc., that is, in what we shall term the "special" spin frame or coordinate system, and it is of interest and importance to note that all their properties can be obtained without the use of any explicit representation. This may be done by defining the g's by the two conditions:

$$\bar{g}^{\mu}{}_{\dot{A} B} = g^{\mu}{}_{\dot{B} A} \tag{9.1a}$$

$$g_{\mu}{}^{\dot{B} A} g_{\nu \dot{B} C} = g_{\mu\nu} \delta^{A}{}_{C} + \tfrac{1}{2} g^{-1/2} \epsilon_{\mu\nu\alpha\beta} g^{\alpha \dot{B} A} g^{\beta}{}_{\dot{B} C} \tag{9.1b}$$
[1]

from which all of our previous results follow, as well as many additional ones of importance. For example, from (9.1b) and its conjugate

$$g_{\mu}{}^{\dot{A} B} g_{\nu \dot{C} B} = g_{\mu\nu} \delta^{\dot{A}}{}_{\dot{C}} - \tfrac{1}{2} g^{-1/2} \epsilon_{\mu\nu\alpha\beta} g^{\alpha \dot{A} B} g^{\beta}{}_{\dot{C} B} \tag{9.1c}$$

[4] The same is true, of course, for each of the matrices obtained from these by raising and lowering the tensor index, respectively. Note however that $\bar{g}^{\mu \dot{A}}{}_{B} = g^{\mu}{}_{\dot{B}}{}^{A}$.

[5] Omitting the unit matrix on the right-hand side.

[1] cf. footnote (4) on the spin-tensors $S_{\mu\nu}{}^{A}{}_{C}$ of § 13b.

we immediately obtain

$$g_{\mu}{}^{\dot B A} g_{\nu \dot B C} + g_{\nu}{}^{\dot B A} g_{\mu \dot B C} = 2 g_{\mu\nu} \delta^{A}{}_{C} \qquad \text{(cf. 8.14)} \tag{9.1d}$$

$$g_{\mu}{}^{\dot B A} g_{\nu \dot B C} - g_{\nu}{}^{\dot B A} g_{\mu \dot B C} = g^{-1/2} \epsilon_{\mu\nu\alpha\beta} g^{\alpha \dot B A} g^{\beta}{}_{\dot B C} \tag{9.1e}$$

and, of course, their respective conjugates and variants obtained by index manipulation. Similarly, we note from (9.1b) that

$$\begin{aligned} \epsilon_{\mu\nu\lambda\kappa} g^{\mu}{}_{\dot B A} g^{\nu \dot B C} g^{\lambda}{}_{\dot D C} &= \tfrac{1}{2} g^{-1/2} \epsilon_{\mu\nu\lambda\kappa} \epsilon^{\mu\nu\alpha\beta} g_{\alpha \dot B A} g_{\beta}{}^{\dot B C} g^{\lambda}{}_{\dot D C} \\ &= g^{-1/2} \delta^{\alpha\beta}_{\lambda\kappa} g_{\alpha \dot B A} g_{\beta}{}^{\dot B C} g^{\lambda}{}_{\dot D C} \\ &= 2 g^{-1/2} (g_{\lambda\kappa} \delta^{C}{}_{A} - g_{\kappa \dot B A} g_{\lambda}{}^{\dot B C}) g^{\lambda}{}_{\dot D C} \\ &= -6 g^{-1/2} g_{\kappa \dot D A} \end{aligned} \tag{9.1f}$$

which relation is needed to establish (8.11b) independently; or, from which we may infer directly that the last two steps above are satisfied by (8.11b). With (9.1d) the above also leads directly to the important identity

$$g^{\mu}{}_{\dot B A} g^{\nu \dot B C} g^{\lambda}{}_{\dot D C} = g^{\mu\nu} g^{\lambda}{}_{\dot D A} - g^{\mu\lambda} g^{\nu}{}_{\dot D A} + g^{\nu\lambda} g^{\mu}{}_{\dot D A} - g^{-1/2} \epsilon^{\mu\nu\lambda\kappa} g_{\kappa \dot D A} \tag{9.1g}$$

Before proceeding to the matter of interpretation we may note the following useful results: From the introductory discussion it follows that the light cone is the locus of points of R_4 which correspond to singular Hermitian matrices. The rows of such a matrix are proportional, $\| X_{\dot A B} \| = \| \chi_{\dot A} \phi_B \|$, so that a vector in R_4 determined by

$$x^{\mu} = \tfrac{1}{2} g^{\mu \dot A B} \chi_{\dot A} \phi_B \tag{9.2a}$$

is a null vector, i.e. using (8.11c) and (6.12b)

$$\begin{aligned} x_{\mu} x^{\mu} &= \tfrac{1}{4} g_{\mu\nu} g^{\mu \dot A B} g^{\nu \dot C D} \chi_{\dot A} \phi_B \chi_{\dot C} \phi_D \\ &= \tfrac{1}{2} (\chi_{\dot A} \chi^{\dot A}) (\phi_B \phi^B) = 0 \end{aligned} \tag{9.2b}$$

Similarly, a vector determined under (8.9) by a positive definite Hermitian matrix is time-like, e.g. for

$$x^{\mu} = \tfrac{1}{2} g^{\mu \dot A B} (\psi_{\dot A} \psi_B + \phi_{\dot A} \phi_B) \tag{9.3a}$$

$$x_{\mu} x^{\mu} = \phi_A \psi^A \phi_{\dot B} \psi^{\dot B} \tag{9.3b}$$

and this is always positive if ϕ_A is not proportional to ψ_A.

§ 10. The Isomorphism between Lorentz Transformations in R_4 and the Group of Linear and Antilinear Geometric Transformations in P_1

(a) *Introduction*

We have seen in § 8 that the determinant of $X_{\dot A B}$ is invariant under the binary unimodular transformation

$$X_{\dot A B}' = \bar{s}^{\dot C}{}_{\dot A} s^{D}{}_{B} X_{\dot C D} \qquad \text{(cf. 8.2)} \tag{10.1}$$

or in matrix form $X' = s^{*} X s$, where $s = \| s^{A}{}_{B} \|$. Equivalently, we may say that under this transformation

$$x_{\mu} x^{\mu} = \tfrac{1}{2} X_{\dot A B} X^{\dot A B} \tag{10.2a}$$

goes into

$$x'_\mu x^{\mu\prime} = \tfrac{1}{2} X_{\dot{A}B}{}' X^{\dot{A}B\prime} \tag{10.2b}$$
$$= \tfrac{1}{2} \bar{s}^C{}_A s^D{}_B \bar{S}^A{}_E S^B{}_F X_{\dot{C}D} X^{\dot{E}F}$$
$$= \tfrac{1}{2} g_{\sigma\dot{C}D} g^{\tau\dot{E}F} \bar{s}^C{}_A s^D{}_B \bar{S}^A{}_E S^B{}_F x^\sigma x_\tau$$
$$= L^\mu{}_\sigma l_\mu{}^\tau x^\sigma x_\tau = x^\sigma x_\sigma$$

where we have introduced

$$L^\mu{}_\sigma l_\mu{}^\tau = \tfrac{1}{2} g_{\sigma\dot{C}D} g^{\tau\dot{E}F} \bar{s}^C{}_A s^D{}_B \bar{S}^A{}_E S^B{}_F = \delta_\sigma{}^\tau \tag{10.3}$$

This relation is satisfied by

$$L^\mu{}_\nu = \tfrac{1}{2} g^{\mu\dot{A}B} \bar{s}^C{}_A s^D{}_B g_{\nu\dot{C}D} \tag{10.4a}$$

or

$$\begin{Vmatrix} L^1{}_1 & L^1{}_2 & L^1{}_3 & L^1{}_4 \\ L^2{}_1 & L^2{}_2 & L^2{}_3 & L^2{}_4 \\ L^3{}_1 & L^3{}_2 & L^3{}_3 & L^3{}_4 \\ L^4{}_1 & L^4{}_2 & L^4{}_3 & L^4{}_4 \end{Vmatrix} = \tfrac{1}{2} \begin{Vmatrix} s^1{}_2 & s^2{}_2 & s^1{}_1 & s^2{}_1 \\ -is^1{}_2 & -is^2{}_2 & is^1{}_1 & is^2{}_1 \\ s^1{}_1 & s^2{}_1 & -s^1{}_2 & -s^2{}_2 \\ s^1{}_1 & s^2{}_1 & s^1{}_2 & s^2{}_2 \end{Vmatrix} \cdot \begin{Vmatrix} \bar{s}^2{}_1 & -i\bar{s}^2{}_1 & \bar{s}^1{}_1 & \bar{s}^1{}_1 \\ \bar{s}^1{}_1 & i\bar{s}^1{}_1 & -\bar{s}^2{}_1 & \bar{s}^2{}_1 \\ \bar{s}^2{}_2 & -i\bar{s}^2{}_2 & \bar{s}^1{}_2 & \bar{s}^1{}_2 \\ \bar{s}^1{}_2 & i\bar{s}^1{}_2 & -\bar{s}^2{}_2 & \bar{s}^2{}_2 \end{Vmatrix} \tag{10.4b}$$

as is readily verified, and as indicated by the notation $L^\mu{}_\nu$ this is a Lorentz transformation. Moreover, it is an element of the proper Lorentz group—which we denote by $\mathscr{L}_4$—since its determinant is unity and $L^4{}_4 = \frac{1}{2}(\bar{s}^1{}_1 s^1{}_1 + \bar{s}^2{}_1 s^2{}_1 + \bar{s}^1{}_2 s^1{}_2 + \bar{s}^2{}_2 s^2{}_2) \geqslant 1$.

On the other hand the matrix $s = \| s^A{}_B \|$ is an element of the binary unimodular (or spinor) group of linear transformations—which we denote by $\mathfrak{S}_4$—and we have the result that the *geometric*[1] spin transformation (10.1), of the type $\phi_A \to \phi_A{}' = s^B{}_A \phi_B$, induces the proper Lorentz transformation $x^\mu \to x^{\mu\prime}$ with matrix given by (10.4).

Thus, to every element a of $\mathfrak{S}_4$ there corresponds an element a', say, of $\mathscr{L}_4$ and it is readily verified that the correspondence $a \to a'$ has the properties: (1) $e \to e'$, where e denotes the identity element; (2) $a^{-1} \to (a')^{-1}$; (3) $ab \to a'b'$. However, the isomorphism is (2 — 1) from $\mathfrak{S}_4$ on $\mathscr{L}_4$ since every proper Lorentz transformation may be represented in the form (10.4*a*) *but* the two matrices s and $-s$ generate the same Lorentz transformation.[2]

(*b*) *Examples of Associated* (*Proper*) *Lorentz and* (*Linear*) *Binary Unimodular Transformation Matrices*

According to (10.4*a*) or (10.4*b*) we find that the Lorentz rotation

$$x^{3\prime} = x^3 \cosh\theta + x^4 \sinh\theta = \frac{x^3}{(1-\beta^2)^{1/2}} - \frac{x^4\beta}{(1-\beta^2)^{1/2}} \tag{10.5a}$$

$$x^{4\prime} = x^3 \sinh\theta + x^4 \cosh\theta = \frac{-x^3\beta}{(1-\beta^2)^{1/2}} + \frac{x^4}{(1-\beta^2)^{1/2}} \qquad \beta = v/c$$

corresponds to the two matrices

$$\| s^A{}_B \| = \pm \begin{Vmatrix} e^{+\theta/2} & 0 \\ 0 & e^{-\theta/2} \end{Vmatrix} = \pm \begin{Vmatrix} 1/\tau & 0 \\ 0 & \tau \end{Vmatrix} \qquad \tau = +\left(\frac{1+\beta}{1-\beta}\right)^{1/4} \tag{10.5b}$$

[1] The reason for this designation here, in contradistinction to coordinate transformation, will be indicated in § 12.

[2] In somewhat technical group-theoretical language, $\mathscr{L}_4$ is isomorphic to the factor group $\mathfrak{S}_4/\mathscr{N}$ where $\mathscr{N}$ is the invariant subgroup of $\mathfrak{S}_4$ consisting of the two elements e and $(-e)$. Cf. v. der Waerden (32); Murnaghan (38); Bargmann (47).

Similarly the pure space rotations

$$x^{1\prime} = x^1 \cos\theta + x^2 \sin\theta \qquad (10.6a)$$
$$x^{2\prime} = -x^1 \sin\theta + x^2 \cos\theta$$

$$x^{2\prime} = x^2 \cos\phi + x^3 \sin\phi \qquad (10.7a)$$
$$x^{3\prime} = -x^2 \sin\phi + x^3 \cos\phi$$

correspond respectively to the two matrices

$$\| s^A{}_B \| = \pm \begin{Vmatrix} e^{-i\theta/2} & 0 \\ 0 & e^{i\theta/2} \end{Vmatrix} \qquad (10.6b)$$

$$\| s^A{}_B \| = \pm \begin{Vmatrix} \cos(\phi/2) & i\sin(\phi/2) \\ i\sin(\phi/2) & \cos(\phi/2) \end{Vmatrix} \qquad (10.7b)$$

and the Lorentz identity corresponds to the two matrices

$$\| s^A{}_B \| = \pm \begin{Vmatrix} 1 & 0 \\ 0 & 1 \end{Vmatrix} \qquad (10.8)$$

We note that the elements of the subgroup $\mathscr{R}$ of $\mathscr{L}_4$, comprising the spatial rotations, correspond to unitary unimodular transformations which define the special linear group $\mathfrak{S}_{\mathscr{R}}$. This follows directly from the observation that $x^4 = \frac{1}{2} g^{4\dot{A}B} X_{\dot{A}B}$ $\left(\text{or equivalently } \| g^{4\dot{A}B} \| = \begin{Vmatrix} 1 & 0 \\ 0 & 1 \end{Vmatrix}\right)$ is invariant under this added restriction, viz.

$$\begin{Vmatrix} 1 & 0 \\ 0 & 1 \end{Vmatrix} = s^* \begin{Vmatrix} 1 & 0 \\ 0 & 1 \end{Vmatrix} s \qquad (10.9)$$

which forces the unitarity of s. Here again the isomorphism is $(2-1)$ from $\mathfrak{S}_{\mathscr{R}}$ on $\mathscr{R}$.[3] Finally, we draw special attention to the fact that under unitary unimodular transformations a contravariant conjugated or dotted spinor, say $\psi^{\dot{A}}$, transforms like ψ_A as is readily verified. This result is of particular importance for our later work.

(c) *Examples of Associated (Improper) Lorentz and (Antilinear) Binary Unimodular Transformation Matrices*

The so-called antilinear *geometric* transformations of the form

$$\phi_A \rightarrow \phi_A{}' = s^{\dot{B}}{}_A \, \phi_{\dot{B}} \qquad (10.10)$$

or

$$\phi_{\dot{A}} \rightarrow \phi_{\dot{A}}{}' = \bar{s}^{B}{}_{\dot{A}} \, \phi_B$$

may similarly be shown to induce an improper Lorentz transformation $x^\mu \rightarrow x^{\mu\prime}$ represented by

$$x^{\mu\prime} = \tfrac{1}{2} g^{\mu\dot{A}B} X'_{\dot{A}B} = L^\mu{}_\nu x^\nu \qquad (10.11a)$$

with

$$L^\mu{}_\nu = \tfrac{1}{2} g^{\mu\dot{A}B} \, \bar{s}^{C}{}_{\dot{A}} \, s^{\dot{D}}{}_B \, g_{\nu\dot{D}C} \qquad (10.11b)$$

[3] $\mathscr{R}$ is isomorphic to $\mathfrak{S}_{\mathscr{R}}/\mathscr{N}$.

i.e.

$$\| L^{\mu}{}_{\nu\ \mathrm{imp}} \| = \tfrac{1}{2} \begin{Vmatrix} s^{\dot 1}{}_2 & s^{\dot 2}{}_2 & s^{\dot 1}{}_1 & s^{\dot 2}{}_1 \\ -is^{\dot 1}{}_2 & -is^{\dot 2}{}_2 & is^{\dot 1}{}_1 & is^{\dot 2}{}_1 \\ s^{\dot 1}{}_1 & s^{\dot 2}{}_1 & -s^{\dot 1}{}_2 & -s^{\dot 2}{}_2 \\ s^{\dot 1}{}_1 & s^{\dot 2}{}_1 & s^{\dot 1}{}_2 & s^{\dot 2}{}_2 \end{Vmatrix} \cdot \begin{Vmatrix} \bar{s}^{\dot 2}{}_1 & i\bar{s}^{\dot 2}{}_1 & \bar{s}^{\dot 1}{}_1 & \bar{s}^{\dot 1}{}_1 \\ \bar{s}^{\dot 1}{}_1 & -i\bar{s}^{\dot 1}{}_1 & -\bar{s}^{\dot 2}{}_1 & s^{\dot 2}{}_1 \\ \bar{s}^{\dot 2}{}_2 & i\bar{s}^{\dot 2}{}_2 & \bar{s}^{\dot 1}{}_2 & \bar{s}^{\dot 1}{}_2 \\ \bar{s}^{\dot 1}{}_2 & -i\bar{s}^{\dot 1}{}_2 & -\bar{s}^{\dot 2}{}_2 & \bar{s}^{\dot 2}{}_2 \end{Vmatrix} \qquad (10.11c)$$

An example of this type of transformation is

$$\phi_A \to \phi_A{}' = \pm\bar{\phi}_A \qquad (10.12a)$$

which corresponds to the improper Lorentz transformation

$$x^{1\prime} = x^1, \qquad x^{2\prime} = -x^2, \qquad x^{3\prime} = x^3, \qquad x^{4\prime} = x^4 \qquad (10.12b)^4$$

i.e.

$$\| s^A{}_B \| = \pm \begin{Vmatrix} 1 & 0 \\ 0 & 1 \end{Vmatrix} \qquad (10.13a)$$

$$\| L^{\mu}{}_{\nu\ \mathrm{imp}} \| = \tfrac{1}{2} \begin{Vmatrix} 0 & \pm 1 & \pm 1 & 0 \\ 0 & \mp i & \pm i & 0 \\ \pm 1 & 0 & 0 & \mp 1 \\ \pm 1 & 0 & 0 & \pm 1 \end{Vmatrix} \cdot \begin{Vmatrix} 0 & 0 & \pm 1 & \pm 1 \\ \pm 1 & \mp i & 0 & 0 \\ \pm 1 & \pm i & 0 & 0 \\ 0 & 0 & \mp 1 & \pm 1 \end{Vmatrix} \qquad (10.13b)$$

Similarly the antilinear transformations specified by

$$\| s^A{}_B \| = \pm \begin{Vmatrix} 0 & -1 \\ 1 & 0 \end{Vmatrix} \qquad (10.14a)$$

correspond to reflection through the origin, viz. from (10.11c)

$$\begin{Vmatrix} -1 & 0 & 0 & 0 \\ 0 & -1 & 0 & 0 \\ 0 & 0 & -1 & 0 \\ 0 & 0 & 0 & 1 \end{Vmatrix} = \tfrac{1}{2} \begin{Vmatrix} \mp 1 & 0 & 0 & \pm 1 \\ \pm i & 0 & 0 & \pm i \\ 0 & \pm 1 & \pm 1 & 0 \\ 0 & \pm 1 & \mp 1 & 0 \end{Vmatrix} \cdot \begin{Vmatrix} \pm 1 & \pm i & 0 & 0 \\ 0 & 0 & \mp 1 & \pm 1 \\ 0 & 0 & \mp 1 & \mp 1 \\ \mp 1 & \pm i & 0 & 0 \end{Vmatrix} \qquad (10.14b)$$

as may be computed directly from

$$\begin{aligned} x^{\mu\prime} &= \tfrac{1}{2} g^{\mu\dot{A}B} X'_{\dot{A}B} \\ X'_{\dot{A}B} &= \bar{s}^{\dot C}{}_{\dot A}\, s^{D}{}_{B}\, X_{\dot{D}C} \end{aligned} \qquad (10.14c)$$

and our choice for the representation of the g's.

(d) The Isomorphism Theorems

The isomorphism theorems may be stated as:

I. For every proper Lorentz transformation $L^{\mu}{}_{\nu}$ in R_4 there are exactly two unimodular matrices $s = \| s^A{}_B \|$ and $-s$ which satisfy

$$L^{\mu}{}_{\nu} = \tfrac{1}{2} g^{\mu\dot{A}B}\, \bar{s}^{\dot C}{}_{\dot A}\, s^{D}{}_{B}\, g_{\nu\dot{C}D} \qquad (10.15a)^5$$

[4] Every improper Lorentz transformation can be written as a product of (10.12b) and a proper one.

[5] Relating transformations of contravariant variables in R_4 and covariant variables in P_1.

II. For every improper Lorentz transformation $L^{\mu}{}_{\nu\,\mathrm{imp}}$ there are exactly two unimodular matrices $s = \| s^{A}{}_{B} \|$ and $-s$ which satisfy

$$L^{\mu}{}_{\nu\,\mathrm{imp}} = \tfrac{1}{2} g^{\mu \dot{A} B} \bar{s}^{\dot{C}}{}_{A}\, s^{\dot{D}}{}_{B}\, g_{\nu \dot{D} C} \tag{10.15b}$$

An equivalent statement refers to the numerical invariance of the g-spin tensors under the simultaneous applications of a linear (antilinear) geometric spin transformation and the proper (improper) Lorentz transformation inverse to it [cf. equations 7.3–7.4 and §§ 11, 12], viz.

I′. For every proper Lorentz transformation $L^{\mu}{}_{\nu}$ there are exactly two unimodular matrices $S = \| S^{A}{}_{B} \|$ and $-S$ which satisfy

$$g^{\mu} = \bar{S} g^{\nu} S^{-1} L^{\mu}{}_{\nu} \tag{10.15a′}$$

where $g^{\mu} = \| g^{\mu \dot{A}}{}_{B} \|$.

II′. For every improper Lorentz transformation $L^{\mu}{}_{\nu}$ there are exactly two unimodular matrices $S = \| S^{A}{}_{B} \|$ and $-S$ which satisfy

$$g^{\mu} = S \bar{g}^{\nu} \bar{S}^{-1} L^{\mu}{}_{\nu} = S \tilde{g}^{\nu} \bar{S}^{-1} L^{\mu}{}_{\nu} \tag{10.15b′}$$ [6]

In the future we shall sometimes refer to a spin transformation associated to a given Lorentz transformation as its spin image.

§ 11. Infinitesimal 2-Spinor Transformations

We wish to consider now an alternative elementary proof of Theorem I′ constructed in terms of infinitesimal spinor transformations. This will serve primarily to illustrate the general technique of working with infinitesimal transformations, which is a very simple procedure, and at the same time provide several results which are basic to the later analysis of various types of spinor wave fields.

The essential problem here is to show that to every proper Lorentz transformation there corresponds a spin transformation such that the application of the two together leaves the spin-tensors $g^{\mu}{}_{\dot{A}B}$ numerically invariant. Clearly it suffices to consider the infinitesimal transformations, viz.

[6] Here we relate transformations on contravariant variables in both R_4 and P_1. Then from (7.4)

$$g^{\mu \dot{A} B''} = g^{\nu \dot{C} D}\, \bar{S}^{A}{}_{C}\, S^{B}{}_{D}\, L^{\mu}{}_{\nu} = g^{\mu \dot{A} B} \tag{1}$$

where we substitute $L^{\mu}{}_{\nu}$ in the form given by (10.15a). This is just the statement of (10.15a') above since from (1) we get

$$g^{\mu \dot{A}}{}_{B} = g^{\nu \dot{C} D}\, \bar{S}^{A}{}_{C}\, S_{BD}\, L^{\mu}{}_{\nu} \tag{2}$$

$$= -g^{\nu \dot{C}}{}_{D}\, \bar{S}^{A}{}_{C}\, S_{B}{}^{D}\, L^{\mu}{}_{\nu} \tag{3}$$

$$= \bar{S}^{A}{}_{C}\, g^{\nu \dot{C}}{}_{D}\, (S^{-1})^{D}{}_{B}\, L^{\mu}{}_{\nu} \tag{4}$$

where we have noted that S being unimodular implies $\epsilon_{AB} S^{A}{}_{C} S^{B}{}_{D} = \epsilon_{CD}$; from which $S_{AB} = -(S^{-1})_{BA}$ and by raising and lowering indices $-S_{A}{}^{B} = (S^{-1})^{B}{}_{A}$, etc. (cf. §§ 4$a$, c and esp. (4.7a) *et seq.*). Alternatively, from (4) we get

$$(\bar{S}^{-1})^{E}{}_{A}\, S^{B}{}_{F}\, g^{\mu \dot{A}}{}_{B} = g^{\nu \dot{E}}{}_{F} L^{\mu}{}_{\nu} \tag{5}$$

i.e.

$$\bar{S}^{-1} g^{\mu} S = g^{\nu} L^{\mu}{}_{\nu}$$

or

$$-(\bar{S}^{-1})^{E}{}_{A}\, S_{B}{}^{F}\, g^{\mu \dot{A} B} = g^{\nu \dot{E} F} L^{\mu}{}_{\nu} \tag{6}$$

$$(\bar{S}^{-1})^{E}{}_{A}\, (S^{-1})^{F}{}_{B}\, g^{\mu \dot{A} B} = g^{\nu \dot{E} F} L^{\mu}{}_{\nu} \tag{7}$$

$$g^{\mu \dot{A} B}\, \bar{s}^{E}{}_{A}\, s^{F}{}_{B}\, g_{\rho \dot{E} F} = 2 L^{\mu}{}_{\rho} \tag{8}$$

Equation (5) also yields the useful relation

$$\bar{S}^{A}{}_{E}\, g^{\mu}{}_{\dot{A}B}\, S^{B}{}_{F} = g^{\nu}{}_{\dot{E}F} L^{\mu}{}_{\nu} \tag{9}$$

Lorentz:

$$x_\mu' = x_\mu + \omega_\mu{}^\nu x_\nu \tag{11.1a}$$

characterized by the infinitesimal skew tensor $\omega_{\mu\nu}$.

Spinor:

$$\begin{aligned} \phi_A' &= \phi_A + \eta^B{}_A \phi_B \\ \phi_{\dot A}' &= \phi_{\dot A} + \eta^{\dot B}{}_{\dot A} \phi_{\dot B} \end{aligned} \tag{11.1b}$$

characterized by the infinitesimal *symmetric*[1] spinor η_{AB}. By the above statement of the problem it suffices to show that for every $\omega^{\mu\nu}$ there exist η_{AB} such that

$$\omega^{\mu\nu} g_{\nu\dot C A} + \eta^B{}_A g^\mu{}_{\dot C B} + \eta^{\dot D}{}_{\dot C} g^\mu{}_{\dot D A} = 0 \tag{11.2}$$

The solution is readily seen to be

$$\eta_{AB} = \tfrac{1}{4}\omega^{\mu\nu} g_{\mu\dot C A} g_\nu{}^{\dot C}{}_B \tag{11.3}$$

and conjugate, as may be verified with the use of (8.11b) and (8.14).[2]

The transformation matrices of (11.1b) are therefore

$$\begin{aligned} &\delta^B{}_A + \tfrac{1}{4}\omega^{\mu\nu} g_{\mu\dot C A} g_\nu{}^{\dot C B} \\ &\delta^{\dot B}{}_{\dot A} + \tfrac{1}{4}\omega^{\mu\nu} g_{\mu\dot A C} g_\nu{}^{\dot B C} \end{aligned} \tag{11.4}$$

and it remains only to prove that considered as a representation of the (proper) Lorentz group the integrability conditions (4.12) are satisfied. Now, as previously noted, the transformation (11.1a) is represented by $1 + \frac{1}{2}\omega^{\mu\nu}\mathfrak{T}_{\mu\nu}$ in any representation of the Lorentz group; hence, in the present case we find on comparison with (11.4)

$$\| 1 + \tfrac{1}{2}\omega^{\mu\nu}\mathfrak{T}_{\mu\nu} \|_A^B = \delta_A^B + \tfrac{1}{4}\omega^{\mu\nu} g_{\mu\dot C A} g_\nu{}^{CB} \tag{11.5}$$

or

$$\| \mathfrak{T}_{\mu\nu} \|_A^B = \tfrac{1}{4}(g_{\mu\dot C A} g_\nu{}^{\dot C B} - g_{\nu\dot C A} g_\mu{}^{\dot C B})$$

where the indices A, B are here considered as matrix indices. From (11.5) and (9.1e and complex conjugate)

$$\| \mathfrak{T}_{\mu\nu} \|_A^B = \tfrac{1}{4} g^{-1/2} \epsilon_{\mu\nu\alpha\beta} g^\alpha{}_{\dot C A} g^{\beta\dot C B} = \tfrac{1}{2} g^{-1/2} \epsilon_{\mu\nu\alpha\beta} \| \mathfrak{T}^{\alpha\beta} \|_A^B \tag{11.6}$$

from which (in the preferred coordinate systems)

$$\begin{aligned} \| [\mathfrak{T}_{\kappa\lambda}, \mathfrak{T}_{\mu\nu}] \|_A^E &= -\tfrac{1}{16}\{\epsilon_{\kappa\lambda\alpha\beta}\epsilon_{\mu\nu\rho\sigma} - \epsilon_{\mu\nu\alpha\beta}\epsilon_{\kappa\lambda\rho\sigma}\} g^\alpha{}_{\dot C A} g^{\beta\dot C B} g^\rho{}_{\dot D B} g^{\sigma\dot D E} \\ &= -\tfrac{1}{16}\epsilon_{\kappa\lambda\alpha\beta}\epsilon_{\mu\nu\rho\sigma}\{g^\alpha{}_{\dot C A} g^{\beta\dot C B} g^\rho{}_{\dot D B} g^{\sigma\dot D E} - g^\sigma{}_{\dot C A} g^{\rho\dot C B} g^\beta{}_{\dot D B} g^{\alpha\dot D E}\} \\ &= -\tfrac{1}{4}\epsilon_{\kappa\lambda\alpha\beta}\epsilon_{\mu\nu\rho\sigma}\{2g^{\beta\rho} \| \mathfrak{T}^{\alpha\sigma} \|_A^E - i\epsilon^{\alpha\beta\rho\tau} \| \mathfrak{T}^\sigma{}_\tau \|_A^E\} \\ &= \| \tfrac{1}{2}\epsilon^\rho{}_{\kappa\lambda\alpha}\epsilon_{\mu\nu\rho\sigma}\mathfrak{T}^{\alpha\sigma} + \tfrac{1}{4} i\epsilon_{\kappa\lambda\beta\alpha}\epsilon_{\mu\nu\rho\sigma}\epsilon^{\beta\alpha\rho\tau}\mathfrak{T}^\sigma{}_\tau \|_A^E \text{ [3]} \end{aligned} \tag{11.7}$$

[1] The symmetry of η_{AB} is prescribed by the condition that the metric spinor be invariant, viz. $\epsilon_{AB} = (\delta^C{}_A + \eta^C{}_A)(\delta^D{}_B + \eta^D{}_B)\epsilon_{CD}$.

[2] $$\begin{aligned} &\omega^{\mu\nu} g_{\nu\dot C A} + \eta^B{}_A g^\mu{}_{\dot C B} + \eta^{\dot D}{}_{\dot C} g^\mu{}_{\dot D A} \\ &= \omega^{\mu\nu}\{g_{\nu\dot C A} + \tfrac{1}{4} g_{\mu\dot E A} g_\nu{}^{\dot E B} g^\mu{}_{\dot C B} + \tfrac{1}{4} g_{\mu\dot C G} g_\nu{}^{\dot D G} g^\mu{}_{\dot D A}\} \\ &= \omega^{\mu\nu}\{g_{\nu\dot C A} + \tfrac{1}{4} g_{\mu\dot E A}(2\delta^\mu{}_\nu \delta^{\dot E}{}_{\dot C} - g^{\mu\dot E\dot B} g_{\nu\dot C B}) + \tfrac{1}{4} g_{\mu\dot C G}(2\delta^\mu{}_\nu \delta^G{}_A - g^{\mu\dot D G} g_{\nu\dot D A})\} = 0 \end{aligned}$$

[3] The transition from the previous step involves only the application of (9.1g) and its conjugate to the first and second terms respectively, and further rearrangement of terms with the use of $\mathfrak{T}_{\mu\nu}$ from (11.5).

With the help of (11.6) and the properties of the ϵ-systems this reduces to the integrability conditions

$$[\mathfrak{T}_{\kappa\lambda}, \mathfrak{T}_{\mu\nu}] = g_{\lambda\mu}\mathfrak{T}_{\kappa\nu} - g_{\kappa\mu}\mathfrak{T}_{\lambda\nu} + g_{\kappa\nu}\mathfrak{T}_{\lambda\mu} - g_{\lambda\nu}\mathfrak{T}_{\kappa\mu} \tag{11.8}$$

For future reference, we note that corresponding to the infinitesimal Lorentz transformation $\omega^{\mu\nu}$ a first-rank spinor undergoes the transformation

$$\begin{aligned}\delta\phi_A &= \eta_A{}^B\,\phi_B \qquad (11.9a)\\ &= -\tfrac{1}{4}\omega^{\mu\nu}\{\tfrac{1}{2}(g_\mu{}^{\dot{C}}{}_A\, g_{\nu\dot{C}}{}^B - g_\nu{}^{\dot{C}}{}_A\, g_{\mu\dot{C}}{}^B)\}\phi_B\\ &= -\tfrac{1}{4}\omega^{\mu\nu}\, S_{\mu\nu A}{}^B\,\phi_B\end{aligned}$$

which equation also serves to define the spin-tensor $S_{\mu\nu A}{}^B$ whose properties will be considered further in § 13*b*. If we express ϕ_A in usual column matrix form, this result may be written as

$$\delta\phi = \tfrac{1}{2}\omega^{\mu\nu}\mathfrak{T}_{\mu\nu\ \mathrm{op}}\,\phi \tag{11.9b}$$

with $\mathfrak{T}_{\mu\nu\ \mathrm{op}} = -\tfrac{1}{2}S_{\mu\nu A}{}^B$ (cf. 11.5).

§ 12. Alternate Modes of Interpretation of Spin Transformations

We wish to consider now the two alternative viewpoints on spin transformations and the fundamental g-spinors. These different interpretations have their logical origin in the spinor theory as developed through projective geometry; this cannot be considered in this text, but can at least be made fairly plausible by intuitive argument.

Consider the fundamental relations

$$x^\mu = \tfrac{1}{2}g^{\mu\dot{A}B}X_{\dot{A}B} \qquad X_{\dot{A}B} = g_{\mu\dot{A}B}x^\mu \tag{12.1}$$

and the spin transformations

$$\begin{aligned}\phi^A &\to \phi^{A\prime} = S^A{}_B\,\phi^B \qquad (12.2)\\ \phi_A &\to \phi_A{}' = s^B{}_A\,\phi_B\end{aligned}$$

According to the *first* interpretation (§ 10) the linear transformation (12.2) is considered as a *geometric* transformation[1] in the spin space and simply establishes another correspondence of the type (12.1) between points of P_1 and R_4, i.e. (12.2) induces the transformation

$$x^{\mu\prime} = \tfrac{1}{2}g^{\mu\dot{A}B}X'_{\dot{A}B} \tag{12.3a}$$

or from (12.1, 12.2)

$$x^{\mu\prime} = L^\mu{}_\nu x^\nu \tag{12.3b}$$

with

$$L^\mu{}_\nu = \tfrac{1}{2}g^{\mu\dot{A}B}\,\bar{s}^{\dot{C}}{}_{\dot{A}}\,s^D{}_B\,g_{\nu\dot{C}D}$$

Thus we obtain the results of the two preceding sections in which this viewpoint was assumed, and in which spin and Lorentz transformations *are not* made independently.

[1] i.e. essentially a permutation of points of the spin space, cf. § 4.

According to the *second* interpretation, the linear transformation (12.2) is regarded as a coordinate transformation or change of spin frame in spin space, which therefore *relabels* the points in spin space but does not alter the association of points in P_1 and R_4, i.e. (12.2) corresponds to

$$x^{\mu} = g^{\mu\dot{A}B'} X_{\dot{A}B}' \tag{12.4a}$$

where

$$g^{\mu\dot{A}B'} = \bar{S}^{\dot{A}}{}_{\dot{C}} S^{B}{}_{D}\, g^{\mu\dot{C}D} \tag{12.4b}$$

and with (12.3a, b) this gives

$$g^{\mu\dot{A}B} = g^{\nu\dot{A}B'} L^{\mu}{}_{\nu} \tag{12.4c}$$

In this *second* interpretation the spin-tensors $g^{\mu\dot{A}B}$, $g_{\mu\dot{A}B}$, etc., transform as the spin tensors of § 7, viz. under constant spin frame transformations they transform as

$$\begin{aligned} g^{\mu\dot{A}B'} &= \bar{S}^{\dot{A}}{}_{\dot{C}} S^{B}{}_{D}\, g^{\mu\dot{C}D} \\ g_{\mu\dot{A}B}{}' &= \bar{s}^{\dot{C}}{}_{\dot{A}}\, s^{D}{}_{B}\, g_{\mu\dot{C}D} \end{aligned} \tag{12.5a}$$

under Lorentz transformations as

$$\begin{aligned} g^{\mu\dot{A}B'} &= L^{\mu}{}_{\nu}\, g^{\nu\dot{A}B} \\ g_{\mu\dot{A}B}{}' &= l^{\nu}{}_{\mu}\, g_{\nu\dot{A}B} \end{aligned} \tag{12.5b}$$

and under combined transformations as

$$\begin{aligned} g^{\mu\dot{A}B''} &= L^{\mu}{}_{\nu} \bar{S}^{\dot{A}}{}_{\dot{C}} S^{B}{}_{D}\, g^{\nu\dot{C}D} \\ g_{\mu\dot{A}B}{}'' &= l^{\nu}{}_{\mu}\, \bar{s}^{\dot{C}}{}_{\dot{A}}\, s^{D}{}_{B}\, g_{\nu\dot{C}D} \end{aligned} \tag{12.5c}$$

Thus, Lorentz and spin transformations are here performed independently; but it should be noted that having subjected the g's to a Lorentz transformation (12.5b) or a spin transformation (12.5a) we can always find a spin or Lorentz transformation respectively which transforms them back into themselves. In other words (cf. Theorems I′, II′, § 10d), given the $L^{\mu}{}_{\nu}$, $l^{\nu}{}_{\mu}$ in (12.5b) we can always find $S^{A}{}_{B}$, $s^{B}{}_{A}$ such that

$$\begin{aligned} g^{\mu\dot{A}B''} &= g^{\mu\dot{A}B} \\ g_{\mu\dot{A}B}{}'' &= g_{\mu\dot{A}B} \end{aligned} \tag{12.6}$$

since

$$g^{\mu\dot{A}B''} = L^{\mu}{}_{\nu}\, g^{\nu\dot{A}B'} = g^{\mu\dot{A}B}$$

according to (12.5b, c) and (12.4c), i.e. if $L^{\mu}{}_{\nu}$ is the Lorentz transformation induced according to the *first* interpretation.

Although it entails a certain loss of generality, still it is sufficient for most purposes (in the special relativistic theory) to look upon the g-spinors as numerically constant,[2] and (denoted usually by σ's) this is the sense in which they appear in most physical literature. For the greater part of our work, except where indicated by the context, they may be thought of in this simpler way.

[2] Lorentz and spin transformations are then *not* performed independently, and a spinor is simply a quantity undergoing the binary unimodular transformation (with a factor of sign) associated with the particular Lorentz transformation to which the coordinates in R_4 are subjected.

§ 13. Special Spinors and their Tensor Equivalents

(a) *Introduction*

From the preceding sections we know that spinors of even rank—with equal numbers of dotted and undotted indices—(e.g. ϕ, $\phi_{A\dot{B}}$, $\phi_{A\dot{B}C\dot{D}}$, . . . etc.) correspond directly to world tensors of half their rank. Among the more useful standard spinors are the equivalents to the gradient, divergence and D'Alembertian which follow directly from the definitions and are simply listed below.[1]

$$\partial_\mu = \tfrac{1}{2}g_{\mu A\dot{B}}\partial^{A\dot{B}} \qquad \partial_{A\dot{B}} = g^\mu{}_{A\dot{B}}\partial_\mu \tag{13.1a}$$

$$\partial_\mu \phi^\mu = \phi^\mu{}_{,\mu} = \tfrac{1}{2}\partial_{A\dot{B}}\phi^{A\dot{B}} \tag{13.1b}$$

$$\partial_\mu \partial^\mu = \Box = \tfrac{1}{2}\partial_{A\dot{B}}\partial^{A\dot{B}} \tag{13.1c}$$

$$\partial_{A\dot{C}}\partial^{A\dot{B}} = \Box\ \delta^{\dot{B}}{}_{\dot{C}}$$

(b) *Symmetric Second-Rank Spinors—The Fundamental S-Spin Tensors*

In addition to the above-mentioned simple correspondence of spinors and tensors, there are special (1—1) correspondences which subsist between spinors of even rank and tensors of (in general) higher rank under certain symmetry restrictions. For most practical purposes the important correspondence of this type is that between second-rank symmetric spinors and self-dual antisymmetric tensors. To establish this correspondence we consider first the properties[2] of the spin tensor $S_{\mu\nu}{}^A{}_C$, which appeared in § 11, defined by the equations

$$\begin{aligned} S_{\mu\nu}{}^A{}_C &= \tfrac{1}{2}(g_\mu{}^{\dot{B}A}g_{\nu\dot{B}C} - g_\nu{}^{\dot{B}A}g_{\mu\dot{B}C}) \\ &= \tfrac{1}{2}(\bar{g}_\nu{}^A{}_{\dot{B}}g_\mu{}^{\dot{B}}{}_C - \bar{g}_\mu{}^A{}_{\dot{B}}g_\nu{}^{\dot{B}}{}_C) \end{aligned} \tag{13.2}$$

This is evidently skew in μ, ν, and from $S_{\mu\nu}{}^A{}_A = 0$ it follows that $S_{\mu\nu AB}$ is also symmetric in A, B. From (13.2) and (8.14) we obtain the useful formula

$$S_{\mu\nu}{}^A{}_C = g_\mu{}^{\dot{B}A}g_{\nu\dot{B}C} - g_{\mu\nu}\,\delta^A{}_C \tag{13.3}$$

and from the above definitions with (8.11*a*, *b*, *c*) we find the useful identities:[3]

$$g^{\mu A\dot{B}}S_{\mu\nu CD} = -\delta^B{}_C g_\nu{}^A{}_D - \delta^B{}_D g_\nu{}^A{}_C \tag{13.4}$$

$$S^{\mu\nu AB}S_{\mu\nu CD} = 4(\delta^A{}_C\,\delta^B{}_D + \delta^A{}_D\,\delta^B{}_C) \tag{13.5}$$

$$\bar{S}^{\mu\nu\dot{A}\dot{B}}S_{\mu\nu CD} = 0 \tag{13.6}$$

$$S^{\lambda\mu}{}_{AB}\bar{S}_{\mu\nu}{}^{\dot{C}\dot{D}} = \tfrac{1}{2}(g^{\lambda\dot{C}}{}_A g_\nu{}^{\dot{D}}{}_B + g^{\lambda\dot{C}}{}_B g_\nu{}^{\dot{D}}{}_A + g^{\lambda\dot{D}}{}_A g_\nu{}^{\dot{C}}{}_B + g^{\lambda\dot{D}}{}_B g_\nu{}^{\dot{C}}{}_A) \tag{13.7}$$

[1] Various other forms are, of course, obtainable by raising and lowering tensor and spinor indices and need not be listed separately.

[2] cf. Harish-Chandra (46); Ruse (37); Taub (39).

[3] The derivations involve, as usual, considerable manipulation—but are essentially straightforward. They may be verified for our particular spin frame, and hence as spinor equations, in general.

Combining (8.8, 8.10) and (13.2) we find for the special coordinate systems:

$$\| S_{12}{}^{A}{}_{B} \| = \begin{Vmatrix} i & 0 \\ 0 & -i \end{Vmatrix} \qquad \| S_{14}{}^{A}{}_{B} \| = \begin{Vmatrix} 0 & -1 \\ -1 & 0 \end{Vmatrix} \tag{13.8}$$

$$\| S_{13}{}^{A}{}_{B} \| = \begin{Vmatrix} 0 & 1 \\ -1 & 0 \end{Vmatrix} \qquad \| S_{24}{}^{A}{}_{B} \| = \begin{Vmatrix} 0 & -i \\ i & 0 \end{Vmatrix}$$

$$\| S_{23}{}^{A}{}_{B} \| = \begin{Vmatrix} 0 & i \\ i & 0 \end{Vmatrix} \qquad \| S_{34}{}^{A}{}_{B} \| = \begin{Vmatrix} -1 & 0 \\ 0 & 1 \end{Vmatrix}$$

and

$$\| S_{12AB} \| = \begin{Vmatrix} 0 & -i \\ -i & 0 \end{Vmatrix} \qquad \| S_{14AB} \| = \begin{Vmatrix} -1 & 0 \\ 0 & 1 \end{Vmatrix} \tag{13.9}$$

$$\| S_{13AB} \| = \begin{Vmatrix} -1 & 0 \\ 0 & -1 \end{Vmatrix} \qquad \| S_{24AB} \| = \begin{Vmatrix} i & 0 \\ 0 & i \end{Vmatrix}$$

$$\| S_{23AB} \| = \begin{Vmatrix} i & 0 \\ 0 & -i \end{Vmatrix} \qquad \| S_{34AB} \| = \begin{Vmatrix} 0 & 1 \\ 1 & 0 \end{Vmatrix}$$

This latter gives

$$S_{12} = S^{12} = -iS_{34} \qquad S_{13} = S^{13} = iS_{24} \qquad S_{14} = -S^{14} = iS_{23} \tag{13.10a}$$

or (cf. 3.1*a*)

$$\check{S}^{\mu\nu} = \tfrac{1}{2} g^{-1/2} \epsilon^{\mu\nu\sigma\tau} S_{\sigma\tau} = S^{\mu\nu} \tag{13.10b}$$ [4]

With the above forms (13.8, 13.9) and (8.8, 8.10) we can establish the further identities:

$$S^{\kappa\lambda A}{}_{B} S_{\mu\nu}{}^{B}{}_{A} = -2\eta^{\kappa\lambda}_{\mu\nu} \tag{13.11}$$

$$S^{\kappa\lambda A}{}_{B} \bar{g}_{\mu}{}^{\dot{B}}{}_{C} = -\eta^{\kappa\lambda}_{\mu\nu} \bar{g}^{\nu A}{}_{C} \tag{13.12}$$

$$g_{\mu}{}^{A}{}_{B} S^{\kappa\lambda B}{}_{C} = \eta^{\kappa\lambda}_{\mu\nu} g^{\nu A}{}_{C} \tag{13.13}$$

where

$$\eta^{\kappa\lambda}_{\mu\nu} = \delta^{\kappa\lambda}_{\mu\nu} + g^{-1/2} \epsilon^{\kappa\lambda\alpha\beta} g_{\alpha\mu} g_{\beta\nu} \tag{13.14}$$

It may also be remarked that

$$S_{\kappa\lambda A}{}^{B} S_{\mu\nu B}{}^{C} - S_{\mu\nu A}{}^{B} S_{\kappa\lambda B}{}^{C}$$
$$= -2\{-g_{\kappa\mu} S_{\lambda\nu A}{}^{C} + g_{\lambda\mu} S_{\kappa\nu A}{}^{C} - g_{\lambda\nu} S_{\kappa\mu A}{}^{C} + g_{\kappa\nu} S_{\lambda\mu A}{}^{C}\} \tag{13.15}$$

as is readily verified in the special spin frame (13.8), and hence as spinor equations, in general. From the connection $\mathfrak{T}_{\mu\nu A}{}^{B} = -\frac{1}{2} S_{\mu\nu A}{}^{B}$ for the infinitesimal spinor

[4] It is of interest to note that the properties of the spin tensor $S^{\mu\nu A}{}_{C}$, etc., are contained in the more general symbolic definition of the g-spinor systems of § 9. This may be seen inversely as follows:

From $$2S_{\mu\nu}{}^{A}{}_{C} = g_{\mu}{}^{\dot{B}A} g_{\nu\dot{B}C} - g_{\nu}{}^{\dot{B}A} g_{\mu\dot{B}C} \qquad (13.2) \tag{1}$$

and $$g_{\mu}{}^{\dot{B}A} g_{\nu\dot{B}C} + g_{\nu}{}^{\dot{B}A} g_{\mu\dot{B}C} = 2g_{\mu\nu} \delta^{A}{}_{C} \qquad (8.14) \tag{2}$$

(which is independently contained in (9.1*b*) *et seq.*) it follows that

$$g_{\mu}{}^{\dot{B}A} g_{\nu\dot{B}C} = g_{\mu\nu} \delta^{A}{}_{C} + S_{\mu\nu}{}^{A}{}_{C} = g_{\mu\nu} \delta^{A}{}_{C} + \check{S}_{\mu\nu}{}^{A}{}_{C} \tag{3a}$$

$$= g_{\mu\nu} \delta^{A}{}_{C} + \tfrac{1}{2} g^{-1/2} \epsilon_{\mu\nu\alpha\beta} S^{\alpha\beta A}{}_{C} \tag{3b}$$

From (3*a*) $S^{\alpha\beta A}{}_{C} = g^{\alpha\dot{B}A} g^{\beta}{}_{\dot{B}C} - g^{\alpha\beta} \delta^{A}{}_{C}$ and substituting in (3*b*) it follows that

$$g_{\mu}{}^{\dot{B}A} g_{\nu\dot{B}C} = g_{\mu\nu} \delta^{A}{}_{C} + \tfrac{1}{2} g^{-1/2} \epsilon_{\mu\nu\alpha\beta} g^{\alpha\dot{B}A} g^{\beta}{}_{\dot{B}C} \tag{4}$$

which is one of the general defining equations (9.1*b*).

transformations (cf. 11.9*b*) this then reads simply as the integrability conditions (11.8).

Because of the self-duality condition (13.10*b*) it follows that if ϕ^{AB} is a symmetric spinor, the tensor defined by

$$A_{\mu\nu} = S_{\mu\nu AB}\,\phi^{AB} \tag{13.16a}$$

is self-dual. The inverse relations follow on noting (13.5), viz.

$$\phi^{CD} = \tfrac{1}{8}A_{\mu\nu}S^{\mu\nu CD} \tag{13.16b}$$

If we compute the determinant of ϕ_{AB} we find (note 13.11, 13.14)

$$|\,\phi_{AB}\,| = \tfrac{1}{2}\phi_{AB}\,\phi^{AB} = \tfrac{1}{16}A_{\mu\nu}\,A^{\mu\nu} \tag{13.17}$$

which means that ϕ_{AB} is singular if, and only if, the " invariant " $A_{\mu\nu}A^{\mu\nu}$ of the associated self-dual 6-vector is null. Moreover, since any symmetric second-rank spinor may be written in the form

$$\phi_{AB} = \tfrac{1}{2}(\chi_A\,\psi_B + \chi_B\,\psi_A) \tag{13.18}$$

it follows that χ_A is proportional to ψ_A *only* when the above invariant is null. Hence, a self-dual antisymmetric tensor *of null invariant* determines a unique simple (first-rank) spinor, and conversely. As particularly emphasized by Whittaker,[5] the consequence of this is that the spinor calculus *is* included in the tensor calculus . . . in the sense that to every spinor there corresponds a tensor, and to every relation between spinors there corresponds a relation between tensors.[6]

In consequence of (13.16*a*, *b*) there are two further correspondences between symmetric second-rank spinors and (real and complex) skew tensors. Thus, according to Chapter I, § 3, there are two self-dual tensors, say $\frac{1}{2}(G_{\mu\nu} + \breve{G}_{\mu\nu})$ and $\frac{1}{2}(\bar{G}_{\mu\nu} + \check{\bar{G}}_{\mu\nu})$, associated with a general antisymmetric tensor $G_{\mu\nu}$, and from (13.16) we have

$$\tfrac{1}{2}(G_{\mu\nu} + \breve{G}_{\mu\nu}) = S_{\mu\nu AB}\,\phi_1{}^{AB} \tag{13.19a}$$

$$\tfrac{1}{2}(\bar{G}_{\mu\nu} + \check{\bar{G}}_{\mu\nu}) = S_{\mu\nu AB}\,\phi_2{}^{AB} \tag{13.19b}$$

Therefore, $\frac{1}{2}(G_{\mu\nu} - \breve{G}_{\mu\nu}) = \bar{S}_{\mu\nu AB}\bar{\phi}_2{}^{AB}$ and

$$G_{\mu\nu} = S_{\mu\nu AB}\,\phi_1{}^{AB} + \bar{S}_{\mu\nu AB}\,\bar{\phi}_2{}^{AB} \tag{13.20a}$$

$$\breve{G}_{\mu\nu} = S_{\mu\nu AB}\,\phi_1{}^{AB} - \bar{S}_{\mu\nu AB}\,\bar{\phi}_2{}^{AB} \tag{13.20b}$$

Inversely, it follows from (13.16*b*) that $\phi_1{}^{AB}$ and $\phi_2{}^{AB}$ are determined by

$$\phi_1{}^{AB} = \tfrac{1}{16}(G_{\mu\nu} + \breve{G}_{\mu\nu})S^{\mu\nu AB} = \tfrac{1}{8}G_{\mu\nu}S^{\mu\nu AB} \tag{13.21a}$$

$$\bar{\phi}_2{}^{AB} = \tfrac{1}{16}(G_{\mu\nu} - \breve{G}_{\mu\nu})\bar{S}^{\mu\nu AB} = \tfrac{1}{8}G_{\mu\nu}\bar{S}^{\mu\nu AB} \tag{13.21b}$$

since $S^{\mu\nu AB}$ is self-dual.

[5] Whittaker (37). Cf. also Ruse (37); Taub (39).

[6] We shall not have occasion to consider these interesting questions which lead to Whittaker's so-called " catalytic " processes. It may be mentioned, however, that this procedure appears to be the simplest for transcribing Dirac's spin $\frac{1}{2}$ equations into tensor form, particularly when we pass over to non-homoloidal spaces. Cf. previous footnote.

A *real* antisymmetric tensor $F_{\mu\nu}$ determines a *single* symmetric spinor by the equation

$$\phi_{AB} = \tfrac{1}{8} S_{\mu\nu AB} F^{\mu\nu} \tag{13.22a}$$

and its inverse

$$F_{\mu\nu} = S_{\mu\nu}{}^{AB} \phi_{AB} + \bar{S}_{\mu\nu}{}^{AB} \bar{\phi}_{AB} \tag{13.22b}$$

§ 14. 4-Spinors

(a) *Introduction—Definition of 4-Spinors*

The Dirac theory of the electron (cf. Chapter IV, §§ 28*b*, 30*a*) first brought spinor analysis to prominence,* but the wave function of his theory is (as originally derived from the matrix-algebraic viewpoint, cf. Chapter V, § 34(*a*)) a four-index symbol called a Dirac- or 4-spinor which is defined by its transformation properties when the coordinates in R_4 are subjected to a Lorentz transformation. These 4-spinors are, as will be seen, simply a combination of two simple 2-spinors; the particular pair of first-rank spinors (as regards variance) depending on which of several equivalent pairs of 2-spinor equations Dirac's is decomposed into. Following Veblen (36) and Taub (39*b*, 40), we take a form which expresses the basic contravariant 4-spinor in terms of one covariant regular spinor and one contravariant conjugated spinor, e.g.

$$\Psi^a = \begin{Vmatrix} \Psi^1 \\ \Psi^2 \\ \Psi^3 \\ \Psi^4 \end{Vmatrix} = \begin{Vmatrix} \psi_1 \\ \psi_2 \\ \phi^{\dot{1}} \\ \phi^{\dot{2}} \end{Vmatrix} \tag{14.1}$$ [1]

These entities have also been extensively investigated by Kramers (37, 41); Belinfante (39*a*, *b*, *c*; 40*a*, *b*); Lubanski (41, 42); *et al.*,[2] who, incidentally, call them " undors " and their analysis " undor calculus ".

The formalism to be developed is intimately related to the chosen form of the Dirac equation (§ 30)—since we do not enter upon the rather involved abstract geometrical bases[3]—and consequently our treatment cannot be as general as might be desirable. Our aim is to maintain as close a connection as possible with the preceding 2-spinor analysis, and to achieve this end we shall adopt certain definitions which are not necessarily convenient for other purposes.[4]

(b) *The γ-Spin Tensors*

For the purposes of orientation, and the procedure suggested thereby, we remark that the Dirac equation (cf. §§ 30(*a*), 34(*a*)) may be written in the form

$$(\partial_\mu \gamma^\mu - i\kappa)\Psi = 0 \tag{14.2}$$

* As regards physical application, the essential mathematical concepts having been early established by various writers, in particular, Cartan (38*a*, *b*); Weyl (25).

[1] N.B. Since 4-spinors are distinguished from 2-spinors both in font and index type, we may for the most part simply refer to " spinors " without risk of confusion.

[2] cf. Majorana (37); Racah (37); Furry (38).

[3] cf. Veblen (33, 35); Schouten (49*a*, *b*).

[4] This procedure does not in itself restrict the generality since it will be seen that other equivalent formalisms may easily be obtained by simple modifications of the theory.

where $\kappa = mc/\hbar$ (m = electron mass), Ψ is a one-column matrix of 4 rows or 4×1 matrix, and the γ's are 4×4 matrices satisfying the commutation relations

$$\gamma^\mu \gamma^\nu + \gamma^\nu \gamma^\mu = 2g^{\mu\nu} \tag{14.3}$$

and as such constitute a matrix realization of the abstract hypercomplex number system satisfying the same relations.[5]

We now note that the γ^μ's may be realized by the definition

$$\gamma^{\mu a}{}_b = \left\| \begin{matrix} 0 & -i \, \| \bar{g}^\mu{}_{AB} \| \\ i \, \| g^{\mu AB} \| & 0 \end{matrix} \right\| \tag{14.4a}$$

where the g's are 2-spin tensors. In the special coordinate system of § 8(c) we have

$$\gamma^{1a}{}_b = \left\| \begin{matrix} 0 & 0 & 0 & i \\ 0 & 0 & i & 0 \\ 0 & i & 0 & 0 \\ i & 0 & 0 & 0 \end{matrix} \right\| \qquad \gamma^{3a}{}_b = \left\| \begin{matrix} 0 & 0 & i & 0 \\ 0 & 0 & 0 & -i \\ i & 0 & 0 & 0 \\ 0 & -i & 0 & 0 \end{matrix} \right\| \tag{14.4b}$$
[6]

$$\gamma^{2a}{}_b = \left\| \begin{matrix} 0 & 0 & 0 & 1 \\ 0 & 0 & -1 & 0 \\ 0 & 1 & 0 & 0 \\ -1 & 0 & 0 & 0 \end{matrix} \right\| \qquad \gamma^{4a}{}_b = \left\| \begin{matrix} 0 & 0 & -i & 0 \\ 0 & 0 & 0 & -i \\ i & 0 & 0 & 0 \\ 0 & i & 0 & 0 \end{matrix} \right\|$$

so that γ^1, γ^2, γ^3 are skew-Hermitian and γ^4 Hermitian. Let us further introduce

$$\gamma_5 = \frac{-g^{+1/2}}{4!} \epsilon_{\mu\nu\sigma\tau} \gamma^\mu \gamma^\nu \gamma^\sigma \gamma^\tau \tag{14.5a}$$

which is seen to satisfy the relations $\gamma_5 \gamma_\mu + \gamma_\mu \gamma_5 = 0$, and takes the form

$$\gamma_5 = \left\| \begin{matrix} 1 & 0 & 0 & 0 \\ 0 & 1 & 0 & 0 \\ 0 & 0 & -1 & 0 \\ 0 & 0 & 0 & -1 \end{matrix} \right\| \tag{14.5b}$$

in the present coordinate system. From the definitions (14.4) we may readily establish the fundamental relations

$$\gamma^{\mu\nu} = \tfrac{1}{2}(\gamma^\mu \gamma^\nu - \gamma^\nu \gamma^\mu) = \left\| \begin{matrix} - \| S^{\mu\nu A}{}_C \| & 0 \\ 0 & \| \bar{S}^{\mu\nu A}{}_C \| \end{matrix} \right\| \quad \text{(cf. 13.2)} \tag{14.6a}$$

and

$$\gamma_5 \gamma^\mu = \left\| \begin{matrix} 0 & -i \, \| \bar{g}^\mu{}_{AB} \| \\ -i \, \| g^{\mu AB} \| & 0 \end{matrix} \right\| \tag{14.6b}$$

[5] What the algebraist terms a semi-simple ring: the order of the system or number of independent elements N, formed by all possible distinct products of the γ's, satisfying according to a general theorem the relation $N = \Sigma \, d_i^{\,2}$, where the d_i are the dimensionalities of the inequivalent irreducible matrix representations. In the present case there is but one such, of dimension 4, which gives 16 elements or the number of independent 4×4 matrices. We shall consider these matters further in Chapter V, and it may be mentioned that familiarity with these questions and the Dirac equation is not essential to the following.

The interested reader may, of course, consult Chapter V as well as the following references: Artin (46); Harish-Chandra (46a, b); Kemmer (43); Pauli (35); v. der Waerden (30, 31).

[6] The notation $\gamma^{\mu a}{}_b$ implies contravariance in the index a and covariance in the index b under 4-spin transformation, or that $\gamma^{\mu a}{}_b$ is a second-rank mixed 4-spinor (cf. *infra*). This agrees with the identification of Ψ in (14.2) with the contravariant 4-spinor Ψ^a of (14.1). Usually, a γ^μ without spinor indices will indicate the $\gamma^{\mu a}{}_b$.

(*c*) 4-*Spinor Transformations*

Considerations on the Lorentz invariance of the Dirac equation (14.2) lead to the necessary condition (cf. Chapter V) that for every Lorentz transformation $L^{\mu}{}_{\nu}$ there should exist a non-singular matrix satisfying the relations

$$\gamma^{\mu\prime} = L^{\mu}{}_{\nu}\gamma^{\nu} = T\gamma^{\mu}T^{-1} \tag{14.7}$$

when the function Ψ undergoes the transformation $\Psi' = T\Psi$. This connection is important but incidental for the moment, where our primary interest is the conditions such that the transformation

$$\Psi^{a\prime} = T^{a}{}_{b}\Psi^{b} \tag{14.8a}$$

be the spin *image* of the Lorentz transformation

$$x^{\mu\prime} = L^{\mu}{}_{\nu}x^{\nu}$$

with Ψ^{a} given by (14.1) *and* the matrix $T^{a}{}_{b}$ determined according to its constituent 2-spinor form and our previous results, viz. (cf. §§ 10*a*–*d*)

$$\| T^{a}{}_{b} \| = T(L) = \begin{Vmatrix} \| \tilde{s}^{A}{}_{B} \| & 0 \\ 0 & \| \bar{S}^{\dot{A}}{}_{\dot{B}} \| \end{Vmatrix} \tag{14.8b}$$

corresponding to a proper Lorentz transformation,[7] and

$$\| T^{a}{}_{b} \| = \begin{Vmatrix} 0 & \| \bar{S}_{\dot{A}B} \| \\ -\| S^{\dot{A}B} \| & 0 \end{Vmatrix} \tag{14.8c}$$

corresponding to an improper Lorentz transformation.[8] The necessary conditions are simply as stated in (14.7). Thus, from the definition (14.4*a*) of the γ's and the above forms of $T^{a}{}_{b}$, this condition states (for a proper transformation) that

$$\bar{g}^{\nu}{}_{\dot{A}B}L^{\mu}{}_{\nu} = s^{C}{}_{A}\bar{g}^{\mu}{}_{\dot{C}D}\bar{s}^{D}{}_{B} \tag{14.8b'}$$

and

$$g^{\nu\dot{A}B}L^{\mu}{}_{\nu} = \bar{S}^{\dot{A}}{}_{\dot{C}}g^{\mu\dot{C}D}S^{B}{}_{D}$$

which are equivalent forms of the spin-image relation (cf. § 10*d*, especially footnote 6) $L^{\mu}{}_{\nu} = \frac{1}{2}g^{\mu\dot{A}B}\bar{S}^{\dot{C}}{}_{\dot{A}}S^{D}{}_{B}g_{\nu\dot{C}D}$. A similar result obtains for the improper transformations corresponding to (14.8*c*).

[7] Under proper Lorentz transformations we have

$$\begin{matrix} \psi^{A} \to \psi^{A\prime} = S^{A}{}_{B}\,\psi^{B} \\ \phi^{\dot{A}} \to \phi^{\dot{A}\prime} = \bar{S}^{\dot{A}}{}_{\dot{B}}\,\phi^{\dot{B}} \end{matrix} \quad \text{or} \quad \begin{matrix} \psi_{A} \to \psi_{A}{}' = -S_{A}{}^{B}\,\psi_{B} = s^{B}{}_{A}\,\psi_{B} \\ \phi^{\dot{A}} \to \phi^{\dot{A}\prime} = \bar{S}^{\dot{A}}{}_{\dot{B}}\,\phi^{\dot{B}} \end{matrix}$$

We should note the inversion of indices (transposed matrix $\| \tilde{s}^{A}{}_{B} \|$) due to the fact that the two constituent 2-spinors transform (conjugate) contragrediently.

[8] The two spinors are chosen such that they transform into one another under improper transformations, so that we have

$$\begin{matrix} \psi^{A} \to \psi^{A\prime} = \bar{S}^{\dot{A}}{}_{B}\,\phi^{\dot{B}} \\ \phi^{A} \to \phi^{A\prime} = \bar{S}^{\dot{A}}{}_{B}\,\psi^{\dot{B}} \end{matrix} \quad \text{or} \quad \begin{matrix} \psi_{A} \to \psi_{A}{}' = \bar{S}_{\dot{A}B}\,\phi^{\dot{B}} \\ \phi^{\dot{A}} \to \phi^{\dot{A}\prime} = -S^{\dot{A}B}\,\psi_{B} \end{matrix}$$

It may be remarked that equation (14.7) may be written in the explicit form

$$L^{\mu}{}_{\nu}\gamma^{\nu a}{}_{b} = T^{a}{}_{c}(T^{-1})^{d}{}_{b}\gamma^{\mu c}{}_{d} = T^{a}{}_{c}t^{d}{}_{b}\gamma^{\mu c}{}_{d} \tag{14.7'}$$

in agreement with the previously noted mixed spinor character of the fundamental γ-matrices.

(*d*) *Behaviour of* 4-*Spinors under Proper Lorentz Transformations*

From our previous examples for 2-spinors (cf. § 10*b*) we find, for example,

$$\begin{aligned} x^{3\prime} &= \frac{x^3}{(1-\beta^2)^{1/2}} - \frac{x^4\beta}{(1-\beta^2)^{1/2}} \\ x^{4\prime} &= \frac{-x^3\beta}{(1-\beta^2)^{1/2}} + \frac{x^4}{(1-\beta^2)^{1/2}} \end{aligned} \rightarrow T = \begin{Vmatrix} 1/\tau & 0 & 0 & 0 \\ 0 & \tau & 0 & 0 \\ 0 & 0 & \tau & 0 \\ 0 & 0 & 0 & 1/\tau \end{Vmatrix} \tag{14.9a}$$

$$\begin{aligned} x^{1\prime} &= x^1 \cos\theta + x^2 \sin\theta \\ x^{2\prime} &= -x^1 \sin\theta + x^2 \cos\theta \end{aligned} \rightarrow T = \begin{Vmatrix} e^{-i\theta/2} & 0 & 0 & 0 \\ 0 & e^{i\theta/2} & 0 & 0 \\ 0 & 0 & e^{-i\theta/2} & 0 \\ 0 & 0 & 0 & e^{i\theta/2} \end{Vmatrix} \tag{14.9b}$$

$$\begin{aligned} x^{2\prime} &= x^2 \cos\phi + x^3 \sin\phi \\ x^{3\prime} &= -x^2 \sin\phi + x^3 \cos\phi \end{aligned} \rightarrow T = \begin{Vmatrix} \cos(\phi/2) & i\sin(\phi/2) & 0 & 0 \\ i\sin(\phi/2) & \cos(\phi/2) & 0 & 0 \\ 0 & 0 & \cos(\phi/2) & i\sin(\phi/2) \\ 0 & 0 & i\sin(\phi/2) & \cos(\phi/2) \end{Vmatrix} \tag{14.9c}$$

in the representation previously chosen for 2-spinors.

(*e*) *Behaviour of* 4-*Spinors under Improper Lorentz Transformations*

Similarly we find (cf. 14.3)

$$x^{2\prime} = -x^2 \quad \rightarrow T = \begin{Vmatrix} 0 & 0 & 0 & 1 \\ 0 & 0 & -1 & 0 \\ 0 & -1 & 0 & 0 \\ 1 & 0 & 0 & 0 \end{Vmatrix} = i\gamma^1\gamma^3\gamma^4 \tag{14.10a}$$

and for reflection through the origin,

$$x^{k\prime} = -x^k, \quad x^{4\prime} = x^4 \quad \rightarrow T = \begin{Vmatrix} 0 & 0 & -1 & 0 \\ 0 & 0 & 0 & -1 \\ 1 & 0 & 0 & 0 \\ 0 & 1 & 0 & 0 \end{Vmatrix} = -i\gamma^4 \tag{14.10b}$$

. . . .

Finally, we note the *essential* difference between the 2-spinor theory and the 4-spinor theory: the transformations of 4-spinors give a representation of the extended Lorentz group. In particular we find with (14.7), and again noting (14.3),

$$x^{k\prime} = x^k, \quad x^{4\prime} = -x^4 \quad \rightarrow T = \gamma^1\gamma^2\gamma^3 \tag{14.11a}$$

$$x^{\mu\prime} = -x^{\mu} \quad \rightarrow T = \gamma_5 \tag{14.11b}$$

(f) The Metric 4-Spinor

As in the 2-spinor calculus the metric 4-spinor, ε_{ab}, will be an antisymmetric matrix; the rules for lowering and raising spin indices

$$\Psi_a = \varepsilon_{ab}\Psi^b \tag{14.12}$$
$$\Psi^a = \varepsilon^{ca}\Psi_c$$

where $\varepsilon_{ab}\varepsilon^{ac} = \delta_b{}^c$ or

$$\| \varepsilon^{ab} \| = \| \tilde{\varepsilon}_{ab} \|^{-1} = - \| \varepsilon_{ab} \|^{-1} = \varepsilon^0 \qquad (14.13)^9$$

However, we cannot assume that ε_{ab} transforms according to

$$\varepsilon_{ab}{}' = t^c{}_a t^d{}_b \varepsilon_{cd} = \varepsilon_{ab} \tag{14.14}$$

or

$$\varepsilon' = \tilde{t}\varepsilon t = \varepsilon$$

Instead we must take

$$\varepsilon' = \tilde{t}\varepsilon t = L^{-1}\varepsilon \tag{14.15a}$$

or equivalently

$$\varepsilon^{0\prime} = T\varepsilon^0\tilde{T} = L\varepsilon^0 \tag{14.15b}$$

(where L is the determinant of the associated Lorentz transformation $\| L^\mu{}_\nu \|$) in order to agree with our specification of the representations of Lorentz transformations by 4-spinors;[10] as may be verified with the explicit forms given in the two preceding sections.[11] The above conditions may be described by the statement that ε_{ab} and ε^{ab} are pseudoscalars (of different weights) relative to Lorentz transformations or, briefly, " pseudospinors ".

The form which we take for ε_{ab} is

$$\varepsilon_{ab} = \left\| \begin{matrix} i \, \| \epsilon^{AB} \| & 0 \\ 0 & -i \, \| \tilde{\epsilon}_{AB} \| \end{matrix} \right\| = \left\| \begin{matrix} 0 & i & 0 & 0 \\ -i & 0 & 0 & 0 \\ 0 & 0 & 0 & -i \\ 0 & 0 & i & 0 \end{matrix} \right\| \qquad (14.16a)^{12}$$

[9] The symbol ^{0} will be used to denote " transposed inverse matrix ".

[10] which are chosen to be in accord with the representations (in particular of improper transformations) afforded by transformation of 2-spinors.

[11] Equivalently, this is a condition restricting the arbitrary factor in the matrix T allowed by the general relation (14.7). It may be remarked that the change from the condition $T\varepsilon^0\tilde{T} = \varepsilon^0$ to (14.15b) corresponds to making the spin-frame transformation

$$\Psi \to \Phi = \frac{1}{\sqrt{2}}(1 + i\gamma_5)\Psi = R\Psi$$

with unitary matrix R, and this amounts to replacing a matrix $P(L)$ satisfying $P\varepsilon^0\tilde{P} = \varepsilon^0$ by $T(L) = RP(L)R^{-1}$ which satisfies (14.15b). This transformation may be considered as the spin image of the (dual) transformations (3.1a–3.3a) since if we transform the γ^μ by it we find

$$\gamma^{\mu\prime} = R\gamma^\mu R^{-1} = i\gamma_5\gamma^\mu = \frac{1}{3!} g^{-1/2}\epsilon^{\mu\nu\sigma\tau}\gamma_\nu\gamma_\sigma\gamma_\tau$$

as may be verified in our special coordinate system.

[12] Cf. Brauer-Weyl (35).

and correspondingly,

$$\varepsilon^{ab} = \left\| \begin{matrix} 0 & -i & 0 & 0 \\ i & 0 & 0 & 0 \\ 0 & 0 & 0 & i \\ 0 & 0 & -i & 0 \end{matrix} \right\| \tag{14.16b}$$

If we raise and lower the (suppressed) spinor indices of the $\gamma^{\mu a}{}_b$ in (14.3) we obtain the relations

$$\gamma^{\mu ab}\gamma^{\nu}{}_{bc} + \gamma^{\nu ab}\gamma^{\mu}{}_{bc} = -2g^{\mu\nu}\,\delta^a{}_c \tag{14.17a}$$

and

$$\gamma^{\mu}{}_a{}^b\gamma^{\nu}{}_b{}^c + \gamma^{\nu}{}_a{}^b\gamma^{\mu}{}_b{}^c = 2g^{\mu\nu}\,\delta_a{}^c \tag{14.17b}$$

From (14.4b) and the above metric forms, these matrices are realized by

$$\gamma^{1ab} = \left\| \begin{matrix} 0 & 0 & 1 & 0 \\ 0 & 0 & 0 & -1 \\ -1 & 0 & 0 & 0 \\ 0 & 1 & 0 & 0 \end{matrix} \right\| \qquad \gamma^{2ab} = \left\| \begin{matrix} 0 & 0 & -i & 0 \\ 0 & 0 & 0 & -i \\ i & 0 & 0 & 0 \\ 0 & i & 0 & 0 \end{matrix} \right\| \tag{14.18a}$$

$$\gamma^{3ab} = \left\| \begin{matrix} 0 & 0 & 0 & -1 \\ 0 & 0 & -1 & 0 \\ 0 & 1 & 0 & 0 \\ 1 & 0 & 0 & 0 \end{matrix} \right\| \qquad \gamma^{4ab} = \left\| \begin{matrix} 0 & 0 & 0 & 1 \\ 0 & 0 & -1 & 0 \\ 0 & 1 & 0 & 0 \\ -1 & 0 & 0 & 0 \end{matrix} \right\|$$

$$\gamma^{1}{}_{ab} = \left\| \begin{matrix} 0 & 0 & -1 & 0 \\ 0 & 0 & 0 & 1 \\ 1 & 0 & 0 & 0 \\ 0 & -1 & 0 & 0 \end{matrix} \right\| \qquad \gamma^{2}{}_{ab} = \left\| \begin{matrix} 0 & 0 & -i & 0 \\ 0 & 0 & 0 & -i \\ i & 0 & 0 & 0 \\ 0 & i & 0 & 0 \end{matrix} \right\| \tag{14.18b}$$

$$\gamma^{3}{}_{ab} = \left\| \begin{matrix} 0 & 0 & 0 & 1 \\ 0 & 0 & 1 & 0 \\ 0 & -1 & 0 & 0 \\ -1 & 0 & 0 & 0 \end{matrix} \right\| \qquad \gamma^{4}{}_{ab} = \left\| \begin{matrix} 0 & 0 & 0 & 1 \\ 0 & 0 & -1 & 0 \\ 0 & 1 & 0 & 0 \\ -1 & 0 & 0 & 0 \end{matrix} \right\|$$

$$\gamma^{1}{}_a{}^b = \left\| \begin{matrix} 0 & 0 & 0 & -i \\ 0 & 0 & -i & 0 \\ 0 & -i & 0 & 0 \\ -i & 0 & 0 & 0 \end{matrix} \right\| \qquad \gamma^{2}{}_a{}^b = \left\| \begin{matrix} 0 & 0 & 0 & 1 \\ 0 & 0 & -1 & 0 \\ 0 & 1 & 0 & 0 \\ -1 & 0 & 0 & 0 \end{matrix} \right\| \tag{14.18c}$$

$$\gamma^{3}{}_a{}^b = \left\| \begin{matrix} 0 & 0 & -i & 0 \\ 0 & 0 & 0 & i \\ -i & 0 & 0 & 0 \\ 0 & i & 0 & 0 \end{matrix} \right\| \qquad \gamma^{4}{}_a{}^b = \left\| \begin{matrix} 0 & 0 & -i & 0 \\ 0 & 0 & 0 & -i \\ i & 0 & 0 & 0 \\ 0 & i & 0 & 0 \end{matrix} \right\|$$

and it follows from the pseudospinor character of ε that the $\gamma^{\mu ab}$ and $\gamma^{\mu}{}_{ab}$ are pseudospinors while the $\gamma^{\mu}{}_a{}^b$, like the $\gamma^{\mu a}{}_b$, are spinors. We note further that $\gamma^{\mu}{}_{ab} = -\gamma^{\mu}{}_{ba}$, which can be expressed in the form

$$\varepsilon\gamma^{\mu} = \tilde{\gamma}^{\mu}\varepsilon \qquad \varepsilon\gamma_5 = \tilde{\gamma}_5\varepsilon \tag{14.19}$$ [13]

[13] This result may in turn be taken as the condition defining the matrix ε (up to a factor). The proof of the existence of an antisymmetric matrix satisfying this condition is given, for example, by Pauli (35) (cf. Chapter V), quite apart from any special representation of the γ's. The role of ε as metric is intimately related to the Dirac equation, and will be considered further in Chapter IV.

We note here the further useful relations:

$$\varepsilon\gamma_5\gamma^\mu = \begin{Vmatrix} 0 & -\|\bar{g}^{\mu\dot{A}}{}_{\dot{B}}\| \\ -\|g^\mu{}_A{}^B\| & 0 \end{Vmatrix} = \begin{Vmatrix} 0 & -\|\bar{g}^{\mu\dot{A}}{}_{\dot{B}}\| \\ -\|\bar{g}^{\mu\dot{A}}{}_{\dot{B}}\| & 0 \end{Vmatrix} \tag{14.20a}$$

$$\varepsilon\gamma_5\gamma^{\mu\nu} = \begin{Vmatrix} i\,\|S^{\mu\nu AB}\| & 0 \\ 0 & i\,\|\bar{S}^{\mu\nu}{}_{\dot{A}\dot{B}}\| \end{Vmatrix}. \tag{14.20b}$$

§ 15. Special 4-Spinors and their Tensor Equivalents

(a) *Introduction*

In the 2-spinor calculus we found that second-rank symmetric spinors correspond in (1–1) fashion to self-dual antisymmetric world-tensors. As might be expected, a similar situation obtains in the 4-spinor calculus, but with the greater number of components the correspondence is now more inclusive.[1] 4-spinors of higher rank are defined as quantities which transform like products of 4-spinors, e.g. a second-rank spinor Ψ^{ab} transforms like $\Psi^a\Phi^b$ and may be regarded as a matrix of 16 rows and 1 column or as a 4×4 matrix

$$\|\Psi^{ab}\| = \begin{Vmatrix} \Psi^{11} & \dots & \Psi^{14} \\ \vdots & & \\ \Psi^{41} & \dots & \Psi^{44} \end{Vmatrix} = \begin{Vmatrix} \|\phi_{1AB}\| & \|\tilde{\phi}_2{}^{\dot{A}}{}_B\| \\ \|\phi_3{}^{\dot{A}}{}_B\| & \|\phi_4{}^{\dot{A}\dot{B}}\| \end{Vmatrix} \tag{15.1}$$

where we have indicated the general 2-spinor composition.

(b) *The γ-Matrices and Second-Rank Spinors*

We obtain the tensors determined by second-rank spinors through the application of the preceding γ-matrix formalism. The 16 matrices γ^μ, $\gamma^{\mu\nu}$, $\gamma_5\gamma^\mu$, γ_5 and 1 are linearly independent and form a basis for all 4-dimensional matrices.[2] From this fact it immediately follows that an arbitrary second-rank spinor may be written as

$$\Psi^{\bullet\bullet} = \|\Psi^{ac}\| = \tfrac{1}{4}[\tfrac{1}{2}iG_{\mu\nu}\gamma_5\gamma^{\mu\nu} + U_\mu\gamma_5\gamma^\mu + \check{\chi}_\mu\gamma^\mu + i\check{\phi} + \rho\gamma_5]\varepsilon^0 \tag{15.2a}$$ [3]

or

$$\Psi^{ac} = \tfrac{1}{4}\{\tfrac{1}{2}iG_{\mu\nu}\gamma_5{}^a{}_b\gamma^{\mu\nu bc} + U_\mu\gamma_5{}^a{}_b\gamma^{\mu bc} + \check{\chi}_\mu\gamma^{\mu ac} + i\check{\phi}\,\delta^{ac} + \rho\gamma_5{}^{ac}\} \tag{15.2b}$$

that is, relative to the Lorentz group it determines an antisymmetrical tensor, a 4-vector, a pseudo 4-vector, a pseudoscalar and a scalar; all in general complex. Apart from numerical factors, the fact that the coefficients must have the tensor character indicated follows directly by subjecting $\Psi^{\bullet\bullet}$ to the transformation T (14.8*a*), viz.

$$\Psi^{ab\prime} = T^a{}_cT^b{}_d\Psi^{cd} \tag{15.3a}$$

or matrically

$$\Psi^{\bullet\bullet\prime} = T\Psi^{\bullet\bullet}\tilde{T} \tag{15.3b}$$

[1] The following development will be required in the illustrative examples from the various meson theories in Chapter IV.

[2] cf. Pauli (35); also Chapter V.

[3] N.B. In the following we shall sometimes use heavy dots to indicate the position of indices not written explicitly.

Then, noting (14.7), (14.15*b*), and the further relation $T\gamma_5 T^{-1} = L\gamma_5$ which follows from (14.7) we find

$$\Psi^{\cdot\cdot\prime} = \tfrac{1}{4}[\tfrac{1}{2}iG'_{\cdot\mu\nu}\gamma_5\gamma^{\mu\nu} + U'_{\mu}\gamma_5\gamma^{\mu} + \check{\chi}'_{\mu}\gamma^{\mu} + i\check{\phi}' + \rho'\gamma_5]\varepsilon^0 \tag{15.4}$$

where the primed quantities are the transforms of the tensor quantities in (15.2*a*) under the associated Lorentz transformation.[4]

From (15.2) we readily obtain the explicit expressions for the tensor quantities determined by a general second-rank spinor Ψ^{ab}, viz.

$$G_{\mu\nu} = -i(\varepsilon\gamma_5\gamma_{\mu\nu})_{ab}\Psi^{ab} = -i\mathscr{S}p(\varepsilon\gamma_5\gamma_{\mu\nu}\Psi^{\cdot\cdot}) \qquad (15.5a)^5$$

$$U_{\mu} = \mathscr{S}p(\varepsilon\gamma_5\gamma_{\mu}\Psi^{\cdot\cdot}) \tag{15.5b}$$

$$\check{\chi}_{\mu} = \mathscr{S}p(\varepsilon\gamma_{\mu}\Psi^{\cdot\cdot}) \tag{15.5c}$$

$$\check{\phi} = -i\mathscr{S}p(\varepsilon\Psi^{\cdot\cdot}) \tag{15.5d}$$

$$\rho = \mathscr{S}p(\varepsilon\gamma_5\Psi^{\cdot\cdot}) \tag{15.5e}$$

In particular if Ψ^{ab} is symmetric, $\check{\phi} = \check{\chi}_{\mu} = 0$; and if antisymmetric, $U_{\mu} = G_{\mu\nu} = 0$. The tensors $G_{\mu\nu}$, U_{μ} are suitable for describing the vector meson, or Proca,[6] field; the pseudo-quantities for describing the pseudoscalar field; and in particular, as will be seen in Chapter IV, the general spinor Ψ^{ab} with the condition $\rho = 0$ provides a synthesis of the two theories.

In Kemmer's (38) classification of meson fields there are two cases dual to these, namely, the so-called pseudovector and scalar fields, respectively, in which the field quantities differ from the preceding cases in their behaviour under improper transformations. This leads us to consider the spinor Φ^{ab}, which is essentially the dual of Ψ^{ab}, defined by

$$\Phi^{\cdot\cdot} = \gamma_5\check{\Psi}^{\cdot\cdot} = \tfrac{1}{4}[\tfrac{1}{2}i\check{G}_{\mu\nu}\gamma^{\mu\nu} + \check{U}_{\mu}\gamma^{\mu} + \chi_{\mu}\gamma_5\gamma^{\mu} + i\phi\gamma_5 + \check{\rho}]\varepsilon^0 \tag{15.6a}$$

which transforms under $\Phi^{\cdot\cdot\prime} = T\Phi^{\cdot\cdot}\tilde{T}$ according to

$$\Phi^{\cdot\cdot\prime} = \tfrac{1}{4}[\tfrac{1}{2}i\check{G}'_{\mu\nu}\gamma^{\mu\nu} + \check{U}'_{\mu}\gamma^{\mu} + \chi'_{\mu}\gamma_5\gamma^{\mu} + i\phi'\gamma_5 + \check{\rho}']\varepsilon^0 \tag{15.6b}$$

and thus determines an antisymmetric pseudotensor, a pseudovector, a vector, a scalar, and a pseudoscalar.[7] Corresponding to (15.5) we now find

$$\check{G}_{\mu\nu} = -i\mathscr{S}p(\varepsilon\gamma_{\mu\nu}\Phi^{\cdot\cdot}) \tag{15.7a}$$

$$\check{U}_{\mu} = \mathscr{S}p(\varepsilon\gamma_{\mu}\Phi^{\cdot\cdot}) \tag{15.7b}$$

$$\chi_{\mu} = -\mathscr{S}p(\varepsilon\gamma_5\gamma_{\mu}\Phi^{\cdot\cdot}) \tag{15.7c}$$

$$\phi = -i\mathscr{S}p(\varepsilon\gamma_5\Phi^{\cdot\cdot}) \tag{15.7d}$$

$$\check{\rho} = \mathscr{S}p(\varepsilon\Phi^{\cdot\cdot}) \tag{15.7e}$$

[4] e.g. $\check{\chi}_{\mu}\gamma^{\mu}\varepsilon^0 \to \check{\chi}_{\mu}T\gamma^{\mu}\varepsilon^0\tilde{T} = \check{\chi}_{\mu}T\gamma^{\mu}T^{-1}T\varepsilon^0\tilde{T} = \check{\chi}_{\mu}L^{\mu}{}_{\nu}L\gamma^{\nu}\varepsilon^0 = \check{\chi}_{\nu}'\gamma^{\nu}\varepsilon^0$.

[5] If, further, we write $\Psi^{\cdot\cdot}$ in the 2-spinor form (15.1), and note (14.20*b*), it follows that

$$-i\mathscr{S}p(\varepsilon\gamma_5\gamma_{\mu\nu}\Psi^{\cdot\cdot}) = (S_{\mu\nu}{}^{AB}\phi_{1AB} + \bar{S}_{\mu\nu AB}\bar{\phi}_4{}^{AB}) = G_{\mu\nu} \quad \text{by (13.20}a\text{)}.$$

[6] Proca (36, 38); cf. Chapters IV, V. Our notation here is chosen to agree with that of these later chapters.

[7] The actual dual or pseudospinor $\check{\Psi}^{ab}$ transforms as

$$\check{\Psi}' = \frac{L}{4}[\tfrac{1}{2}i\check{G}'_{\mu\nu}\gamma_5\gamma^{\mu\nu} + \check{U}_{\mu}'\gamma_5\gamma^{\mu} + \chi_{\mu}'\gamma^{\mu} + i\phi' + \check{\rho}'\gamma_5]\varepsilon^0$$

where L is the determinant of the associated Lorentz transformation, so that its reflection properties are different from those of a proper spinor. As previously noted, multiplication by an odd number of factors γ_5 changes a pseudospinor into a spinor and conversely.

and with the condition $\check{\rho} = 0$ we can again obtain a synthesis of the pseudovector and scalar meson fields in terms of Φ^{ab}. However, the individual cases, pseudovector and scalar, are now not obtained by the restrictions to symmetry and antisymmetry, respectively, on Φ^{ab}. Instead the conditions are

$$\Psi^{\cdot\cdot} = \pm\tilde{\Psi}^{\cdot\cdot} \tag{15.8a}$$

or in terms of Φ^{ab}

$$\Phi^{\cdot\cdot} = \pm\gamma_5\tilde{\Phi}^{\cdot\cdot}\tilde{\gamma}_5 \tag{15.8b}$$

The explanation of this behaviour involves the detailed examination of the corresponding tensor and spinor field equations, and must therefore be deferred to Chapter IV.

§ 16. Infinitesimal 4-Spinor Transformations

As previously noted, the study of the infinitesimal transformation properties of field functions is essential to the discussion of the associated spin angular momentum; in particular for the determination of the eigenvalues. The behaviour of 4-spinors under infinitesimal Lorentz transformations follows immediately from their relation to 2-spinors and the results of § 11 for infinitesimal transformations of the latter.

Thus, corresponding to $\delta x^\mu = \omega^{\mu\nu} x_\nu$, we have

$$\delta\Psi^{\cdot} = \left\| \begin{matrix} \delta\psi_B \\ \delta\phi^{\dot{B}} \end{matrix} \right\| = \left\| \begin{matrix} \eta^A{}_B\psi_A \\ -\eta_{\dot{A}}{}^{\dot{B}}\phi^{\dot{A}} \end{matrix} \right\| = \tfrac{1}{4}\omega^{\mu\nu} \left\| \begin{matrix} g_{\mu\dot{C}}{}^A g_\nu{}^{\dot{C}}{}_B \psi_A \\ -\bar{g}_{\mu\dot{C}A}\bar{g}_\nu{}^{\dot{C}B}\,\phi^{\dot{A}} \end{matrix} \right\| \tag{16.1a}$$

by equation (11.3). We write the right-hand side in the matrix form

$$\frac{\omega^{\mu\nu}}{4} \left\| \begin{matrix} g_{\mu\dot{C}}{}^A g_\nu{}^{\dot{C}}{}_B & 0 \\ 0 & -\bar{g}_{\mu\dot{C}A}\bar{g}_\nu{}^{\dot{C}B} \end{matrix} \right\| \cdot \left\| \begin{matrix} \psi_A \\ \phi^{\dot{A}} \end{matrix} \right\| \tag{16.1b}$$

or

$$\frac{\omega^{\mu\nu}}{4} \left\| \begin{matrix} -\tfrac{1}{2}(g_\mu{}^{\dot{C}A} g_{\nu\dot{C}B} - g_\nu{}^{\dot{C}A} g_{\mu\dot{C}B} & 0 \\ 0 & \tfrac{1}{2}(\bar{g}_{\nu\dot{C}A}\bar{g}_\mu{}^{\dot{C}B} - \bar{g}_{\mu\dot{C}A}\bar{g}_\nu{}^{\dot{C}B}) \end{matrix} \right\| \Psi^{\cdot}$$

by virtue of the antisymmetry of $\omega^{\mu\nu}$. From the two defining forms of the spin-tensors $S_{\mu\nu AB}$ given by equation (13.2), this is

$$\frac{\omega^{\mu\nu}}{4} \left\| \begin{matrix} -\| S_{\mu\nu}{}^A{}_B \| & 0 \\ 0 & \bar{S}_{\mu\nu\dot{A}}{}^{\dot{B}} \| \end{matrix} \right\| \Psi^{\cdot} = \tfrac{1}{4}\omega^{\mu\nu}\gamma_{\mu\nu}\Psi \tag{16.1c}$$

the last result by (14.6a). Thus we have finally

$$\delta\Psi^{\cdot} = \tfrac{1}{2}\omega^{\mu\nu}\mathfrak{T}_{\mu\nu}{}^{\cdot}{}_{\cdot}\Psi^{\cdot} \tag{16.2a}$$

with

$$\mathfrak{T}_{\mu\nu} = \tfrac{1}{2}\gamma_{\mu\nu} = \tfrac{1}{2}\gamma_\mu\gamma_\nu \qquad (\mu \neq \nu) \tag{16.2b}$$

which satisfies the C.R.

$$\gamma_\mu\mathfrak{T}_{\lambda\nu} - \mathfrak{T}_{\lambda\nu}\gamma_\mu = g_{\mu\lambda}\gamma_\nu - g_{\mu\nu}\gamma_\lambda \tag{16.3}$$

and the integrability conditions (4.12), by virtue of the basic anticommutation rules (14.3).

The calculation of the infinitesimal transformations of higher-rank 4-spinors may again be based on the results for 2-spinors by direct generalization of the above procedure, although they follow immediately from the fact that the higher-rank 4-spinors transform as the *direct* products of simple 4-spinors (cf. *infra*). If we follow the first procedure for a second-rank spinor, for example, we first write it as a matrix of 16 rows and 1 column; taking the first row of Ψ^{ab} in its 2-spinor form (15.1) as the first four components, the second row as the next four, etc.,

$$\| \Psi^{ab} \| = \left\| \begin{matrix} \phi_{1\ 11} \\ \phi_{1\ 21} \\ \phi_3{}^{\dot 1}{}_1 \\ \phi_3{}^{\dot 2}{}_1 \\ \phi_{1\ 12} \\ \vdots \end{matrix} \right\| \tag{16.4}$$

Then, in computing $\delta\Psi^{\cdot\cdot}$, we use the results obtained for 2-spinors (cf. § 11) in the more general form appropriate to this case, viz.

$$\begin{aligned} \delta\phi_{AC} &= \delta^D{}_C\,\eta^B{}_A\,\phi_{BD} + \delta^B{}_A\,\eta^D{}_C\,\phi_{BD} \\ \delta\phi^{\dot A}{}_C &= -\delta^D{}_C\,\eta_{\dot B}{}^{\dot A}\,\phi^{\dot B}{}_D + \delta^{\dot A}_{\dot B}\,\eta^D{}_C\,\phi^{\dot B}{}_D \end{aligned} \tag{16.5}$$

etc., and proceeding in this way we find that the transformation matrices $\mathfrak{T}_{\mu\nu}$ are now given by

$$\mathfrak{T}_{\mu\nu} = \beta_\mu\beta_\nu - \beta_\nu\beta_\mu = \tfrac{1}{2}[(I \times \gamma_{\mu\nu}) + (\gamma_{\mu\nu} \times I)] \tag{16.6}$$

that is, the matrices β_μ are obtained from the matrices γ_μ by

$$\beta_\mu = \tfrac{1}{2}[(I \times \gamma_\mu) + (\gamma_\mu \times I)] \tag{16.7}$$

where the cross denotes the direct, or Kronecker, matrix product[1] of the 4-dimensional matrices I (unit) and γ_μ. Thus,

$$(I \times \gamma_\mu) = \left\| \begin{matrix} \gamma_\mu & & & \\ & \gamma_\mu & & \\ & & \gamma_\mu & \\ & & & \gamma_\mu \end{matrix} \right\| \tag{16.8a}$$

and

$$(\gamma_\mu \times I) = \left\| \begin{matrix} \gamma_\mu{}^1{}_1 I & \gamma_\mu{}^1{}_2 I \ldots \\ \gamma_\mu{}^2{}_1 I & \gamma_\mu{}^2{}_2 I \ldots \\ \vdots & \vdots \end{matrix} \right\| \tag{16.8b}$$

which are transforms of one another by a permutation matrix, i.e. $(I \times \gamma_\mu) = P(\gamma_\mu \times I)P^{-1}$.

It should also be noted that, as mentioned above, the infinitesimal second-rank transformation $\delta\Psi^{\cdot\cdot} = \frac{1}{2}\omega^{\mu\nu}\mathfrak{T}_{\mu\nu}\Psi^{\cdot\cdot}$ with $\mathfrak{T}_{\mu\nu}$ given by (16.6) follows directly from the fact that $\Psi^{\cdot\cdot}$ transforms according to the product representation—which in this infinitesimal case, noting (16.2), is given by (16.6). Finally, it may be mentioned that the matrices β_μ satisfy the so-called Duffin-Kemmer commutation rules (cf. Chapter V),

$$\beta_\mu\beta_\nu\beta_\rho + \beta_\rho\beta_\nu\beta_\mu = g_{\mu\nu}\beta_\rho + g_{\rho\nu}\beta_\mu \tag{16.9}$$

[1] cf. Murnaghan (38); van d. Waerden (32).

(as is readily verified from (16.7) and (14.3)) and the representation of the β-algebra given by (16.7) is not irreducible.* It can be reduced into irreducible constituents which are 10-, 5-, and 1-dimensional. We shall defer the discussion of these matters until Chapter V, which considers the general basis of the matrix-algebraic aspect of particle theory, but from the last sections it is already evident how close is the inter-relation of matrix-algebraic and field aspects.

§ 17. (*a*) Irreducible Tensors and Spinors

The irreducibility of a group representation guarantees by definition that the linear space spanned by the vectors (in the general sense) constituting the basis of the matrix representation is the entire carrier space; reducibility of a representation being equivalent to the existence of a common proper invariant subspace of the collection of matrices constituting the representation. Therefore, the components of the so-called irreducible quantities (with respect to a given group) span the entire carrier space of the irreducible representation to which they are said to belong—and there exists no invariant subspace other than the trivial zero space. In other perhaps more descriptive words, the components transform among themselves under the elements of the group and do not separate into two or more sets of components which, transforming only among themselves, would span two or more invariant subspaces.[1]

A tensor or spinor which transforms irreducibly under the Lorentz group is characterized by two positive integral or half-integral numbers (k, l), the so-called associated quantum numbers,[2] which also label the corresponding irreducible representation $\mathfrak{D}(k, l)$ of the proper Lorentz group which is induced in the carrier space $\mathfrak{R}(k, l)$. We denote by $\psi(k, l)$, for example, the corresponding irreducible quantities with $(2k + 1)(2l + 1)$ independent components ordered, in general, into a column matrix. The general irreducible quantity is in fact (cf. *infra*) a spinor with $2k$ undotted and $2l$ dotted indices, symmetric in each set of indices, and therefore having just the indicated number of independent components.

The basis for these conclusions is quite simple and essentially well known—though perhaps more familiar in the somewhat disguised form of the analysis of angular momentum operators in quantum mechanics.[3]

It would probably be more logical to obtain first the irreducible representations of the unitary-unimodular (spinor) group $\mathfrak{S}_{\mathscr{R}}$ which, according to the results of § 10, will be *at most* double-valued representations of the 3-dimensional rotation group; but we start with the rotation group $\mathscr{R}$ in order to make fuller use of the more

* Equivalently, an arbitrary second-rank 4-spinor is not irreducible under the proper Lorentz group; the same remark applies to the preceding first-rank spinors. Cf. § 17 for the classification of quantities irreducible under the *proper* Lorentz group and the Lorentz group.

[1] cf. Murnaghan (38); van der Waerden (32); Cartan (38*a*, *b*); Weyl (31).

[2] It is often stated that in the relativistic theory an angular momentum, spin in particular, is associated with two quantum numbers. However, it may be noted here that, as will be seen in Chapter IV, it is the sum $k + l$ of these numbers which enters the definition of the spin so that the characterization by k *and* l is of rather formal significance in this connection.

[3] cf. for example, van der Waerden (32); Weyl (31).

familiar results of quantum mechanics. In this case the integrability conditions (4.12) take the simple form

$$\mathfrak{T}_x\mathfrak{T}_y - \mathfrak{T}_y\mathfrak{T}_x = \mathfrak{T}_z \qquad \text{(cyclically)} \tag{17.1}$$

or the familiar C.R. form

$$\begin{aligned} [\mathscr{J}_z, \mathscr{J}^+] &= \mathscr{J}^+ \\ [\mathscr{J}_z, \mathscr{J}^-] &= -\mathscr{J}^- \\ [\mathscr{J}^+, \mathscr{J}^-] &= 2\mathscr{J}_z \end{aligned} \tag{17.2}$$

with $\mathscr{J}_x = i\mathfrak{T}_x$, etc., and $\mathscr{J}^{\pm} = \mathscr{J}_x \pm i\mathscr{J}_y$. In the usual way we find the simultaneous eigenvectors $V_m(m = j, j-1, \ldots -j)$ of $\mathscr{J}^2$, $\mathscr{J}_z$ which give the non-vanishing matrix elements

$$\begin{aligned} (m+1 \mid \mathscr{J}^+ \mid m) &= [(j-m)(j+m+1)]^{1/2} \\ (m-1 \mid \mathscr{J}^- \mid m) &= [(j+m)(j-m+1)]^{1/2} \\ (m \mid \mathscr{J}_z \mid m) &= m \end{aligned} \tag{17.3}$$

with $\mathscr{J}^2 V_m = j(j+1)V_m$.[4] This $(2j+1)$-dimensional vector space we denote by $\mathfrak{R}(j)$, and the equations (17.3) state that it is exhausted by the operations $\mathscr{J}_z$, $\mathscr{J}^+$, $\mathscr{J}^-$; or equivalently, that it is transformed into itself by the infinitesimal transformations $\mathfrak{T}_x$, $\mathfrak{T}_y$, $\mathfrak{T}_z$ (and hence all transformations of the representation) and is therefore invariant. By the same token there is no invariant subspace—which is to say that $\mathfrak{R}(j)$ is irreducible. We now turn to the question of the explicit realization of a system of quantities, that is basis vectors, transforming according to the irreducible representation defined by (17.3), which we denote by $\mathfrak{D}(j)$.

Let us look at the particular representative matrices of the $\mathfrak{T}$'s arising from the transformations of 3-vectors, i.e. the infinitesimal transformations in the representation of $\mathscr{R}$ *by itself*. From § 5 we have

$$\delta x^\alpha = \tfrac{1}{2}\omega^{\mu\nu}(\delta^\alpha_\mu g_{\beta\nu} - \delta^\alpha_\nu g_{\beta\mu})x^\beta = \tfrac{1}{2}\omega^{\mu\nu}\mathfrak{T}_{\mu\nu}{}^\alpha{}_\beta x^\beta \tag{17.4}$$

and specializing to the case of the rotation subgroup $\mathscr{R}$ we find

$$\mathfrak{T}_x = \mathfrak{T}_{23}{}^i{}_j = \begin{Vmatrix} 0 & 0 & 0 \\ 0 & 0 & -1 \\ 0 & 1 & 0 \end{Vmatrix} \tag{17.5}$$

$$\mathfrak{T}_y = \mathfrak{T}_{31}{}^i{}_j = \begin{Vmatrix} 0 & 0 & -1 \\ 0 & 0 & 0 \\ 1 & 0 & 0 \end{Vmatrix}$$

$$\mathfrak{T}_z = \mathfrak{T}_{12}{}^i{}_j = \begin{Vmatrix} 0 & -1 & 0 \\ 1 & 0 & 0 \\ 0 & 0 & 0 \end{Vmatrix}$$

[4] The statement that $\mathscr{J}^2$ commutes with $\mathscr{J}_x$, $\mathscr{J}_y$, $\mathscr{J}_z$ is, group-theoretically, simply the statement that it commutes with the infinitesimal transformations $\mathfrak{T}_x$, $\mathfrak{T}_y$, $\mathfrak{T}_z$ and hence with *all* transformations of the representation; so that by Schur's Lemma it is represented by a scalar matrix (multiple of the unit matrix) in the irreducible representation generated in this manifold of states. Note that in the irreducible representation $\mathfrak{D}(j)$, $-(\mathfrak{T}_x{}^2 + \mathfrak{T}_y{}^2 + \mathfrak{T}_z{}^2) = j(j+1)$ as illustrated by (17.5) and (17.8) for $\mathfrak{D}(1)$ and $\mathfrak{D}(\tfrac{1}{2})$, respectively.

These infinitesimal rotations serve as a basis for all infinitesimal rotations[5] of $\mathscr{R}$, and we shall find it convenient to indicate the content of (17.4) in the form

$$\mathfrak{T}_x \to \delta x = 0, \quad \delta y = -z, \quad \delta z = y \tag{17.6}$$
$$\mathfrak{T}_y \to \delta x = -z, \quad \delta y = 0, \quad \delta z = x$$
$$\mathfrak{T}_z \to \delta x = -y, \quad \delta y = x, \quad \delta z = 0$$

On the other hand we found in § 11 that the infinitesimal (spinor) transformations of $\mathfrak{S}_{\mathscr{R}}$ are

$$\delta\phi_A = -\tfrac{1}{4}\omega^{\mu\nu} S_{\mu\nu A}{}^B \phi_B = \tfrac{1}{2}\omega^{\mu\nu}\mathfrak{T}_{\mu\nu A}{}^B \phi_B \tag{17.7}$$

where, by (13.8),

$$\mathfrak{T}_{23A}{}^B = \tfrac{1}{2}\begin{Vmatrix} 0 & -i \\ -i & 0 \end{Vmatrix} = \frac{1}{2i}\sigma_x \tag{17.8}$$[6]

$$\mathfrak{T}_{31A}{}^B = \tfrac{1}{2}\begin{Vmatrix} 0 & -1 \\ 1 & 0 \end{Vmatrix} = \frac{1}{2i}\sigma_y$$

$$\mathfrak{T}_{12A}{}^B = \tfrac{1}{2}\begin{Vmatrix} -i & 0 \\ 0 & i \end{Vmatrix} = \frac{1}{2i}\sigma_z$$

and again

$$\frac{1}{2i}\sigma_x \to \delta\phi_1 = \frac{1}{2i}\phi_2 \qquad \delta\phi_2 = \frac{1}{2i}\phi_1 \tag{17.9}$$

$$\frac{1}{2i}\sigma_y \to \delta\phi_1 = -\frac{1}{2}\phi_2 \qquad \delta\phi_2 = \frac{1}{2}\phi_1$$

$$\frac{1}{2i}\sigma_z \to \delta\phi_1 = \frac{1}{2i}\phi_1 \qquad \delta\phi_2 = -\frac{1}{2i}\phi_2$$

Under the correspondence $\mathfrak{S}_{\mathscr{R}} \to \mathscr{R}$, the three infinitesimal transformations (17.8) correspond, respectively, to the three infinitesimal transformations (17.5); since according to the results toward the end of § 10*b* we have under the rotation group

$$x \sim \tfrac{1}{2}(-\phi_1^2 + \phi_2^2) \tag{17.10}$$

$$y \sim \frac{1}{2i}(\phi_2^2 + \phi_1^2)$$

$$z \sim \phi_1\phi_2$$

and, for example,

$$\delta x = 0 \qquad \delta y = -z \qquad \delta z = y \tag{17.11a}$$

corresponding to $\mathscr{J}_x$ according to (17.6),

$$\delta x = -\phi_1\,\delta\phi_1 + \phi_2\,\delta\phi_2 = 0 \tag{17.11b}$$

$$\delta y = \frac{1}{i}(\phi_2\,\delta\phi_2 + \phi_1\,\delta\phi_1) = -z$$

$$\delta z = \phi_1\,\delta\phi_2 + \phi_2\,\delta\phi_1 = y$$

corresponding to $\frac{1}{2i}\sigma_x$ according to (17.9). It is now readily seen that if instead of

[5] They may be described as corresponding to unit angular velocity. Cf. Cartan (38*a*, *b*); Weyl (31).

[6] We use the σ's to distinguish from the preceding $\mathscr{J}$'s. They will be recognized as the Pauli matrices.

the transformations (17.8) we take the set $\sigma^+ = \frac{1}{2}(\sigma_x + i\sigma_y)$, $\sigma^- = \frac{1}{2}(\sigma_x - i\sigma_y)$, σ_z, then in the carrier space with basic vectors

$$V_m = \frac{\phi_1{}^r \phi_2{}^s}{\sqrt{r!s!}} \qquad (r + s = 2j,\ r - s = 2m) \tag{17.12}$$

these three infinitesimal transformations of $\mathfrak{S}_{\mathscr{R}}$ induce the transformations (substituting in δV_m from (17.9))

$$\begin{aligned} \sigma^+ &\to \delta V_m = \sqrt{r(s+1)}\,V_{m-1} = [(j+m)(j-m+1)]^{1/2} V_{m-1} \\ \sigma^- &\to \delta V_m = \sqrt{s(r+1)}\,V_{m+1} = [(j-m)(j+m+1)]^{1/2} V_{m+1} \\ \sigma_z &\to \delta V_m = \tfrac{1}{2}(r-s)V_m = mV_m \end{aligned} \tag{17.13}$$

From this result, the equations (17.3) follow directly and it is clear that the monomials $\dfrac{\phi_1{}^r\,\phi_2{}^s}{\sqrt{r!\,s!}}$ coincide with the vectors V_m as indicated. The irreducible representations of $\mathfrak{S}_{\mathscr{R}}$ defined by (17.12, 17.13) coincide with the $\mathfrak{D}(j)$, and the $\mathfrak{D}(j)$ with half-integral j are double-valued representations of $\mathscr{R}$ since a rotation θ about the x^3-axis is represented (cf. § 10*b*)) by the spinor transformations $\| s^A{}_B \| = \pm \begin{Vmatrix} e^{-i\theta/2} & 0 \\ 0 & e^{i\theta/2} \end{Vmatrix}$ corresponding to which the vectors V_m are multiplied by $e^{-i(r-s)\theta/2} = e^{-im\theta}$ so that, in particular, V_j is multiplied by $e^{-ij\theta}$ which gives -1 for $\theta = 2\pi$ and j half-integral.

An arbitrary linear combination of the V_m may be written as $c^{AB\cdots}\ \phi_A \phi_B \ldots$ where the spinor $c^{AB\cdots}$ is *symmetric* in its $2j+1$ indices. These coefficients are therefore the components of a general vector of the space $\mathfrak{R}(j)$ transforming according to the irreducible representation $\mathfrak{D}(j)$ when a spinor transformation corresponding to a given rotation is applied to the variables ϕ_1, ϕ_2; that is, the components of an irreducible quantity under the group $\mathfrak{S}_{\mathscr{R}}$, and hence $\mathscr{R}$—which we denote by $\psi(j)$, for example; the $\psi(j)$ with integral (half-integral) j transforming under $\mathscr{R}$ according to single- (double-)valued representations.

In the analogous analysis of the proper Lorentz group $\mathscr{L}_4$ we note that the transformations $\mathfrak{T}^{\mu\nu}$ (cf. 4.12) and their C.R. can be realized by introducing two 3-vector operators $\mathfrak{K}$, $\mathfrak{L}$, for example, defined by

$$\begin{aligned} \mathfrak{K}_x &= \tfrac{1}{2}(i\mathfrak{T}^{23} + \mathfrak{T}^{41}) & \mathfrak{K}_y &= \tfrac{1}{2}(i\mathfrak{T}^{31} + \mathfrak{T}^{42}) & \mathfrak{K}_z &= \tfrac{1}{2}(i\mathfrak{T}^{12} + \mathfrak{T}^{43}) \\ \mathfrak{L}_x &= \tfrac{1}{2}(i\mathfrak{T}^{23} - \mathfrak{T}^{41}) & \mathfrak{L}_y &= \tfrac{1}{2}(i\mathfrak{T}^{31} - \mathfrak{T}^{42}) & \mathfrak{L}_z &= \tfrac{1}{2}(i\mathfrak{T}^{12} - \mathfrak{T}^{43}) \end{aligned} \tag{17.14}$$

Thus, the $\mathfrak{K}$'s and $\mathfrak{L}$'s form two sets of matrices which commute with each other while the members of each set satisfy (17.2). Here again we find the simultaneous eigenvectors $V_{m_k m_l}$ of $\mathfrak{K}^2$, $\mathfrak{K}_z$, $\mathfrak{L}^2$, $\mathfrak{L}_z$ which define the matrix elements

$$\begin{aligned} (m_k m_l \mid \mathfrak{K}_z \mid m_k m_l) &= m_k \\ (m_k + 1\ m_l \mid \mathfrak{K}^+ \mid m_k m_l) &= [(k - m_k)(k + m_k + 1)]^{1/2} \\ (m_k - 1\ m_l \mid \mathfrak{K}^- \mid m_k m_l) &= [(k + m_k)(k - m_k + 1)]^{1/2} \end{aligned} \tag{17.15}$$

with similar expressions for the representations of the $\mathfrak{L}$'s. The vectors $V_{m_k m_l}$ constitute the $(2k+1)(2l+1)$-dimensional space $\mathfrak{R}(k, l)$, and determine the irreducible representation $\mathfrak{D}(k, l)$ of $\mathfrak{S}_4$ and hence $\mathscr{L}_4$. It is readily shown that a system transforming according to $\mathfrak{D}(k, l)$ is realized, for example, by the basis vectors

$$V_{m_k m_l} = \frac{(\phi^1)^{k+m_k}(\phi^2)^{k-m_k}}{\sqrt{(k+m_k)!\,(k-m_k)!}} \cdot \frac{(\phi^{\dot{1}})^{l+m_l}(\phi^{\dot{2}})^{l-m_l}}{\sqrt{(l+m_l)!\,(l-m_l)!}} \tag{17.16}$$

An arbitrary linear combination of these basis vectors is given by the expression $c_{AB\ldots\dot{R}\dot{S}\ldots}\phi^A\phi^B\ldots\phi^{\dot{R}}\phi^{\dot{S}}\ldots$ where the spinor c is symmetric in its $2k$ undotted and $2l$ dotted indices. These are the prototype irreducible quantities denoted by $\psi(k, l)$.

In the following table we summarize these conclusions regarding the proper Lorentz group, giving the dimensions d_4 of the irreducible representations and corresponding irreducible spinors for the first few values of k and l.

TABLE I

k	l	d_4	Typical irreducible spinors	
0	0	1	ψ	*
$\frac{1}{2}$	0	2 }	ψ_A	
0	$\frac{1}{2}$	2 }	$\psi^{\dot{A}}$	
$\frac{1}{2}$	$\frac{1}{2}$	4	$\psi^{\dot{A}}{}_B$	**
1	0	3 }	ψ_{AB}	***
0	1	3 }	$\psi^{\dot{A}\dot{B}}$	
1	$\frac{1}{2}$	6 }	$\psi_{AB}{}^{\dot{C}}$	
$\frac{1}{2}$	1	6 }	$\psi^{\dot{A}\dot{B}}{}_C$	
$\frac{3}{2}$	0	4 }	ψ_{ABC}	
0	$\frac{3}{2}$	4 }	$\psi^{\dot{A}\dot{B}\dot{C}}$	
1	1	9	$\psi^{\dot{A}\dot{B}}{}_{CD}$	****
$\frac{3}{2}$	$\frac{1}{2}$	8 }	$\psi_{ABC}{}^{\dot{D}}$	
$\frac{1}{2}$	$\frac{3}{2}$	8 }	$\psi^{\dot{A}\dot{B}\dot{C}}{}_D$	
2	0	5 }	ψ_{ABCD}	
0	2	5 }	$\psi^{\dot{A}\dot{B}\dot{C}\dot{D}}$	

These indicate some of the 4-tensor equivalents—when they exist. (*) scalar; (**) vector; (***) self-dual second-rank skew-tensor; (****) second-rank symmetric tensor with zero spur.

The bracketed pairs of numbers in this table correspond to the fact that a reflection turns the representation $\mathfrak{D}(k, l)$ into the representation $\mathfrak{D}(l, k)$ so that they must appear together in a representation of the full Lorentz group. This irreducible representation of the Lorentz group is therefore the direct sum $\mathfrak{D}(k, l) \oplus \mathfrak{D}(l, k)$ of dimension $2(2k+1)(2l+1)$ if $k \neq l$, and simply $\mathfrak{D}(k, k)$ of dimension $(2k+1)^2$ if $k = l$. A typical example of the former instance is that of the 4-dimensional space of 4-spinors Ψ^a which we call the γ-space $\mathfrak{R}(\gamma)$. This is the direct sum of two subspaces $\mathfrak{R}(\frac{1}{2}, 0)$ and $\mathfrak{R}(0, \frac{1}{2})$ transforming respectively according to the irreducible representations $\mathfrak{D}(\frac{1}{2}, 0)$ and $\mathfrak{D}(0, \frac{1}{2})$ of the proper Lorentz group.

As a further illustration of the interrelation of the representations and associated quantities from the above table, we may re-examine the 16-dimensional β-space $\mathfrak{R}(\beta)$ which occurred in § 16 as the direct product space $\mathfrak{R}(\gamma) \times \mathfrak{R}(\gamma)$ of arbitrary second-rank 4-spinors. As previously remarked, these are not irreducible under the proper Lorentz group, and the space $\mathfrak{R}(\gamma) \times \mathfrak{R}(\gamma)$ can be reduced with respect to the proper Lorentz group in the usual way by reducing each product in the product representation, viz.

$$\{\mathfrak{D}(\tfrac{1}{2}, 0) \oplus \mathfrak{D}(0, \tfrac{1}{2})\} \times \{\mathfrak{D}(\tfrac{1}{2}, 0) \oplus \mathfrak{D}(0, \tfrac{1}{2})\} = \mathfrak{D}(1, 0) \oplus 2\mathfrak{D}(\tfrac{1}{2}, \tfrac{1}{2}) \oplus \mathfrak{D}(0, 1) \oplus 2\mathfrak{D}(0, 0) \quad (17.17)$$

with the check by degrees, $16 = 3 + 8 + 3 + 2$. Thus, the space $\mathfrak{R}(\beta)$ is the direct sum of four subspaces $\mathfrak{R}(1, 0)$, $\mathfrak{R}(0, 1)$, $\mathfrak{R}(\tfrac{1}{2}, \tfrac{1}{2})$ and $\mathfrak{R}(0, 0)$, the last two occurring twice. We may write this result as

$$\begin{aligned} \mathfrak{R}(\beta) &= \mathfrak{R}(\beta_{10}) \oplus \mathfrak{R}(\beta_5) \oplus \mathfrak{R}(\beta_1) \\ &= \mathfrak{R}(0, 1) \oplus \mathfrak{R}(\tfrac{1}{2}, \tfrac{1}{2}) \oplus \mathfrak{R}(1, 0) \\ &\quad \oplus \mathfrak{R}(\tfrac{1}{2}, \tfrac{1}{2}) \oplus \mathfrak{R}(0, 0) \\ &\quad \oplus \mathfrak{R}(0, 0) \end{aligned} \quad (17.18)$$

the three separate sets of terms corresponding to the 10-, 5-, and 1-dimensional representations mentioned in § 16. From the above table these are all tensor representations and, in particular, in accord with the general composition of second-rank 4-spinors obtained in §15(*b*).

From the work of § 10(*b*) it follows that if we restrict ourselves to the group of spatial rotations (i.e. the unitary unimodular transformations) $\psi_{AB\ldots}^{\dot{R}\dot{S}\ldots}$ transforms like $\psi_{AB\ldots,\,RS\ldots}$, i.e. $\psi(k, l)$ say, behaves like the direct product $\psi(k) \times \psi(l)$ of two quantities irreducible under this group, transforming according to the, in general, reducible representation $\mathfrak{D}(k) \times \mathfrak{D}(l)$ of $\mathscr{R}$. The product space $\mathfrak{R}(k) \times \mathfrak{R}(l)$ can be reduced with respect to the rotation group with the familiar result

$$\mathfrak{D}(k) \times \mathfrak{D}(l) = \mathfrak{D}(k + l) \oplus \mathfrak{D}(k + l - 1) \oplus \ldots \oplus \mathfrak{D}(|k - l|) \quad (17.19)$$

so that $\psi(k, l)$ reduces into the direct sum of several irreducible quantities $\psi(j)$, each of $2j + 1$ components with $j = k + l,\ k + l - 1, \ldots |k - l|$. From this and a previous result, it follows that unreduced quantities $\phi(k, l)$ with integral (half-integral) $k + l$ belong to single- (double-)valued representations of the rotation group.

§ 17. (*b*) **The Irreducible Representations of the Lorentz Group**[1]

Before proceeding to the consideration of the physical principles, it will be advantageous to consider, in more detail, the *general* irreducible representation of the homogeneous Lorentz group. This analysis will also serve to introduce, in a natural and simple way, the important concepts of the Dirac spinor matrices and expinors, and illustrate the process of reduction of a representation.

[1] Harish-Chandra (47*b*). Cf. Bargmann, V. (47); Gelfand, I. M., and Neumark, M. (46); Gelfand, I. M., and Yaglon, A. M. (48*a*, *b*, *c*).

In essence, we shall find that the general irreducible representation $\mathfrak{D}(\kappa, \kappa^\dagger)$ of the proper Lorentz group can be characterized by an ordered pair of complex numbers $(\kappa, \kappa^\dagger)$ such that $2(\kappa - \kappa^\dagger)$ is an integer, positive, negative, or zero, and if $\kappa, \kappa^\dagger$ are of the form $\frac{1}{4}n(n \neq -2)$, there exist two irreducible representations $\mathfrak{D}^+(\kappa, \kappa^\dagger)$ and $\mathfrak{D}^-(\kappa, \kappa^\dagger)$ corresponding to the pair $(\kappa, \kappa^\dagger)$. The *finite* representations correspond to $\mathfrak{D}^-(\kappa, \kappa^\dagger)$ with $\kappa, \kappa^\dagger$ both of the form $\frac{1}{2}n(n \neq -1)$, the connection between the usual characterization[2] (p, q) and $(\kappa, \kappa^\dagger)$ being given by $p = |\kappa + \frac{1}{2}| - \frac{1}{2}$, $q = |\kappa^\dagger + \frac{1}{2}| - \frac{1}{2}$. The *finite* representations are not unitary, and in general $\mathfrak{D}(\kappa, \kappa^\dagger)$ is unitary only if either

$$\kappa = -\tfrac{1}{2} + i\nu + \tfrac{1}{2}n, \quad \kappa^\dagger = -\tfrac{1}{2} + i\nu - \tfrac{1}{2}n \tag{17.20a}$$

or

$$\kappa = \kappa^\dagger = -\tfrac{1}{2} + \nu, \quad |\nu| \leq \tfrac{1}{2} \tag{17.20b}$$

where ν is an arbitrary real number. However, in the particular case for which $\kappa, \kappa^\dagger$ are both of the form $\frac{1}{4}n(n \neq -2)$, $\mathfrak{D}^-(\kappa, \kappa^\dagger)$ is unitary only if $\frac{1}{2} \geqslant |\kappa + \frac{1}{2}| = |\kappa^\dagger + \frac{1}{2}|$, while $\mathfrak{D}^+(\kappa, \kappa^\dagger)$ is unitary whenever $|\kappa + \frac{1}{2}| = |\kappa^\dagger + \frac{1}{2}|$.

(i) *The General Representation*

For the purposes of the present work, we introduce, instead of the previous $\mathfrak{K}$, $\mathfrak{L}$ of (17.14), vectors $\boldsymbol{K}$ and $\boldsymbol{L}$ defined by

$$K_1 = i\mathfrak{T}^{23}, \quad K_2 = i\mathfrak{T}^{31}, \quad K_3 = i\mathfrak{T}^{12} \tag{17.21}$$
$$L_1 = i\mathfrak{T}^{14}, \quad L_2 = i\mathfrak{T}^{24}, \quad L_3 = i\mathfrak{T}^{34}$$

with which the commutation rules (4.12) of the infinitesimal transformations (or nucleus) take the form (cyclically)

$$[K_1, K_2]_- = iK_3 \tag{17.22a}$$
$$[K_1, L_2]_- = [L_1, K_2] = iL_3 \tag{17.22b}$$
$$[L_1, L_2]_- = -iK_3 \tag{17.22c}$$

We consider an irreducible representation of the proper Lorentz group, denoting the representation space by $\mathfrak{R}$, and reduce this space with respect to the subgroup of spatial rotations $\mathscr{R}$ for which it is known that every representation is completely reducible into a direct sum of irreducible representations, each of finite degree and unitary. As we have seen, every irreducible representation of $\mathscr{R}$ is characterized by an integral or half-integral number $k \geqslant 0$, so that $\mathfrak{R}$ is decomposed into a direct sum of subspaces which are all irreducible under $\mathscr{R}$; and we assume that there occurs at most one subspace $\mathfrak{R}(k)$ corresponding to any k. The coordinate system in $\mathfrak{R}(k)$ can always be adapted to the representation of the matrices of K_1, K_2, K_3, according to

$$(m \mid K_3 \mid m) = m \tag{17.23}$$
$$(m \mid K^+ \mid m-1) = (m-1 \mid K^- \mid m) = [(k+m)(k-m+1)]^{1/2}$$
$$\boldsymbol{K}^2 = k(k+1)$$

[2] For convenience here we use (p, q) instead of the previous notation (k, l).

The vector character of $\boldsymbol{L}$ determines its matrix elements completely, apart from multiplying factors a which may depend on k (cf. for example, Weyl (31))

$$\left.\begin{aligned}(k, m \mid L^+ \mid k-1, m) &= -a^-(k)\,[(k+m)(k+m-1)]^{1/2}\\ (k, m-1 \mid L^- \mid k-1, m) &= a^-(k)\,[(k-m+1)(k-m)]^{1/2}\\ (k, m \mid L_3 \mid k-1, m) &= a^-(k)\,[(k+m)(k-m)]^{1/2}\end{aligned}\right\} \tag{17.24a}$$

$$\left.\begin{aligned}(k, m \mid L^+ \mid k, m-1) &= a(k)\,[(k+m)(k-m+1)]^{1/2}\\ (k, m-1 \mid L^- \mid k, m) &= a(k)\,[(k-m+1)(k+m)]^{1/2}\\ (k, m \mid L_3 \mid k, m) &= a(k)m\end{aligned}\right\} \tag{17.24b}$$

$$\left.\begin{aligned}(k, m \mid L^+ \mid k+1, m-1) &= a^+(k)\,[(k-m+1)(k-m+2)]^{1/2}\\ (k, m-1 \mid L^- \mid k+1, m) &= -a^+(k)\,[(k+m)(k+m+1)]^{1/2}\\ (k, m \mid L_3 \mid k+1, m) &= a^+(k)\,[(k+m+1)(k-m+1)]^{1/2}\end{aligned}\right\} \tag{17.24c}$$

Now, the quantities $J^2 = -\frac{1}{2}\mathfrak{T}^{\kappa\lambda}\mathfrak{T}_{\kappa\lambda}$ and $\mathfrak{T} = -\frac{1}{2}\epsilon^{\kappa\lambda\mu\nu}\mathfrak{T}_{\kappa\lambda}\mathfrak{T}_{\mu\nu}$ are invariants of the proper Lorentz group (cf. Casimir (31)) and hence may be used to characterize the irreducible representations of $\mathscr{L}_4$, in which they are obviously represented by numerical constants.

Explicitly,

$$\begin{aligned}J^2 &= \boldsymbol{K}^2 - \boldsymbol{L}^2 = k(k+1) - (L^+L^- + L_3{}^2 + K_3)\\ \mathfrak{T} &= (\boldsymbol{K}\cdot\boldsymbol{L})\end{aligned} \tag{17.25}$$

and substituting from (17.23, 24) we find

$$\begin{aligned}J^2 = k(k+1)\{1-a^2(k)\} - m\{1+a^2(k)\} &- a^+(k)a^-(k+1)(k-m+1)(2k+3)\\ &- a^-(k)a^+(k-1)(k+m)(2k-1)\end{aligned} \tag{17.26a}$$

$$\mathfrak{T} = k(k+1)a(k) \tag{17.26b}$$

where $a^-(k) \equiv 0$ if $k = 0, \frac{1}{2}$. For $k \neq 0$ there are at least two distinct values of m for which (17.26a) holds, so the coefficient of m must vanish, i.e.

$$\{1 + a^2(k)\}k = a^+(k)a^-(k+1)k(2k+3) - a^-(k)a^+(k-1)k(2k-1) \tag{17.27a}$$

$$\begin{aligned}J^2 = k(k+1)\{1-a^2(k)\} &- a^+(k)a^-(k+1)(k+1)(2k+3)\\ &- a^-(k)a^+(k-1)k(2k-1)\end{aligned} \tag{17.27b}$$

the factor k being introduced in (17.27a) so as to include the case $k = 0$. From (17.27) it follows that

$$a^+(k)a^-(k+1) = \frac{k(k+2) - k^2a^2(k) - J^2}{(2k+1)(2k+3)} \tag{17.28}$$

If the smallest value of k appearing in the decomposition of $\mathfrak{R}$ into $\mathfrak{R}(k)$ is $j \neq 0$, then, noting that $a^-(j) = 0$, we find

$$J^2 = j^2 - \left(\frac{\mathfrak{T}}{j}\right)^2 - 1 \tag{17.29}$$

from (17.26b, 27) with $k = j$, and from (17.28)

$$a^+(k)a^-(k+1) = \frac{(k-j+1)(k+j+1)}{(2k+1)(2k+3)}\left\{1 + \frac{\mathfrak{T}^2}{(k+1)^2j^2}\right\}(k \geqslant j) \qquad (17.30)$$

The same form holds for $j = 0$ provided $\mathfrak{T}$ is considered as μj with $\mu = \mathfrak{T}/j$ a definitely given number, so that $\mathfrak{T}$ vanishes automatically for $j = 0$. Similarly we find from (17.26b)

$$a(k) = \frac{\mu j}{k(k+1)} \qquad (k \geqslant j) \qquad (17.31)$$

with the understanding that $a(0) \equiv 0$.

By multiplying all the eigenvectors in each subspace $\mathfrak{R}(k)$ by a suitable normalization factor, it can always be arranged that

$$a^+(k) = a^-(k+1) = \left[\frac{(k-j+1)(k+j+1)}{(2k+1)(2k+3)}\right]^{1/2}\left[1 + \frac{\mu^2}{(k+1)^2}\right]^{1/2} \qquad (17.32)$$

In particular, if the representation of $\mathscr{L}_4$ is unitary, $\boldsymbol{K}$ and $\boldsymbol{L}$ are Hermitian, so that $a(k)$ is real and $a^+(k) = \bar{a}^-(k+1)$. The normalization factors are then unimodular and the corresponding transformation of $\mathfrak{R}$ is unitary. Therefore, every unitary representation of the Lorentz group can, by suitable unitary transformation, be brought to a form for which (17.32) holds. Then equations (17.30, 31, 32) show that, apart from equivalence, an irreducible representation is completely fixed by the two numbers j, μ, and all unitary representations can be transformed to the form (17.32) without affecting the unitarity. Hence, all irreducible representations which are equivalent to a unitary representation are those for which the right-hand sides of (27.31, 32) are real—so that the necessary and sufficient condition for a representation to be unitary is that $\mathfrak{T} = \mu j$ be real and $\mu^2 > -1$.

(ii) *The Spinor-Matrices u_A, v_A*

To proceed farther we must introduce spinors and the Dirac (36) spinor matrices, which arise essentially in the following way: A totally symmetric spinor is completely defined if we know the number of dotted or undotted indices which have a given value, say 1. Therefore such a spinor, say $\phi^{AB\cdots}_{KL\cdots}$ with $2k$ undotted and $2l$ dotted indices can be represented by a quantity $\phi^{\dot r}_s$ whose indices $\dot r$, s denote how many of the dotted and undotted spinor indices are equal to 1. Then s has $2k+1$ values, $\dot r$ has $2l+1$ values, and it remains to find a matrix which relates spinor indices to the indices $\dot r$, s in order to establish a correspondence between the quantity $\phi^{\dot r}_s$ and the symmetric spinor $\phi^{A\dot B\cdots}_{KL\cdots}$. We can, of course, also relate such a quantity to a spinor without special symmetry properties, but the correspondence so obtained is not $(1-1)$, since $\phi^{\dot r}_s$ only defines the *symmetric* part of the related spinor. Now, since on the one hand the quantity $\phi^{\dot r}_s$ is irreducible, while on the other a spinor without special properties transforms according to a product representation of the Lorentz group which contains $\mathfrak{D}(\frac{1}{2}, 0)$ $2k$-times and $\mathfrak{D}(0, \frac{1}{2})$ $2l$-times, the correlation of such a quantity to an unsymmetric spinor is equivalent to the attainment of an irreducible subspace—and precisely that subspace of the original representation with the largest number of dimensions. Consequently, the entity which relates the spinor indices to the indices $\dot{r}$, s can be found by application of the theory of reduction of representations, as noted by Fierz (39).

Consider, therefore, the simple product representation $\mathfrak{D}(\frac{1}{2}) \times \mathfrak{D}(k)$ of the spatial rotation group. $\mathfrak{D}(\frac{1}{2})$ and $\mathfrak{D}(k)$ are characterized by the infinitesimal transformations

$$\boldsymbol{K}(\tfrac{1}{2}) = \tfrac{1}{2}\boldsymbol{\sigma} \quad \text{and} \quad \boldsymbol{K}(k)$$

and we note that the matrix

$$A(k) = \sum_{s=1}^{3} \sigma_s \times K_s(k) \tag{17.33}$$

by its definition, commutes with the infinitesimal transformations of $\mathfrak{D}(\frac{1}{2}) \times \mathfrak{D}(k)$ and therefore with all matrices of the product representation. Consequently, if $A(k)$ is diagonalized, the representation $\mathfrak{D}(\frac{1}{2}) \times \mathfrak{D}(k)$ is necessarily decomposed into the direct sum of its irreducible parts $\mathfrak{D}(k + \frac{1}{2})$ and $\mathfrak{D}(k - \frac{1}{2})$, so that the reduction of $\mathfrak{D}(\frac{1}{2}) \times \mathfrak{D}(k)$ is referred back to the problem of diagonalizing[3] $A(k)$.

A matrix which diagonalizes $A(k)$ has been given by Dirac, and we write it in the form

$$U = (2k+1)^{-1/2} \begin{Vmatrix} u^1(k+\frac{1}{2}) & u^2(k+\frac{1}{2}) \\ v^1(k) & v^2(k) \end{Vmatrix} \tag{17.34}$$

$$U^{-1} = (2k+1)^{-1/2} \begin{Vmatrix} v_1(k+\frac{1}{2}) & u_1(k) \\ v_2(k+\frac{1}{2}) & u_2(k) \end{Vmatrix} (-1)^{2k+1}$$

where $u^A(k)$ is a rectangular spinor-matrix of $2k + 1$ rows and $2k$ columns and $v^A(k)$ is a rectangular spinor-matrix of $2k$ rows and $2k + 1$ columns. Correspondingly,

$$A(k) = \begin{Vmatrix} K_3 & K^- \\ K^+ & -K_3 \end{Vmatrix} = \begin{Vmatrix} K_1{}^1 & K_1{}^2 \\ K_2{}^1 & K_2{}^2 \end{Vmatrix} \tag{17.35a}$$

where

$$K_{12} = K_{21} = -K_3, \quad K_{11} = K^-, \quad K_{22} = -K^+ \tag{17.35b}$$

are the components of the spinor K_{AB} associated with the vector $\boldsymbol{K}$. From (17.33) or (17.35*a*) it is readily seen that $A(k)$ satisfies the equation $A(A + 1) = k(k + 1)$ so that its eigenvalues are k and $-(k + 1)$, the first of which must occur $(2k + 2)$ times and the second $2k$ times.

From the condition that the u^A, v^B form a matrix U which diagonalizes $A(k)$:

$$A(k) = U^{-1} \begin{Vmatrix} k & 0 \\ 0 & -(k+1) \end{Vmatrix} U,$$

we find

$$u_A(k)v^B(k) = \{k\, \delta_A{}^B - K_A{}^B(k)\}(-1)^{2k+1} \tag{17.36a}$$

$$v_A(k+\tfrac{1}{2})u^B(k+\tfrac{1}{2}) = \{K_A{}^B(k) + (k+1)\, \delta_A{}^B\}(-1)^{2k+1} \tag{17.36b}$$

Similarly, from the conditions $UU^{-1} = U^{-1}U = I$:

$$u^A(k+\tfrac{1}{2})v_A(k+\tfrac{1}{2}) = v^A(k)u_A(k) = (2k+1)(-1)^{2k+1} \tag{17.36c}$$

$$u^A(k+\tfrac{1}{2})u_A(k) = v^A(k)v_A(k+\tfrac{1}{2}) = 0 \tag{17.36d}$$

[3] cf. Casimir, H.—van der Waerden, B. L. (35).

For our immediate purposes the equations (17.36d) are particularly important, since they show that the rectangular matrices $u^A(k+\frac{1}{2})u^B(k)$ and $v^A(k)v^B(k+\frac{1}{2})$ are spinors, symmetric in A and B. From the $u(k)$, $v(k)$, we can form the single-column matrices P and the single-row matrices R:

$$P^{A_t}(k) = P^{A_1\dots A_{2k}}(k) = u^{A_1}(k)u^{A_2}(k-\tfrac{1}{2})\dots u^{A_{2k}}(\tfrac{1}{2}) \qquad (17.37a)$$

$$R^{A_t}(k) = R^{A_1\dots A_{2k}}(k) = v^{A_1}(\tfrac{1}{2})v^{A_2}(1)\dots v^{A_{2k}}(k) \qquad (17.37b)$$

which are symmetric in all spinor indices and satisfy the relations

$$P^{A_t}(k) \times R_{A_t}(k) = (2k)!(-1)^{k(2k-1)} \qquad (17.38a)$$

$$R_{A_t}(k)P^{B_r}(k) = \{\Sigma(\text{Perm } A_t)\ \delta^{B_1}_{A_1}\delta^{B_2}_{A_2}\dots\}(-1)^{k(2k-1)} \qquad (17.38b)$$

The P, R can be written as $P_s{}^{AB\cdots}$ and $R_s{}^{AB\cdots}$ where the index s takes on $2k+1$ values and with respect to which these two forms are matrices of one column and one row, respectively. By means of the operators $P_s{}^{AB\cdots}(k)$ we can relate to a spinor $\phi_{AB\dots}$ of rank $2k$ a quantity ϕ_s which has but one index with a range of $2k+1$ values. Conversely, by means of $R_{s\,AB\dots}(k)$ a symmetric spinor can be related to every ϕ_s:

$$P_s{}^{AB\cdots}(k)\phi_{AB\dots} = \sqrt{(2k)!}\ \phi_s \qquad (17.39a)$$

$$R_{s\,AB\dots}(k)\phi_s = \sqrt{(2k)!}\ \phi_{AB} \qquad (17.39b)$$

If we apply the operator P to a non-symmetric spinor, then the operation is no longer (1–1) reversible (cf. (17.38)); only the symmetrized spinor is recovered, so that we may use $P_s{}^{AB\cdots}$ as the symmetrizer.

Analogous operators can be defined in the dotted spinors, e.g.

$$P^{\dot s}{}_{\dot A\dot B\dots}(l), \quad R^{\dot s\,\dot A\dot B\cdots}(l)$$

which can be built up from the $u^{\dot A}$, $v^{\dot B}$ and satisfy the same relations as the operators with undotted indices, if in equations (17.36–39) we replace each spinor by the conjugate-contragredient form. We can also construct quantities of the type

$$v^A(k)v^B(k+\tfrac{1}{2})\dots v^R(r)$$

$$u^A(k)u^B(k-\tfrac{1}{2})\dots u^M(m)$$

which convert only a part of the indices of a spinor into an index s, so that we obtain quantities labelled by both kinds of indices. In particular, we can build quantities $\phi_s{}^A$, $\phi^{\dot t B}$ which have only one index of each type, and these play an important role in formulating the Dirac equations for higher spin (cf. § 28(b)).

An explicit realization of the spinor matrices u^A, v^A is readily constructed; in particular, satisfying the unitarity condition $v^A(k) = (-)^{2k+1}u_A{}^*(k)$ on $U(k)$ which is convenient for later applications, viz.

$$\| u_1(k) \|_{sr} = \sqrt{s}\ \delta_{s-1,\,r} \qquad \| v^1(k) \| = (-)^{2k+1}\sqrt{r+1}\ \delta_{r+1,\,s} \qquad (17.39')$$

$$\| u_2(k) \|_{sr} = \sqrt{2k-s}\ \delta_{s,\,r} \qquad \| v^2(k) \| = (-)^{2k+1}\sqrt{2k-r}\ \delta_{r,\,s}$$

where r denotes an index which gives the number of indices 1 of a spinor and ranges from 0 to $2k - 1$, and s is a similar index with the range 0 to $2k$. The first three sets are then (listing in the same order)

$$\begin{Vmatrix} 0 \\ 1 \end{Vmatrix} \qquad \begin{Vmatrix} 0 & 1 \end{Vmatrix} \tag{17.39''}$$

$$\begin{Vmatrix} 1 \\ 0 \end{Vmatrix} \qquad \begin{Vmatrix} 1 & 0 \end{Vmatrix}$$

. . . .

$$\begin{Vmatrix} 0 & 0 \\ 1 & 0 \\ 0 & \sqrt{2} \end{Vmatrix} \qquad -\begin{Vmatrix} 0 & 1 & 0 \\ 0 & 0 & \sqrt{2} \end{Vmatrix}$$

$$\begin{Vmatrix} \sqrt{2} & 0 \\ 0 & 1 \\ 0 & 0 \end{Vmatrix} \qquad -\begin{Vmatrix} \sqrt{2} & 0 & 0 \\ 0 & 1 & 0 \end{Vmatrix}$$

. . . .

$$\begin{Vmatrix} 0 & 0 & 0 \\ 1 & 0 & 0 \\ 0 & \sqrt{2} & 0 \\ 0 & 0 & \sqrt{3} \end{Vmatrix} \qquad \begin{Vmatrix} 0 & 1 & 0 & 0 \\ 0 & 0 & \sqrt{2} & 0 \\ 0 & 0 & 0 & \sqrt{3} \end{Vmatrix}$$

$$\begin{Vmatrix} \sqrt{3} & 0 & 0 \\ 0 & \sqrt{2} & 0 \\ 0 & 0 & 1 \\ 0 & 0 & 0 \end{Vmatrix} \qquad \begin{Vmatrix} \sqrt{3} & 0 & 0 & 0 \\ 0 & \sqrt{2} & 0 & 0 \\ 0 & 0 & 1 & 0 \end{Vmatrix}$$

. . . .

(iii) *The Spinor-Matrices* U_A, V_A

To return to the problem originally posed, we shall find it convenient to make a transformation, amounting to a phase change, to new spinor matrices u_A, v_A which satisfy the conditions $u_A(k) = \bar{v}^A(k)$. The defining conditions (17.36) then take the more convenient form

$$K_{AB}(k) = u_A(k)v_B(k) - k\epsilon_{AB} = v_A(k + \tfrac{1}{2})u_B(k + \tfrac{1}{2}) + (k + 1)\epsilon_{AB} \tag{17.40a}$$

$$u_A(k)v^A(k) = 2k \qquad v^A(k)u_A(k) = 2k + 1 \tag{17.40b}$$

$$u^A(k + \tfrac{1}{2})u_A(k) = v^A(k)v_A(k + \tfrac{1}{2}) = 0 \tag{17.40c}$$

and

$$v_A(k + \tfrac{1}{2})u_B(k + \tfrac{1}{2}) - u_B(k)v_A(k) = \epsilon_{BA} \tag{17.40d}$$

or,

$$K_{AB} = U_A V_B - k\epsilon_{AB} = V_A U_B + (k + 1)\epsilon_{AB} \tag{17.41a}$$

$$U_A V^A = 2k \qquad V^A U_A = 2(k + 1) \tag{17.41b}$$

$$U_A U^A = V^A V_A = 0 \tag{17.41c}$$

$$V_A U_B - U_B V_A = \epsilon_{BA} \tag{17.41d}$$

in terms of the two matrices U^A, V^A whose non-zero elements are defined by

$$(k \mid U^A \mid k - \tfrac{1}{2}) = u^A(k) \qquad (k - \tfrac{1}{2} \mid V^A \mid k) = v^A(k) \tag{17.42a}$$

In this latter formulation, k now represents the diagonal matrix whose component in $\mathfrak{R}(k')$ is k' times the unit matrix, and we note that the V^A, U^A have the *step* property

$$\begin{aligned} V^A f(k) &= f(k + \tfrac{1}{2}) V^A \\ U^A f(k) &= f(k - \tfrac{1}{2}) U^A \end{aligned} \tag{17.42b}$$

To the vector $\boldsymbol{L}$ there again corresponds a spinor L_{AB}: $L_{12} = L_{21} = -L_3$, $L_{11} = L^-$, $L_{22} = -L^+$ with the commutation relations (cf. (17.22))

$$[K_{AB}, K_{CD}]_- = \tfrac{1}{2}\{\epsilon_{CA} K_{DB} + \epsilon_{DA} K_{CB} + \epsilon_{CB} K_{DA} + \epsilon_{DB} K_{CA}\} \tag{17.43a}$$

$$[K_{AB}, L_{CD}]_- = [L_{AB}, K_{CD}]_- = \tfrac{1}{2}\{\epsilon_{CA} L_{DB} + \dots\} \tag{17.43b}$$

$$[L_{AB}, L_{CD}]_- = -\tfrac{1}{2}\{\epsilon_{CA} K_{DB} + \dots\} \tag{17.43c}$$

$$-\tfrac{1}{2}K^{AB}K_{AB} = \boldsymbol{K}^2 \qquad -\tfrac{1}{2}L^{AB}L_{AB} = \boldsymbol{L}^2 \qquad -\tfrac{1}{2}K^{AB}L_{AB} = (\boldsymbol{K} \cdot \boldsymbol{L}) = \mathfrak{T} \tag{17.43d}$$

and we introduce

$$\mathfrak{T}_{AB} = \tfrac{1}{2}(K_{AB} - iL_{AB}) \tag{17.44}$$

$$[\mathfrak{T}_{AB}, \mathfrak{T}_{CD}]_- = \tfrac{1}{2}\{\epsilon_{CA} \mathfrak{T}_{DB} + \dots\}$$

$$-\tfrac{1}{2}\mathfrak{T}^{AB}\mathfrak{T}_{AB} = \tfrac{1}{4}(J^2 - 2i\mathfrak{T}) = \tfrac{1}{4}(j - i\mu)^2 - \tfrac{1}{4}$$

From equations (17.24, 30, 31) and (17.41), it follows that

$$\begin{aligned} L_{AB} = \frac{\mathfrak{T}}{k(k+1)} K_{AB} &+ i\sqrt{\left(\frac{(k-j)(k+j)}{(2k-1)(2k+1)}\right)}\sqrt{\left(1 + \frac{\mu^2}{k^2}\right)} U_A U_B \\ &+ V_A V_B i\sqrt{\left(\frac{(k-j)(k+j)}{(2k-1)(2k+1)}\right)}\sqrt{\left(1 + \frac{\mu^2}{k^2}\right)} \end{aligned} \tag{17.45}$$

where, as before, the *now* matrix square root is to be taken positive; and this relation, together with (17.24, 41), gives directly the matrix elements of the u_A, v_A.

In the product space $\mathfrak{R} \times \mathfrak{R}(\tfrac{1}{2}, 0)$, where again $\mathfrak{R}(\tfrac{1}{2}, 0)$ is a two-dimensional space transforming according to the fundamental spinor representation $\mathfrak{D}(\tfrac{1}{2}, 0)$ of the Lorentz group, the components of a vector may be written as ψ_A with ψ_1, ψ_2 forming one-column matrices. We follow essentially the procedure outlined in the preceding section, defining

$$\begin{aligned} \psi &= V^A \frac{1}{\sqrt{(2k+1)}} \psi_A \\ \psi^\dagger &= U^A \frac{1}{\sqrt{(2k+1)}} \psi_A \end{aligned} \tag{17.46a}$$

(and conversely (cf. 17.41))

$$\psi_A = \frac{1}{\sqrt{(2k+1)}} \{U_A \psi - V_A \psi^\dagger\} \tag{17.46b}$$

and consider the transformation

$$\psi_A{}' = \mathfrak{T}_A{}^B \psi_B \tag{17.47a}$$

where

$$\mathfrak{T}_{AB} = b(k)K_{AB} + c(k)U_A U_B + V_A V_B c(k) \tag{17.47b}$$

$$b(k) = \tfrac{1}{2}\left(1 - \frac{ij\mu}{k(k+1)}\right) \qquad c(k) = \tfrac{1}{2}\sqrt{\left(\frac{(k-j)(k+j)}{(2k-1)(2k+1)}\right)}\sqrt{\left(1+\frac{\mu^2}{k^2}\right)}$$

according to (17.44, 45). This induces a linear transformation on the variables ψ and $\psi^\dagger$, viz. (cf. 17.41, 42, 46, 47),

$$\left\|\begin{matrix}\psi' \\ \psi^{\dagger\prime}\end{matrix}\right\| \equiv \left\|\begin{matrix} V^A \dfrac{1}{\sqrt{(2k+1)}} & \psi_A{}' \\ U^A \dfrac{1}{\sqrt{(2k+1)}} & \psi_A{}' \end{matrix}\right\| = \left\|\begin{matrix} b(k+\frac{1}{2})(k+\frac{3}{2}) & c(k+\frac{1}{2})\sqrt{2k(2k+2)} \\ -c(k+\frac{1}{2})\sqrt{2k(2k+2)} & -b(k-\frac{1}{2})(k-\frac{1}{2}) \end{matrix}\right\| \cdot \left\|\begin{matrix}\psi \\ \psi^\dagger\end{matrix}\right\| \tag{17.48a}$$

and the eigenvalues of this transformation are readily found to be $\frac{1}{2}(1 \pm \gamma)$, where γ is given, apart from sign, by $\gamma^2 = (j - i\mu)^2$. If we introduce $\gamma_{k+\frac{1}{2}} = k + \frac{1}{2} - \dfrac{i\mathfrak{T}}{k+\frac{1}{2}}$, and the transformation

$$\left\|\begin{matrix}\psi_{\gamma-1} \\ \psi_{\gamma+1}\end{matrix}\right\| = \frac{1}{\sqrt{2}}\left\|\begin{matrix}\sqrt{(\gamma_{k+\frac{1}{2}}+\gamma)} & \sqrt{(\gamma_{k+\frac{1}{2}}-\gamma)} \\ \sqrt{(\gamma_{k+\frac{1}{2}}-\gamma)} & \sqrt{(\gamma_{k+\frac{1}{2}}+\gamma)}\end{matrix}\right\| \cdot \left\|\begin{matrix}\psi \\ \psi^\dagger\end{matrix}\right\| \tag{17.48b}$$

the square roots being chosen so that

$$\sqrt{(\gamma_{k+\frac{1}{2}}+\gamma)}\sqrt{(\gamma_{k+\frac{1}{2}}-\gamma)} = \sqrt{\{(k+\tfrac{1}{2}-j)(k+\tfrac{1}{2}+j)\}}\sqrt{\left(1+\frac{\mu^2}{(k+\frac{1}{2})^2}\right)},$$

then (17.48*a*) is brought to diagonal form:

$$\left\|\begin{matrix}\psi_{\gamma-1}{}' \\ \psi_{\gamma+1}{}'\end{matrix}\right\| = \frac{1}{\sqrt{2}}\left\|\begin{matrix}\sqrt{(\gamma_{k+\frac{1}{2}}+\gamma)} & \sqrt{(\gamma_{k+\frac{1}{2}}-\gamma)} \\ \sqrt{(\gamma_{k+\frac{1}{2}}-\gamma)} & \sqrt{(\gamma_{k+\frac{1}{2}}+\gamma)}\end{matrix}\right\| \cdot \left\|\begin{matrix}\psi' \\ \psi^{\dagger\prime}\end{matrix}\right\|$$

$$= \tfrac{1}{2}\left\|\begin{matrix}(1+\gamma) & 0 \\ 0 & (1-\gamma)\end{matrix}\right\| \cdot \left\|\begin{matrix}\psi_{\gamma-1} \\ \psi_{\gamma+1}\end{matrix}\right\| \tag{17.48c}$$

Corresponding to γ, we introduce a $\gamma^\dagger$ defined by $\gamma^{\dagger 2} = (j + i\mu)^2$, and write $\gamma = 2\kappa + 1$, $\gamma^\dagger = 2\kappa^\dagger + 1$. In this notation:

$$\gamma_{k+\frac{1}{2}} = k + \tfrac{1}{2} - \frac{i\mathfrak{T}}{k+\frac{1}{2}} = \frac{(k+\frac{1}{2})^2 - (\kappa^\dagger+\frac{1}{2})^2 + (\kappa+\frac{1}{2})^2}{(k+\frac{1}{2})}$$

$$\gamma_{k+\frac{1}{2}} + \gamma = \frac{(k+\kappa+1)^2 - (\kappa^\dagger+\frac{1}{2})^2}{(k+\frac{1}{2})}$$

$$\gamma_{k+\frac{1}{2}} - \gamma = \frac{(k-\kappa)^2 - (\kappa^\dagger+\frac{1}{2})^2}{(k+\frac{1}{2})}$$

$$i\mathfrak{T} = \tfrac{1}{4}(\gamma^{\dagger 2} - \gamma^2) = (\kappa^\dagger - \kappa)(\kappa^\dagger + \kappa + 1)$$

and if we put $\psi_{\gamma-1} \equiv \psi_{\kappa-\frac{1}{2}}$, $\psi_{\gamma+1} \equiv \psi_{\kappa+\frac{1}{2}}$, then it follows from (17.46, 48) that

$$\left.\begin{aligned}\psi_{\kappa-\frac{1}{2}} &= \mathfrak{V}^A(\kappa, \kappa^\dagger)\psi_A \\ \psi_{\kappa+\frac{1}{2}} &= \mathfrak{U}^A(\kappa+\tfrac{1}{2}, \kappa^\dagger)\psi_A\end{aligned}\right\} \tag{17.49}$$

where

$$\mathfrak{B}^A(\kappa, \kappa^\dagger) = \tfrac{1}{2}V^A\sqrt{\left\{\frac{(k + \frac{1}{2} - \kappa + \kappa^\dagger)(k - \frac{1}{2} - \kappa - \kappa^\dagger)}{k(k + \frac{1}{2})}\right\}} + \tfrac{1}{2}\sqrt{\left\{\frac{(k + 1 + \kappa + \kappa^\dagger)(k + \kappa - \kappa^\dagger)}{k(k + \frac{1}{2})}\right\}}\, U^A \tag{17.50}$$

$$\mathfrak{U}^A(\kappa, \kappa^\dagger) = \tfrac{1}{2}V^A\sqrt{\left\{\frac{(k + 1 + \kappa + \kappa^\dagger)(k + \kappa - \kappa^\dagger)}{k(k + \frac{1}{2})}\right\}} + \tfrac{1}{2}\sqrt{\left\{\frac{(k + \frac{1}{2} - \kappa + \kappa^\dagger)(k - \frac{1}{2} - \kappa - \kappa^\dagger)}{k(k + \frac{1}{2})}\right\}}\, U^A$$

With (17.46*b*) and (17.48*b*) we also find the connection

$$(2\kappa + 1)\psi_A = \mathfrak{U}_A(\kappa, \kappa^\dagger)\psi_{\kappa-\frac{1}{2}} - \mathfrak{B}_A(\kappa + \tfrac{1}{2}, \kappa^\dagger)\psi_{\kappa+\frac{1}{2}} \tag{17.51}$$

and when $\gamma = 2\kappa + 1 \neq 0$, the two eigenvalues $\frac{1}{2}(1 \pm \gamma)$ of (17.48*a*) are distinct, so that the two eigenvectors $\psi_{\kappa\pm\frac{1}{2}}$ are linearly independent. We shall not consider the many additional relations of interest which may be derived (cf. Harish-Chandra, *loc. cit.*), except to note the final results which contain all of them:

$$\mathfrak{U}_A(\kappa, \kappa^\dagger)\mathfrak{B}^A(\kappa, \kappa^\dagger) = 2\kappa \qquad \mathfrak{B}^A(\kappa, \kappa^\dagger)\mathfrak{U}_A(\kappa, \kappa^\dagger) = 2\kappa + 1 \tag{17.52a}$$

$$\mathfrak{U}_A(\kappa + \tfrac{1}{2}, \kappa^\dagger)\mathfrak{U}^A(\kappa, \kappa^\dagger) = \mathfrak{B}^A(\kappa, \kappa^\dagger)\mathfrak{B}_A(\kappa + \tfrac{1}{2}, \kappa^\dagger) = 0 \tag{17.52b}$$

$$\mathfrak{B}_A(\kappa + \tfrac{1}{2}, \kappa^\dagger)\mathfrak{U}_B(\kappa + \tfrac{1}{2}, \kappa^\dagger) - \mathfrak{U}_B(\kappa + \tfrac{1}{2}, \kappa^\dagger)\mathfrak{B}_A(\kappa + \tfrac{1}{2}, \kappa^\dagger) = \epsilon_{BA} \tag{17.52c}$$

$$\begin{aligned}\mathfrak{T}_{AB} &= \mathfrak{U}_A(\kappa, \kappa^\dagger)\mathfrak{B}_B(\kappa, \kappa^\dagger) - \kappa\epsilon_{AB} \\ &= \mathfrak{B}_A(\kappa + \tfrac{1}{2}, \kappa^\dagger)\mathfrak{U}_B(\kappa + \tfrac{1}{2}, \kappa^\dagger) + (\kappa + 1)\epsilon_{AB}\end{aligned} \tag{17.52d}$$

The transformation (17.48*b*) is non-singular when $\gamma = 2\kappa + 1 \neq 0$, and the correspondence between ψ_A and the pair $(\psi_{\kappa-\frac{1}{2}}, \psi_{\kappa+\frac{1}{2}})$ is a (1–1) correspondence. Of course, an entirely similar reduction of the product space $\mathfrak{R} \times \mathfrak{R}(0, \frac{1}{2})$, corresponding to dotted spinors, can be carried out. The procedure and results are virtually identical, except for the appropriate changes to ψ_A, $\mathfrak{U}_A$, $\mathfrak{B}^A$, etc. Corresponding to (17.51), for example, we find

$$(2\kappa^\dagger + 1)\psi_A = \mathfrak{U}_A(\kappa, \kappa^\dagger)\psi_{\kappa^\dagger-\frac{1}{2}} - \mathfrak{B}_A(\kappa, \kappa^\dagger + \tfrac{1}{2})\psi_{\kappa^\dagger+\frac{1}{2}} \tag{17.53}$$

and corresponding to (17.52)

$$\mathfrak{U}_A(\kappa, \kappa^\dagger)\mathfrak{B}^A(\kappa, \kappa^\dagger) = 2\kappa^\dagger \qquad \mathfrak{B}^A(\kappa, \kappa^\dagger)\mathfrak{U}_A(\kappa, \kappa^\dagger) = 2\kappa^\dagger + 1 \tag{17.54a}$$

$$\mathfrak{U}_A(\kappa, \kappa^\dagger + \tfrac{1}{2})\mathfrak{U}^A(\kappa, \kappa^\dagger) = \mathfrak{B}^A(\kappa, \kappa^\dagger)\mathfrak{B}_A(\kappa, \kappa^\dagger + \tfrac{1}{2}) = 0 \tag{17.54b}$$

$$\mathfrak{B}_A(\kappa, \kappa^\dagger + \tfrac{1}{2})\mathfrak{U}_B(\kappa, \kappa^\dagger + \tfrac{1}{2}) - \mathfrak{U}_B(\kappa, \kappa^\dagger)\mathfrak{B}_A(\kappa, \kappa^\dagger) = \epsilon_{\dot{B}A} \tag{17.54c}$$

$$\begin{aligned}\mathfrak{T}_{A\dot{B}} &= \mathfrak{U}_A(\kappa, \kappa^\dagger)\mathfrak{B}_B(\kappa, \kappa^\dagger) - \kappa^\dagger\epsilon_{A\dot{B}} \\ &= \mathfrak{B}_A(\kappa, \kappa^\dagger + \tfrac{1}{2})\mathfrak{U}_B(\kappa, \kappa^\dagger + \tfrac{1}{2}) + (\kappa^\dagger + 1)\epsilon_{A\dot{B}}\end{aligned} \tag{17.54d}$$

The interpretation of these results is quite analogous to that of our work in the preceding section. Recalling the result (17.29) and the definitions of γ, $\gamma^\dagger$, we have

$$\gamma^2 = J^2 + 1 - 2i\mathfrak{T} \qquad \gamma^{\dagger 2} = J^2 + i + 2i\mathfrak{T} \tag{17.55}$$

which may be used to define γ and $\gamma^\dagger$ as operators in terms of $\boldsymbol{K}$ and $\boldsymbol{L}$ even for reducible representations in which J^2 and $\mathfrak{T}$ are no longer represented by multiples of the unit matrix. In the *product* space, the infinitesimal transformations corresponding to $\boldsymbol{K}$ and $\boldsymbol{L}$ are

$$\mathfrak{K} = \boldsymbol{K} + \tfrac{1}{2}\boldsymbol{\sigma} \qquad \mathfrak{L} = \boldsymbol{L} + \tfrac{1}{2}i\boldsymbol{\sigma} \tag{17.56a}$$

and the operators corresponding to $\mathfrak{T}$ and J^2 are

$$\mathscr{I} = (\mathfrak{K}\mathfrak{L}) = \mathfrak{T} + \tfrac{3}{4}i + i\mathscr{A} \tag{17.56b}$$

$$\mathscr{J}^2 = \mathfrak{K}^2 - \mathfrak{L}^2 = J^2 + \tfrac{3}{2} + 2\mathscr{A} \tag{17.56c}$$

where

$$\mathscr{A} = \tfrac{1}{2}(\boldsymbol{K\sigma}) - \tfrac{1}{2}i(\boldsymbol{L\sigma}) \tag{17.56d}$$

$\mathscr{A}$ commutes with $\mathfrak{K}$ and $\mathfrak{L}$, since it is an invariant given by $-\frac{1}{2}\mathfrak{T}^{\kappa\lambda}\mathfrak{T}_{\kappa\lambda}(\frac{1}{2})$, where $\mathfrak{T}^{\kappa\lambda}$ and $\mathfrak{T}_{\kappa\lambda}(\frac{1}{2})$ are the infinitesimal transformations of the two spaces, and hence it must be scalar in an irreducible subspace of the product representation. From (17.56*d*)

$$(\mathscr{A} + \tfrac{1}{2})^2 = \tfrac{1}{4}(J^2 - 2i\mathfrak{T} + 1) = \tfrac{1}{4}\gamma^2$$

so that its eigenvalues are $-\frac{1}{2}(1 \pm \gamma)$, and $-\mathscr{A}$ is precisely the transformation $\psi_A \to \psi_A{}' = \mathfrak{T}_A^{\ B}\psi_B$ (17.47).

If Γ and $\Gamma^\dagger$ denote the operators in the product space corresponding to γ and $\gamma^\dagger$ respectively, then

$$\Gamma^{\dagger 2} = \mathscr{J}^2 + 2i\mathscr{I} + 1 = J^2 + 1 + 2i\mathfrak{T} = \gamma^{\dagger 2} \tag{17.57a}$$

$$\Gamma^2 = \mathscr{J}^2 - 2i\mathscr{I} + 1 = J^2 + 1 - 2i\mathfrak{T} + 4\mathscr{A} + 3 = \gamma^2 + 4\mathscr{A} + 3 \tag{17.57b}$$

and $\Gamma^2 = (\gamma \mp 1)^2$ for the two irreducible subspaces corresponding to the eigenvalues $-\frac{1}{2}(1 \pm \gamma)$ of $\mathscr{A}$. Hence we may choose $\Gamma^\dagger = \gamma^\dagger$, and $\Gamma = \gamma \mp 1$, the choice of sign depending on the corresponding choice in $\mathscr{A}$.

Thus, in fact, the vectors $\psi_{\kappa\pm\frac{1}{2}}$, $\psi_{\kappa\dagger\pm\frac{1}{2}}$ transform according to the representations $(\kappa \pm \frac{1}{2}, \kappa^\dagger)$, $(\kappa, \kappa^\dagger \pm \frac{1}{2})$ respectively, and should therefore be labelled $\psi(\kappa \pm \frac{1}{2}, \kappa^\dagger)$ and $\psi(\kappa, \kappa^\dagger \pm \frac{1}{2})$. The j values of these representations are given by $|\kappa \pm \frac{1}{2} - \kappa^\dagger|$ and $|\kappa - \kappa^\dagger \mp \frac{1}{2}|$ respectively, if that of the original representation $\mathfrak{D}(\kappa, \kappa^\dagger)$ is $|\kappa - \kappa^\dagger|$.

From the definitions of γ, $\gamma^\dagger$ we have

$$j = \frac{\varepsilon\gamma + \varepsilon^\dagger\gamma^\dagger}{2} \qquad i\mu = \frac{\varepsilon^\dagger\gamma^\dagger - \varepsilon\gamma}{2} \qquad i\mathfrak{T} = \frac{\gamma^{\dagger 2} - \gamma^2}{4} \tag{17.58}$$

where ε, $\varepsilon^\dagger$ may take the possible values ± 1, independently. If the numbers γ, $\gamma^\dagger$ are *given* for a particular representation, then it follows that at least one of the quantities $\gamma \pm \gamma^\dagger$ must be an integer, since $2j$ is integral and greater than zero. In the event that only one of these quantities is integral, the value of j is completely determined by the values of γ and $\gamma^\dagger$. This determines ε and $\varepsilon^\dagger$ and therefore μ, except for the case $j = 0$; but even in this case $\mathfrak{T}$ and μ^2 are completely determined. Since only j, $\mathfrak{T}$ and μ^2 enter in the representation, the latter is completely determined by the value pair $(\gamma, \gamma^\dagger)$. However, when both $\gamma + \gamma^\dagger$ and $\gamma - \gamma^\dagger$ are integral and distinct, there are two possible values of j, say j^+ and j^-, such that $j^+ > j^-$. Therefore, there are, in this case, two distinct representations corresponding to γ, $\gamma^\dagger$ $(= 2\kappa + 1,\ 2\kappa^\dagger + 1)$, and they may be denoted by $\mathfrak{D}^+(\kappa, \kappa^\dagger)$ and $\mathfrak{D}^-(\kappa, \kappa^\dagger)$ corresponding to the j values j^+ and j^-, respectively. Thus, a given pair $(\kappa, \kappa^\dagger)$ completely determines the irreducible representation unless κ, $\kappa^\dagger$ are both of the form $\frac{1}{4}n\,(n \neq -2)$. It is readily seen that $\mathfrak{D}^+(\kappa, \kappa^\dagger)$ is always an *infinite* representation. $\mathfrak{D}^-(\kappa, \kappa^\dagger)$ is finite only if j^- and j^+ differ by an integer, as is the case when both κ and $\kappa^\dagger$ are of the form $\frac{1}{2}n\,(n \neq -1)$. Since reversing the sign of γ or $\gamma^\dagger$ does not alter the representation, the four pairs $(\kappa, \kappa^\dagger)$, $(-\kappa - 1, \kappa^\dagger)$, $(\kappa, -\kappa^\dagger - 1)$ and $(-\kappa - 1, -\kappa^\dagger - 1)$ determine the same representation. It follows from (17.58) that at least one of the quantities $\kappa - \kappa^\dagger$ and $\kappa + \kappa^\dagger + 1$ is of the form $\frac{1}{2}n$, and for convenience it may be fixed that κ, $\kappa^\dagger$ are to be so chosen that $\kappa - \kappa^\dagger$ is an integer or half-integer. This is the only condition to be imposed on the choice of κ, $\kappa^\dagger$.

(iv) *Unitary Representations*

Since $\mathfrak{T} = \mu j$ is real for a unitary representation, it follows from the condition $i\mathfrak{T} = \frac{1}{4}(\gamma^{\dagger 2} - \gamma^2) = (\kappa^\dagger - \kappa)(\kappa^\dagger + \kappa + 1)$ that either $(\kappa^\dagger + \kappa + 1)$ is pure imaginary or $\kappa^\dagger - \kappa = \pm j = 0$. In the first case the condition $\mu^2 > -1$ is automatically satisfied and κ, $\kappa^\dagger$ are given by

$$\kappa = -\tfrac{1}{2} + i\nu + \tfrac{1}{2}n \qquad \kappa^\dagger = -\tfrac{1}{2} + i\nu - \tfrac{1}{2}n$$

In the second case it follows that $\kappa = \kappa^\dagger$ is real and

$$1 \geqslant (\kappa + \kappa^\dagger + 1)^2 = (2\kappa + 1)^2$$

so that $\frac{1}{2} \geqslant |\kappa + \frac{1}{2}|$, that is, $0 \geqslant \kappa = \kappa^\dagger \geqslant -1$, in accordance with the previously stated condition

$$\kappa = \kappa^\dagger = -\tfrac{1}{2} + \nu \qquad |\nu| \leqslant \tfrac{1}{2}$$

In particular, if κ, $\kappa^\dagger$ are both of the form $\frac{1}{4}n(n \neq -2)$, then $j^+ = |\kappa + \frac{1}{2}| + |\kappa^\dagger + \frac{1}{2}|$ and $\pm j^- = |\kappa + \frac{1}{2}| - |\kappa^\dagger + \frac{1}{2}|$. Therefore, $\mathfrak{D}^-(\kappa, \kappa^\dagger)$ is unitary only if $\frac{1}{2} \geqslant |\kappa + \frac{1}{2}| = |\kappa^\dagger + \frac{1}{2}|$ while $\mathfrak{D}^+(\kappa, \kappa^\dagger)$ is unitary whenever $|\kappa + \frac{1}{2}| = |\kappa^\dagger + \frac{1}{2}|$, since in the latter case

$$\pm i\mathfrak{T} = \{|\kappa + \tfrac{1}{2}| - |\kappa^\dagger + \tfrac{1}{2}|\}\{|\kappa + \tfrac{1}{2}| + |\kappa^\dagger + \tfrac{1}{2}|\} = 0$$

and $j = j^+ \neq 0$.

By reflection $\mathfrak{T}$ goes over into $-\mathfrak{T}$, so that a representation is reflection invariant only if $\mathfrak{T}$ is zero, i.e. $\kappa + \frac{1}{2} = \pm(\kappa^\dagger + \frac{1}{2})$. Thus, of the two main cases above, the representations under the first condition are reflection invariant only if $\nu = 0$, while the representations under the second are always reflection invariant.

(v) *Expinors*

As for the finite representations previously considered, that is the $\mathfrak{D}^-(\kappa, \kappa^\dagger)$ with κ, $\kappa^\dagger$ both of the form $\frac{1}{2}n(n \neq -1)$, we can construct a series in the variables ϕ_1, ϕ_2 and $\dot\phi_1$, $\dot\phi_2$ such that the coefficients of this series transform according to the *general* irreducible representation $\mathfrak{D}(\kappa, \kappa^\dagger)$ when a spinor transformation, corresponding to a given Lorentz transformation, is applied to the ϕ_A and $\phi_{\dot{A}}$. These coefficients form the components of a vector of the space $\mathfrak{R}(\kappa, \kappa^\dagger)$, and such vectors have been called " expinors ".[4] For the expinor transforming according to $\mathfrak{D}(\kappa, \kappa^\dagger)$, the series is homogeneous in each of the two sets of variables (ϕ_1, ϕ_2) and $(\phi_{\dot{1}}, \phi_{\dot{2}})$ of degree 2κ and $2\kappa^\dagger$, respectively, and it is therefore referred to as a homogeneous expinor. In case both 2κ and $2\kappa^\dagger$ are integral and greater than zero, the corresponding homogeneous expinor is the same as the symmetric spinor with 2κ undotted and $2\kappa^\dagger$ dotted indices.

To obtain the explicit expinor form, put

$$\dot{\phi}_1 = -\bar{\phi}_2 \qquad \dot{\phi}_2 = \bar{\phi}_1$$

so that $\dot{\phi}_1$, $\dot{\phi}_2$ transform in the same way as ϕ_1, ϕ_2 under spatial rotation, and consider terms of the type

$$\frac{\phi_1{}^p \phi_2{}^q \phi_{\dot{1}}{}^r \phi_{\dot{2}}{}^s}{\Delta^{n-\lambda}} = (-1)^s \frac{\phi_1{}^p \phi_2{}^q \dot{\phi}_1{}^s \dot{\phi}_2{}^r}{\Delta^{n-\lambda}} \tag{17.59}$$

where

$$\Delta = \phi_1\phi_{\dot{1}} + \phi_2\phi_{\dot{2}} = \phi_1\dot{\phi}_2 - \phi_2\dot{\phi}_1 = |\,\phi_1\,|^2 + |\,\phi_2\,|^2 \tag{17.60}$$

and p, q, r, s, n are integers $\geqslant 0$ while λ is a fixed complex number. On account of (17.60) the various terms for different values of p, q, r, s, n are, in general, not linearly independent. To get around this, we may proceed as follows. Put

$$\Phi_M^{K,j} = \sum_{M=m+m'} \frac{2k\,!}{(k-m)\,!\,(k+m)\,!}\,\frac{2k'\,!}{(k'-m')\,!\,(k'+m')\,!}\,\phi_1{}^{k-m}\phi_2{}^{k+m}\dot{\phi}_1{}^{k'-m'}\dot{\phi}_2{}^{k'+m'} \tag{17.61}$$

where $K = k + k'$, $j = k' - k$, and $2k$, $2k'$ are integers $\geqslant 0$. M takes on the usual values K, $K - 1, \ldots -K$ and similarly for m, m' with respect to k, k', and the summation is over all m, m' such that $m + m' = M$. As shown by van der Waerden (32), every polynomial $P^{k,k'}$ which is homogeneous of degree $2k$ and $2k'$ in each of the sets of variables ϕ and $\dot{\phi}$ respectively, can be expressed uniquely in terms of $\Phi_M^{K,j}$ as

$$P^{k,k'} = \sum_{o\leq\nu\leq 2k,\,2k'} \; \sum_{M=-K+\nu}^{K-\nu} A_{M,\nu}\,\Delta^\nu \Phi_M^{K-\nu,j} \tag{17.62}$$

with constant coefficients $A_{M,\nu}$, the index ν taking on all integral values in the range $0 \leq \nu \leq 2k$, $2k'$; and since every polynomial in ϕ, $\dot{\phi}$ is decomposable into a sum of homogeneous polynomials, this expansion is always possible and unique. Therefore, express the numerator of (17.59) in the form (17.62) and arrange the result according to descending powers of Δ.

Now consider a series of terms of the type (17.59), homogeneous of degree $\lambda - j$ and $\lambda + j$ in ϕ and $\dot{\phi}$ respectively—with $2j = r + s - p - q$ an integer. According to (17.62) it can be put in the form

$$\sum_{K\geq|j|}^{\infty} \; \sum_{M=-K}^{K} a_M^{K,j} \frac{\Phi_M^{K,j}}{\Delta^{K-\lambda}} \tag{17.63}$$

[4] cf. Dirac (45); Harish-Chandra (47b).

in which K ranges over $|j|, |j|+1, \ldots$ and the $a_M^{K,j}$ are constants. If now we apply a spinor transformation

$$\phi_A = s_A{}^B \phi_B' \qquad \dot\phi_A = s_{\dot A}{}^{\dot B} \dot\phi_{\dot B}' \tag{17.64}$$

and again put $\dot\phi_1' = -\bar\phi_2'$, $\dot\phi_2' = \bar\phi_1'$, then $\Phi_M^{K,j}(\phi, \dot\phi)$ goes over into a polynomial in ϕ' and $\dot\phi'$ of degree $K-j$, $K+j$ respectively. By (17.62) this can again be expressed linearly in terms of $\Delta'^{\nu}\Phi'^{K-\nu,j}_M$ where $\Delta' = |\phi_1'|^2 + |\phi_2'|^2$. Further,

$$\Delta = |\phi_1|^2 + |\phi_2|^2 = m(s)\{|\phi_1'|^2 + |\phi_2'|^2\} - \{|s_1{}^1\dot\phi_1' + s_1{}^2\dot\phi_2'|^2 + |s_2{}^1\dot\phi_1' + s_2{}^2\dot\phi_2'|^2\} \tag{17.65a}$$

$$m(s) = |s_1{}^1|^2 + |s_1{}^2|^2 + |s_2{}^1|^2 + |s_2{}^2|^2$$

or

$$\Delta = m(s)\Delta'\left\{1 - \frac{D'}{m(s)\Delta'}\right\} \tag{17.65b}$$

and $0 < D'/m(s)\Delta' < 1$, since $\Delta, \Delta' > 0$. Therefore we can expand $1/\Delta^{K-\lambda}$ in (17.63) by the binomial theorem, in ascending powers of $D'/m(s)\Delta'$. Again, D' is a polynomial in ϕ' and $\dot\phi'$ so that the product of its powers with $\Phi'^{K,j}_M$ can also be expressed in the form (17.62). Collecting together the coefficients of $\Phi'^{K,j}_M/\Delta'^{K-\lambda}$, this gives

$$\sum_{K,M}^{\infty} a_M^{K,j} \frac{\Phi_M^{K,j}}{\Delta^{K-\lambda}} = \sum_{K,M}^{\infty} a'^{K,j}_M \frac{\Phi'^{K,j}_M}{\Delta'^{K-\lambda}} \tag{17.66}$$

and the linear transformation of the coefficients $a_M^{K,j} \to a'^{K,j}_M$ constitutes a representation of the Lorentz transformation associated with (17.64).

The important question of convergence has been entirely ignored here, and for this we refer the interested reader to the complete development given in the original paper of Harish-Chandra (*loc. cit.*). However, for our later work we shall require a more explicit formulation of the properties of the basic vectors in $\mathfrak{D}(\kappa, \kappa^\dagger)$. Moreover, the following brief considerations serve to show more clearly the meaning of the results of the several preceding sections.

If we make the substitutions

$$\left.\begin{aligned} \phi_1 &= \rho\cos\tfrac{1}{2}\theta e^{\frac{1}{2}i\chi}e^{\frac{1}{2}i\phi} \\ \phi_2 &= \rho\sin\tfrac{1}{2}\theta e^{\frac{1}{2}i\chi}e^{-\frac{1}{2}i\phi} \end{aligned}\right\} \begin{aligned} &\rho, \theta, \chi, \phi \text{ real} \\ &\rho > 0,\ \pi \geqslant \theta \geqslant 0,\ 2\pi > \phi \geqslant 0,\ 4\pi > \chi \geqslant 0 \end{aligned} \tag{17.67}$$

$$y = \cos\theta$$

$$\frac{1}{\sqrt{2\pi}} A_M^{K,j} = \sqrt{\frac{(K-j)!\,(K+j)!}{(K-M)!\,(K+M)!}}\left(\frac{2}{2K+1}\right) a_M^{K,j}$$

then (17.63) takes the form

$$\rho^{2\lambda} e^{-ij\chi} \sum_{K \geqslant |j|}^{\infty} A_M^{K,j} \frac{1}{\sqrt{2\pi}} P_M^{K,j}(y) e^{-iM\phi} \tag{17.68a}$$

where

$$P_M^{K,j}(y) = \frac{(-1)^{K+j}}{2^K}\sqrt{\frac{(K-M)!\,(K+\frac{1}{2})}{(K+M)!\,(K+j)!\,(K-j)!}} \times (1-y)^{\frac{1}{2}(M-j)}(1+y)^{\frac{1}{2}(M+j)} \frac{d^{K+M}}{dy^{K+M}}\{(1-y)^{K+j}(1+y)^{K-j}\} \tag{17.68b}$$

and the functions $Q_M^{K,j} = \frac{1}{\sqrt{2\pi}} P_M^{K,j}(y)e^{-iM\phi}$ form a complete orthonormal set. To obtain the same matrix representation as given in 17*b*(i), the basic vectors must be chosen as

$$i^K \sqrt{\left(\frac{(K-\lambda-1)\,!}{(K+\lambda+1)\,!}\right)} Q_M^{K,j} e^{-ij\chi} \rho^{2\lambda} \tag{17.68c}$$

the factorial being, in general, interpreted through the gamma function, which, through the relation $\Gamma(z)\Gamma(1-z) = \pi/\sin z\pi$, permits alternative forms in the exceptional case when λ is an integer and either $(K-\lambda-1)!$ or $(K+\lambda+1)!$ becomes infinite.

We may also consider functions on the three variables y, ϕ, χ, viz.

$$S_M^{K,j} = \frac{1}{\sqrt{4\pi}} Q_M^{K,j} e^{-ij\chi} = \frac{(-1)^{K+j}}{\sqrt{8\pi}} \sqrt{\frac{(K-M)\,!\,(K+M)\,!}{(K-j)\,!\,(K+j)\,!}} (K+\tfrac{1}{2}) \frac{\Phi_M^{K,j}}{\Delta^K} \tag{17.69a}$$

which are known to form a complete set, such that for $f(y, \phi, \chi)$ continuous in the interval $-1 \leq y \leq 1$, $0 \leq \phi \leq 2\pi$, $0 \leq \chi \leq 4\pi$,

$$f = \sum_\nu \mathfrak{A}_\nu S_\nu \qquad \mathfrak{A}_\nu = (S_\nu, f) = \int \bar{S}_\nu f \, dy \, d\phi \, d\chi \tag{17.69b}$$

In the more general series

$$\sum_{j=-\infty}^{\infty} \sum_{K \geqslant j}^{\infty} A_M^{K,j} S_M^{K,j} = \sum_\nu \mathfrak{A}_\nu S_\nu = \mathfrak{A}$$

which we denote by $\mathfrak{A}$; the summation over j is for all integral and half-integral values. We consider the formal defining relation

$$\psi(\phi, \dot{\phi}) = \sum_{j=-\infty}^{\infty} \sum_{K \geqslant j}^{\infty} a_M^{K,j} \frac{\Phi_M^{K,j}}{\Delta^{K-\lambda}} = \rho^{2\lambda}\mathfrak{A} \tag{17.70}$$

which, if there exists a continuous function satisfying (17.69*b*) and $f(y, \phi, 0) = f(y, \phi, 4\pi)$, $f(y, 0, \chi) = f(y, 2\pi, \chi \pm 2\pi)$, is said to be a continuous function of $\phi, \dot{\phi}$; or, continuously differentiable if it can be differentiated term by term.

Now introduce the operators ϕ_A, $\partial/\partial\phi_A$, and conjugates, where $\partial/\partial\phi_A$ is defined by

$$\frac{\partial}{\partial\phi_A} = \frac{\partial\rho}{\partial\phi_A}\frac{\partial}{\partial\rho} + \frac{\partial\chi}{\partial\phi_A}\frac{\partial}{\partial\chi} + \frac{\partial y}{\partial\phi_A}\frac{\partial}{\partial y} + \frac{\partial\phi}{\partial\phi_A}\frac{\partial}{\partial\phi}$$

and $\partial\xi/\partial\phi_A$ is $\left\{\frac{\partial(\phi_1, \phi_2, \dot{\phi}_1, \dot{\phi}_2)}{\partial(\rho, \chi, y, \phi)}\right\}^{-1}$ times the minor of $\partial\phi_A/\partial\xi$ in the Jacobian

$$\frac{\partial(\phi_1, \phi_2, \dot{\phi}_1, \dot{\phi}_2)}{\partial(\rho, \chi, y, \phi)} \qquad (\xi = \rho, \chi, y \text{ or } \phi)$$

and consider the infinitesimal spinor transformation

$$\begin{aligned} \phi_A &= \phi_A' + \eta_A{}^B \phi_B' \\ \dot{\phi}_A &= \dot{\phi}_A' + \dot{\eta}_A{}^{\dot{B}} \dot{\phi}_{\dot{B}}' \end{aligned} \tag{17.71}$$

applied to the expinor space. Let $\psi(\phi, \dot{\phi})$ be a series which is homogeneous and continuously differentiable so that $\psi = \rho^{2\lambda}\mathfrak{A}$. Corresponding to (17.71) ψ goes over into $\psi' = \rho^{2\lambda}\mathfrak{A}'$, so that in each Lorentz frame there exist continuously differentiable functions $f(y, \phi, \chi)$ and $f'(y', \phi', \chi')$ of required properties. If we write $\rho^{2\lambda}f(y, \phi, \chi) = \Psi(\phi, \dot{\phi})$ and similarly $\rho'^{2\lambda}f'(y', \phi', \chi') = \Psi'(\phi', \dot{\phi}')$, then $\Psi'(\phi', \dot{\phi}') = \Psi(\phi, \dot{\phi})$. From (17.71) it follows that, to quantities of the first order,

$$\Psi'(\phi, \dot{\phi}) = \Psi'(\phi', \dot{\phi}') + \eta_A^{\ B}\phi_B' \frac{\partial}{\partial \phi_A'}\Psi' + \eta_{\dot{A}}^{\ \dot{B}}\phi_{\dot{B}}' \frac{\partial}{\partial \phi_{\dot{A}}'}\Psi' \tag{17.72a}$$
$$= \Psi(\phi, \dot{\phi}) + (\eta_A^{\ B}\phi_B\partial^A + \eta_{\dot{A}}^{\ \dot{B}}\phi_{\dot{B}}\partial^{\dot{A}})\Psi$$
$$= \{1 - \eta_{AB}\mathfrak{T}^{AB} - \eta_{\dot{A}\dot{B}}\mathfrak{T}^{\dot{A}\dot{B}}\}\Psi(\phi, \dot{\phi})$$

$$\mathfrak{T}^{AB} = \tfrac{1}{2}(\phi^A\partial^B + \phi^B\partial^A) \qquad \left(\partial^A \equiv \frac{\partial}{\partial \phi_A}\right) \tag{17.72b}$$

and, since ψ is continuously differentiable, the functions $\mathfrak{T}^{AB}\Psi(\phi, \dot{\phi})$ and $\mathfrak{T}^{\dot{A}\dot{B}}\Psi(\phi, \dot{\phi})$ are *represented* by the corresponding series $\mathfrak{T}^{AB}\psi$ and $\mathfrak{T}^{\dot{A}\dot{B}}\psi$.

According to results of § 11, the spinor transformation (17.71) is associated with the Lorentz transformation

$$x_\mu = x_\mu' + \omega_\mu^{\ \nu}x_\nu' \tag{17.73a}$$

through the relations (cf. 11.3)

$$\eta_{AB} = \tfrac{1}{4}\omega^{\mu\nu}g_{\mu\dot{C}A}g_\nu^{\ \dot{C}}{}_B \tag{17.73b}$$
$$= -\tfrac{1}{4}\omega^{\mu\nu}S_{\mu\nu AB}$$

and by (13.22*a*, *b*) we have inversely

$$\omega_{\mu\nu} = -\tfrac{1}{2}(S_{\mu\nu AB}\eta^{AB} + \bar{S}_{\mu\nu \dot{A}\dot{B}}\eta^{\dot{A}\dot{B}}) \tag{17.73c}$$

The transformation operator representing (17.73*a*) is $1 - \frac{1}{2}\omega_{\mu\nu}\mathfrak{T}^{\mu\nu}$, and therefore the connection between $\mathfrak{T}^{\mu\nu}$ and $\mathfrak{T}_{AB}$ is given by

$$\mathfrak{T}_{AB} = -\tfrac{1}{4}\mathfrak{T}^{\mu\nu}S_{\mu\nu AB} = -\tfrac{1}{4}\mathfrak{T}^{\mu\nu}g_\mu^{\ \dot{C}}{}_A g_{\nu\dot{C}B} \tag{17.74}$$

which is readily shown to be just the form $\frac{1}{2}(K_{AB} - iL_{AB})$ of (17.44).

Further, if ψ is homogeneous of degree $2\kappa = \lambda - j$ and $2\kappa^\dagger = \lambda + j$ in the undotted and dotted variables respectively,

$$\phi_A\partial^A\psi = 2\kappa\psi \tag{17.75}$$
$$\phi_{\dot{A}}\partial^{\dot{A}}\psi = 2\kappa^\dagger\psi$$

by term-by-term differentiation, and the ϕ_A, ∂^A are seen to satisfy the commutation rules

$$[\partial^A, \phi^B]_- = \epsilon^{AB} = [\phi^A, \partial^B]_- \tag{17.76}$$

The properties of ϕ_A, ∂_A are the same as those of the spinor matrices $\mathfrak{U}_A$, $\mathfrak{V}_A$ respectively, and, in fact, these matrix representations may be shown to be identical, provided the basic vectors are suitably chosen. This then gives the underlying meaning of the equations (17.49, 51), since if ψ_A is a vector of $\mathfrak{R}(\kappa, \kappa^\dagger) \times \mathfrak{R}(\frac{1}{2}, 0)$ such that the corresponding pair of series $\psi_A(\phi, \dot{\phi})$ are of the type (17.75), then obviously $\phi^A\psi_A$ and $\partial^A\psi_A$ are two series of degrees $(2\kappa + 1, 2\kappa^\dagger)$ and $(2\kappa - 1, 2\kappa^\dagger)$ respectively, and therefore transform according to $\mathfrak{D}(\kappa + \frac{1}{2}, \kappa^\dagger)$ and $\mathfrak{D}(\kappa - \frac{1}{2}, \kappa^\dagger)$.

CHAPTER III

General Field Theory

§ 18. Introduction

The classical theory of fields is essentially a generalization of classical point mechanics to continuously many variables; the field functions ψ^r (cf. *infra*) and their time derivatives $\dot{\psi}^r$, at each space-time point, playing the roles of the familiar canonically conjugate generalized coordinates and momenta, or q's and p's, of Hamiltonian mechanics. As indicated by this correspondence, the entire Hamiltonian-Lagrange theory of point mechanics can be taken over in the field theory with but minor changes in the mathematical formalism. The basic change lies in the fact that the primary quantities with which we operate are now termed *densities*, e.g. momentum density, energy density, etc., since the physical quantities of interest are obtained from them by integration over the space-like coordinates of the continuum, and correspondingly the ordinary *partial* derivative is replaced by its natural generalization the *variational* derivative.

We shall consider first the derivation of the field equations from the variation principle which is simply an extension of Hamilton's principle of stationary action—$\delta\int_{t_0}^{t_1} L\,dt = 0$—with the Lagrangian L now given by the space integral of a Lagrangian density $\mathscr{L}$. From elementary considerations of invariance we easily obtain the derived field density tensors representing the energy momentum, angular momentum, etc., and their respective continuity equations which express the conservation laws. These results hold quite generally for arbitrary fields whether associated with charge or not, i.e. whose quantized form describes charges or neutral quanta, respectively. Proceeding to the consideration of complex fields, which are suitable for describing charged particles, we find that the further condition of gauge invariance of first and second kinds permits the definition of a current-charge density 4-vector which satisfies the usual continuity equation; and with which Maxwell's equations obtain in the presence of an external electromagnetic field if the conservation laws for combined matter (m) and electromagnetic (e.m.) fields, including interaction, are satisfied. Herein we find the intimate connection between the conservation of charge and the gauge invariance of the electromagnetic field which is equivalent to the non-existence of a photon rest mass (cf. Chapter IV, §§ 27iv, 29).

Finally, the whole theory will be cast in canonical Hamiltonian form in which the consideration of infinitesimal contact transformations as generators of the Lorentz group affords virtually self-evident proof of the Lorentz covariance of the formalism, and from which the canonical quantized theory of wave fields follows in the familiar passage from classical to quantal Poisson Brackets. It is to be emphasized that the fields considered in the c-number theory are to be viewed as classical fields whether they originate from the Klein-Gordon, Maxwell, Dirac or any other equations.* Insofar as the classical fields are described by equations already obtained by a first quantization of relativistic point mechanics, it is true that the subsequent transition to the quantized field amounts to a second- or hyper-quantization. Nonetheless we disregard the origin of the equations in this sense so that, for example, in the Dirac electron theory considered as a classical field, the charge (e) and the action ($\hbar$) do not, strictly speaking, assume the roles of quanta of charge and action until the transition to the quantized field formalism in which, as a many-particle theory, this significance is manifest.

§ 19. The Lagrangian Density and the Variation Principle

Before proceeding with the programme briefly outlined above it seems desirable to comment on the role of the Hamiltonian variation principle[1] in the underlying formulation of special-relativistic field theories. This is perhaps particularly appropriate since the field equations, which within the framework of present physical theories and space-time concepts are differential equations, have not been derived from a variation principle requiring that the space-time integral of an invariant $\mathscr{L}$, the Lagrangian density, taken over any fixed region be stationary:

$$\delta \int_{\substack{\text{arbitrary region} \\ \Omega}} \mathscr{L}\sqrt{-g}\, d\omega = 0 \qquad d\omega = dx^1 dx^2 dx^3 dx^4 \qquad (19.1)^2$$

This principle is, indeed, a very convenient way, in theory, of seeking the correct field equations since we have *only* (a euphemism, at best) to find the correct $\mathscr{L}$ as a functional of a selected class of field variables and their derivatives. This has been well described by Eddington (37) as the search for an action "whose sole *raison d'être* is to be varied—and, moreover, *varied in such a way as to defy the laws of nature!*" Yet, there are good and general reasons for adopting this procedure in view of the fact that, if the field equations are equivalent to a variation principle, the conservation laws are an automatic consequence of the general invariance, as we shall see.[3] Of the same nature is the fact that theories which cannot be cast in a form equivalent to a variation principle are heir to special difficulties, in particular in connection with the introduction of interaction with an electromagnetic field (cf. §§ 22, 31), over and beyond those which already plague all field theories.[4]

Granting the importance and implicit power of this principle, we may properly

* That is, they are classical in the sense that the field variables are, as yet, commuting quantities; not, in general, in the sense of this description as applied to the electromagnetic and gravitational fields.

[1] cf., for example, Eddington (37); Pauli (22, 41).

[2] N.B. As is well known, if the action is to be an absolute integral invariant, the integrand must be a relative scalar of weight one, or what is often termed a scalar density; hence the factor $\sqrt{-g}$. Cf., for example, Weitzenbock (23); de Donder (35).

[3] However, the argument that the consistency of a theory is ensured simply because it is derived from variation of a Lagrangian is fallacious—since the Lagrangian need not have a minimum at all unless it is a positive definite form.

[4] We refer here to problems of self-energy, etc., which are properly discussed in the quantized formalism and are essentially beyond the scope of the present treatment (cf. Wentzel (43)).

ask why it is used largely in an *a posteriori* appeal to its implications. A partial answer to this lies in the fact that, within the framework of the special theory of relativity, the choice of possible invariants is well-nigh unlimited, and to this extent it is perhaps fortunate that the connection between relativistic point mechanics and field theory is sufficiently close to suggest various forms of field equations. We may note that in the general relativity the principle begins to come into its own: the choice is not so broad since there are several intrinsic tensors, suggested by physical and geometrical considerations, from which simple invariants can be constructed. The various attempts to combine the gravitational and electromagnetic fields have been concerned largely with the delimitation of the number of possibilities. Here, too, we may remark that when we derive the so-called canonical energy-momentum-density tensor for a given special-relativistic field, whether directly or in the retrospective manner suggested above, it is in a sense vacant of intrinsic meaning as regards localization of energy and momentum in space. The latter is uniquely determined only in the gravitational theory where the generation of the gravitational field gives the matter energy-momentum tensor a direct physical content.[5] It is, therefore, especially important that the so-called symmetrized energy-momentum-density tensor is the limiting special-relativistic form of the expression defined in the gravitational theory.

We shall consider first a matter field defined by the field variables $\psi^r(x^1, x^2, x^3, x^4)$ where the general-purpose label r may stand for a set of tensor, 2-spinor, or 4-spinor indices—and eventually some index labelling field quantities defining different (for the present, non-interacting) fields. The summation convention is now extended to expressions involving repeated indices $r, s, \ldots$; the sum ranging over all matter fields present and their individual components.[6]

The Lagrangian density will be assumed to be a function of the field components and their first derivatives,

$$\mathscr{L} = \mathscr{L}\big(\psi^r(x);\ \psi^r{}_{,\mu}(x)\big) \tag{19.2}$$

such that it is a real scalar with respect to all transformations admitted in the restricted relativity. Further, since we consider here a conservative system, $\mathscr{L}$ does not depend explicitly on the time nor therefore on the space coordinates for reasons of invariance. The field equations in the present theory are assumed to follow from the variation principle, without accessory or supplementary conditions,[7] for arbitrary variations of the field variables ψ^r, vanishing on the integration boundaries, viz.

$$\frac{\delta I}{\delta \psi^r} \equiv \frac{\partial \mathscr{L}}{\partial \psi^r} - \partial_\mu \left(\frac{\partial \mathscr{L}}{\partial \psi^r{}_{,\mu}}\right) \tag{19.3}$$

$\delta I/\delta\psi^r$ denotes the *functional* derivative of $I[\psi^r]$ with respect to ψ^r, which may be expressed[8] as: $\lim \dfrac{I[\psi^r + \delta\psi^r] - I[\psi^r]}{\int \delta\psi^r\, d\omega}$. It should be noted that if $\mathscr{L}$ is linear in the

[5] cf. Hilbert (15); Weyl (23); Rosenfeld (40); Belinfante (39*c*, 40*b*).

[6] However, we must use caution in raising and lowering the indices $r, s, \ldots$ unless the transformation character of the field variables is explicitly indicated.

[7] That is, all equations of motion are assumed to flow from the variation alone. Cf., however, § 31.

[8] cf. de Donder (35); Heisenberg-Pauli (29, 30); Volterra (31).

gradients $\psi^r{}_{,\mu} = \partial_\mu \psi^r$ the field equations are always of the first order; whereas if $\mathscr{L}$ is allowed to be quadratic in the gradients, the field equations will, in general, be of second order.[9]

(*a*) *Conservation Theorems—The Transformation Properties of Field Variables*

The assumed invariance of $\mathscr{L}$ implies

$$I = \int_\Omega \mathscr{L}\, d\omega = \text{invariant} \tag{19.4}$$

since we now confine our attention to Galilean coordinates; but it should be noted that the region of integration which is Lorentz-invariant refers, *inter alia*, to the volume between two surfaces of constant t. Hence, though all equations and conclusions flowing from the variation principle are guaranteed to be Lorentz-covariant, they carry with them the stigma, so to speak, of the explicit reference to *particular* Lorentz frames—which is not an invariant *concept*. This may be avoided by replacing the concept of a surface of constant t by the invariant concept of a space-like surface specified by $g_{\mu\nu}(y^\mu y^\nu - x^\mu x^\nu) < 0$, i.e. such that no two of its points $\mathrm{P}(y)$, $\mathrm{P}(x)$ can be connected by a light signal. This was first emphasized by Weiss (36, 38) and is essentially the starting point of the more recent considerations of Tomonaga (46), Schwinger (48, 49*a*, *b*), *et al.*

For an arbitrary infinitesimal coordinate transformation $x^\mu \to x^\mu + \delta x^\mu$ the invariance property takes the form

$$\begin{aligned}\delta I &= \int_{\Omega'} \mathscr{L}'(x')\, d\omega' - \int_\Omega \mathscr{L}(x)\, d\omega \\ &= \int_{\Omega'} \{\mathscr{L}'(x') - \mathscr{L}(x')\}\, d\omega' + \int_{\Omega'} \mathscr{L}(x')\, d\omega' - \int_\Omega \mathscr{L}(x)\, d\omega \\ &= 0\end{aligned} \tag{19.5}$$

Then, noting that $\left|\dfrac{\partial x'}{\partial x}\right| = 1 + \partial_\nu\, \delta x^\nu$,

$$\begin{aligned}\int_{\Omega'} \mathscr{L}(x')\, d\omega' &= \int_\Omega \mathscr{L}(x^\mu + \delta x^\mu)(1 + \partial_\nu \delta^\nu)\, d\omega \\ &= \int_\Omega \{\mathscr{L}(x) + \partial_\nu(\mathscr{L}\delta x^\nu)\}\, d\omega\end{aligned} \tag{19.6}$$

with neglect of higher-order corrections; so that the general statement of the invariance property takes the form

[9] Unless $\dfrac{\partial^2 \mathscr{L}}{\partial\psi^r{}_{,\mu}\,\partial\psi^s{}_{,\nu}} + \dfrac{\partial^2 \mathscr{L}}{\partial\psi^r{}_{,\nu}\,\partial\psi^s{}_{,\mu}} = 0$, in which case they are again of first order. Cf. Belinfante (40*b*).

$$\delta I = \int_{\Omega'} \{\mathscr{L}'(x') - \mathscr{L}(x')\} \, d\omega' + \int_{\Omega} \partial_\nu(\mathscr{L}\, \delta x^\nu) \, d\omega \tag{19.7}$$

$$= \int_{\Omega} \{\mathscr{L}'(x) - \mathscr{L}(x) + \partial_\nu(\mathscr{L}\, \delta x^\nu)\} \, d\omega$$

$$= \int_{\Omega} \{\delta^* \mathscr{L} + \partial_\nu(\mathscr{L}\, \delta x^\nu)\} \, d\omega = 0$$

where $\delta^*\mathscr{L}$ is the substantial variation of $\mathscr{L}$ (cf. § 5).

Now for the *substantial* variation

$$\delta^* \psi^r = \psi^{r\prime}(x) - \psi^r(x) \tag{19.8a}$$

we have

$$\delta^*(\psi^r{,}_\lambda) = (\delta^* \psi^r){,}_\lambda \tag{19.8b}$$

i.e. δ^* and ∂ commute, since all quantities are functions of the same variables. This simple relation does not hold for the *local* variation

$$\delta\psi^r = \psi^{r\prime}(x') - \psi^r(x) = \delta^* \psi^r + \psi^r{,}_\mu \, \delta x^\mu \tag{19.9a}$$

since

$$\delta(\psi^r{,}_\lambda) = \psi^{r\prime}{,}_\lambda(x + \delta x) - \psi^r{,}_\lambda(x) \tag{19.9b}$$

$$= \psi^{r\prime}{,}_\lambda + \psi^r{,}_{\lambda\mu} \, \delta x^\mu - \psi^r{,}_\lambda$$

$$= \delta^* \psi^r{,}_\lambda + \psi^r{,}_{\lambda\mu} \, \delta x^\mu$$

and

$$(\delta\psi^r){,}_\lambda = \{\psi^{r\prime}(x + \delta x) - \psi^r(x)\}{,}_\lambda \tag{19.9c}$$

$$= \psi^{r\prime}{,}_\lambda + \psi^r{,}_{\lambda\mu} \, \delta x^\mu + \psi^r{,}_\mu \, \delta x^\mu{,}_\lambda - \psi^r{,}_\lambda$$

Therefore

$$\delta(\psi^r{,}_\lambda) = (\delta\psi^r){,}_\lambda - \psi^r{,}_\mu \, \delta x^\mu{,}_\lambda \tag{19.9d}$$

which may also be obtained directly from (19.9*a*, *b*).

For an infinitesimal *Lorentz* transformation $x^\mu \to x^{\mu\prime} = x^\mu + \omega^{\mu\nu} x_\nu$ the components of ψ^r transform (at a given space-time point) according to

$$\delta\psi^r = \tfrac{1}{2}\omega^{\mu\nu} \mathfrak{T}_{\mu\nu}{}^{(rs)} \psi_s \tag{19.10a}$$

or, symbolically,

$$\delta\psi = \tfrac{1}{2}\omega^{\mu\nu} \mathfrak{T}_{\mu\nu\ \mathrm{op}}\, \psi \tag{19.10b}$$

which is the general form, of which our previous results for tensors (§ 5), 2-spinors (§ 11), and 4-spinors (§ 16) are special cases. From (19.9*d*) the gradients transform as

$$\delta(\psi^r{,}_\lambda) = \tfrac{1}{2}\omega^{\mu\nu} \mathfrak{T}_{\mu\nu}{}^{rs} \psi_{s,\lambda} - \psi^r{,}_\mu g_{\lambda\nu} \omega^{\mu\nu} \tag{19.11a}$$

or

$$\delta(\partial_\lambda \psi) = \tfrac{1}{2}\omega^{\mu\nu}(\mathfrak{T}_{\mu\nu\ \mathrm{op}}\, \partial_\lambda + g_{\lambda\mu}\partial_\nu - g_{\lambda\nu}\partial_\mu)\psi \tag{19.11b}$$

and by definition the corresponding substantial variation is

$$\delta^* \psi^r = \tfrac{1}{2}\omega^{\mu\nu} \mathfrak{T}_{\mu\nu}{}^{rs} \psi_s - \omega^{\mu\nu} \psi^r{,}_\mu x_\nu \tag{19.12a}$$

or

$$\delta^* \psi = \tfrac{1}{2}\omega^{\mu\nu}(\mathfrak{T}_{\mu\nu\ \mathrm{op}} + x_\mu \partial_\nu - x_\nu \partial_\mu)\psi \tag{19.12b}$$

from which $\delta^*\psi{,}_\lambda$ follows by inspection.

(b) Conservation of Energy-Momentum—The Canonical Energy-Momentum Tensor

We return to the general expression of the invariance property (19.7) which takes the form

$$\int_{\Omega} d\omega\, \partial_\nu \left\{ \frac{\partial \mathscr{L}}{\partial \psi^r{}_{,\nu}}\, \delta^* \psi^r + \mathscr{L}\, \delta x^\nu \right\} = 0 \tag{19.13}$$

on noting that $\delta^* \mathscr{L} = \frac{\partial \mathscr{L}}{\partial \psi^r}\, \delta^* \psi^r + \frac{\partial \mathscr{L}}{\partial \psi^{r,\nu}}\, \delta^* \psi^{r,\nu}$ and using the field equations (19.3) to simplify. Let us consider an infinitesimal translation $x^\mu \to x^\mu + \delta x^\mu$, i.e. a rigid displacement of the whole wave field. Then

$$\delta \psi^r = \psi^{r\prime}(x') - \psi^r(x) = 0 \tag{19.14}$$

$$\delta^* \psi^r = \psi^{r\prime}(x) - \psi^r(x) = -\psi^{r,\mu}\, \delta x_\mu$$

and (19.13) becomes

$$0 = \int d\omega\, \partial_\nu \left\{ \frac{\partial \mathscr{L}}{\partial \psi^r{}_{,\nu}}\, \psi^{r,\mu}\, \delta x_\mu - g^{\mu\nu}\, \mathscr{L}\, \delta x_\mu \right\} \tag{19.15}$$

$$= \int d\omega\, \partial_\nu (\mathscr{T}^{\mu\nu}\, \delta x_\mu)$$

where

$$\mathscr{T}^{\mu\nu} = \frac{\partial \mathscr{L}}{\partial \psi^r{}_{,\nu}}\, \psi^{r,\mu} - g^{\mu\nu}\, \mathscr{L} \tag{19.16a}$$

defines the canonical energy-momentum tensor; and since the δx^μ are arbitrary constants we conclude that

$$\mathscr{T}^{\mu\nu}{}_{,\nu} = 0 \tag{19.16b}$$

This fundamental result constitutes the statement of the conservation laws for energy and momentum as is seen by making the following identification between components of $\mathscr{T}^{\mu\nu}$ and the Hamiltonian- (energy) density $\mathfrak{H}$, momentum-density $\mathfrak{G}^k$, energy flux-density $\mathfrak{S}^k$, and the stress tensor $\mathfrak{Z}^{kl}$, viz.

$$\mathfrak{H} = \mathscr{T}^{44} = \frac{\partial \mathscr{L}}{\partial \psi^r{}_{,4}}\, \psi^{r,4} - \mathscr{L} \tag{19.17a}$$

$$c\mathfrak{G}^k = \mathscr{T}^{k4} = \frac{\partial \mathscr{L}}{\partial \psi^r{}_{,4}}\, \psi^{r,k} \tag{19.17b}$$

$$\mathfrak{S}^k = c\mathscr{T}^{4k} = \frac{\partial \mathscr{L}}{\partial \psi^r{}_{,k}}\, \psi^{r,4} \tag{19.17c}$$

$$\mathfrak{Z}^{kl} = \mathscr{T}^{kl} \tag{19.17d}$$

The vanishing of the divergence (19.16b) expresses the conservation laws embodied in

$$\frac{\partial \mathfrak{H}}{\partial t} + \operatorname{div} \mathfrak{S} = 0 \tag{19.18a}$$

$$\frac{\partial \mathfrak{G}}{\partial t} + \operatorname{div} \mathfrak{Z} = 0 \tag{19.18b}$$

and we shall show later (cf. § 24*e*) that the volume-integrated momentum-energy transforms as a 4-vector as required. It is important to bear in mind that, in general, the canonical density tensor $\mathscr{T}^{\mu\nu}$ is not symmetric, nor is the energy-density positive definite. The latter point, which we shall consider further in the next chapter, corresponds to the fact that the fundamental condition that the total (volume-integrated) energy be positive in the absence of interaction can in many cases (cf. Chapter IV, § 32) only be fulfilled by transcription to the quantized or q-number theory. On the other hand the lack of symmetry is immediately pertinent, since it has bearing on the conservation law for angular momentum which we consider next and whose validation requires symmetry of the energy-momentum tensor.

(*c*) *Conservation of Angular Momentum—The Symmetrical Energy-Momentum Tensor*[10]

With the help of equations (19.9*a*) and (19.16*a*), the general result (19.13) takes the form

$$\int_{\Omega} d\omega\, \partial^{\nu}\left(\frac{\partial \mathscr{L}}{\partial \psi^{r,\nu}}\,\delta\psi^{r} - \mathscr{T}_{\mu\nu}\,\delta x^{\mu}\right) = 0 \tag{19.19a}$$

and for an infinitesimal Lorentz rotation we find (cf. 19.10*b*)

$$\int_{\Omega} d\omega\, \partial^{\nu}\left(\frac{\partial \mathscr{L}}{\partial \psi^{,\nu}}\,\mathfrak{T}_{\lambda\mu\ \mathrm{op}}\,\psi + x_{\lambda}\mathscr{T}_{\mu\nu} - x_{\mu}\mathscr{T}_{\lambda\nu}\right)\omega^{\lambda\mu} = 0 \tag{19.19b}$$

Since the $\omega^{\lambda\mu}$ are arbitrary we conclude that

$$\partial^{\nu}\mathscr{J}_{\lambda\mu\nu} = 0 \tag{19.20a}$$

where

$$\mathscr{J}_{\lambda\mu\nu} = \frac{\partial \mathscr{L}}{\partial \psi^{,\nu}}\,\mathfrak{T}_{\lambda\mu\ \mathrm{op}}\,\psi + x_{\lambda}\mathscr{T}_{\mu\nu} - x_{\mu}\mathscr{T}_{\lambda\nu} \tag{19.20b}$$

defines the *total* angular-momentum-density tensor. The vanishing of the divergence of $\mathscr{J}_{\lambda\mu\nu}$ is a necessary condition for the conservation of angular momentum since it guarantees the constancy in time of the volume-integrated skew-tensor $\int d\tau\, \mathscr{J}_{\lambda\mu 4}$, and in particular the integrated *total* angular-momentum components

$$\mathscr{J}_{ij} = \frac{1}{c}\int d\tau\, \mathscr{J}_{ij4} = \int d\tau\left(x_i\mathfrak{G}_j - x_j\mathfrak{G}_i + \frac{1}{c}\frac{\partial \mathscr{L}}{\partial \psi^{,4}}\,\mathfrak{T}_{ij\ \mathrm{op}}\,\psi\right) \tag{19.21}$$

We may write $\mathscr{J}_{\lambda\mu\nu}$ as

$$\mathscr{J}_{\lambda\mu\nu} = \mathscr{M}_{\lambda\mu\nu} + \mathscr{S}_{\lambda\mu\nu} \tag{19.20c}$$

that is, as the sum of an *orbital*- and *spin*-angular momentum-density tensor. The appropriateness of this division may be seen from the well-known fact that in the case of a scalar field (and its dual)—which describes particles without spin—and only in this case (cf. § 26*a*), the canonical tensor $\mathscr{T}_{\mu\nu}$ is symmetric so that the *orbital* density $\mathscr{M}_{\lambda\mu\nu}$ which is the *moment* of $\mathscr{T}_{\mu\nu}$ is divergenceless *ab initio*; the term $\mathscr{S}_{\lambda\mu\nu}$ being absent. However, it is evident that this division is not unique, nor even physically meaningful, unless it is possible to refer the particle to a " rest " system.

[10] cf. Rosenfeld (40); Belinfante (39, 40); Pauli (41).

Hence, the preceding applies directly only to the case of non-zero rest mass, and special consideration is necessary in the contrary case. We leave to the next chapter the necessary detailed analysis of both cases.

We now wish to find a symmetric energy-momentum tensor which can be considered as giving the distribution of energy and momentum in the field. The primary problem is then to modify $\mathscr{T}_{\mu\nu}$ by adding to it a divergenceless tensor $t_{\mu\nu}$, which can be interpreted as the spin-energy-momentum tensor; such that the total energy-momentum tensor defined by

$$\Theta_{\mu\nu} = \mathscr{T}_{\mu\nu} + t_{\mu\nu} \tag{19.22}$$

is again divergenceless, preserving thereby the previously obtained conservation laws. To achieve this we introduce a quantity $\mathsf{S}_{\lambda\mu\nu}$ by writing (19.20a) in the form

$$\begin{aligned}\mathscr{T}_{\lambda\mu} - \mathscr{T}_{\mu\lambda} &= \partial^{\nu}\left(\frac{\partial \mathscr{L}}{\partial \psi^{\cdot\nu}} \mathfrak{T}_{\lambda\mu\ \mathrm{op}}\, \psi\right) = \partial^{\nu} \mathscr{S}_{\lambda\mu\nu} \\ &= \partial^{\nu}(\mathsf{S}_{\mu\lambda\nu} - \mathsf{S}_{\lambda\mu\nu})\end{aligned} \tag{19.23a}$$

from which $\Theta_{\lambda\mu}$ defined by

$$\Theta_{\lambda\mu} = \mathscr{T}_{\lambda\mu} + \mathsf{S}_{\lambda\mu\nu}{}^{\cdot\nu} \tag{19.23b}$$

is symmetric. The condition that $\Theta_{\lambda\mu}$ be divergenceless requires $\mathsf{S}_{\lambda\mu\nu}{}^{\cdot\mu\nu} = 0$ which implies that $\mathsf{S}_{\lambda\mu\nu}$ is skew in μ and ν:

$$\mathsf{S}_{\lambda\mu\nu} = -\mathsf{S}_{\lambda\nu\mu} \tag{19.24a}$$

This condition together with the defining equation

$$\mathsf{S}_{\mu\lambda\nu} - \mathsf{S}_{\lambda\mu\nu} = \mathscr{S}_{\lambda\mu\nu} \tag{19.24b}$$

determine $\mathsf{S}_{\lambda\mu\nu}$ uniquely, and we readily obtain

$$\begin{aligned}\mathsf{S}_{\lambda\mu\nu} &= \tfrac{1}{2}(\mathscr{S}_{\mu\lambda\nu} + \mathscr{S}_{\lambda\nu\mu} + \mathscr{S}_{\mu\nu\lambda}) \\ &= \tfrac{1}{2}\left(\frac{\partial \mathscr{L}}{\partial \psi^{\cdot\nu}} \mathfrak{T}_{\mu\lambda\ \mathrm{op}} + \frac{\partial \mathscr{L}}{\partial \psi^{\cdot\mu}} \mathfrak{T}_{\lambda\nu\ \mathrm{op}} + \frac{\partial \mathscr{L}}{\partial \psi^{\cdot\lambda}} \mathfrak{T}_{\mu\nu\ \mathrm{op}}\right)\psi\end{aligned} \tag{19.24c}$$

The striking fact to be noted (and one which throws considerable light on the role of spin) is that the divergenceless (spin) energy-momentum tensor $t_{\lambda\mu} = \mathsf{S}_{\lambda\mu\nu}{}^{\cdot\nu}$ gives no contribution to the total energy and momentum since

$$\int \Theta_{\lambda 4}\, d\tau = \int \mathscr{T}_{\lambda 4}\, d\tau \tag{19.25}$$

but does contribute to the total angular-momentum density in such a way as to ensure conservation of angular momentum. It should also be mentioned that $\mathscr{J}_{\lambda\mu\nu}$ is *not* the moment of the symmetrized energy-momentum-density tensor since

$$\begin{aligned}x_{\lambda}\Theta_{\mu\nu} - x_{\mu}\Theta_{\lambda\nu} &= \mathscr{M}_{\lambda\mu\nu} + (\mathsf{S}_{\mu\lambda\nu} - \mathsf{S}_{\lambda\mu\nu}) + \partial^{\kappa}(x_{\lambda}\mathsf{S}_{\mu\nu\kappa} - x_{\mu}\mathsf{S}_{\lambda\nu\kappa}) \\ &= \mathscr{J}_{\lambda\mu\nu} + \partial^{\kappa}(x_{\lambda}\mathsf{S}_{\mu\nu\kappa} - x_{\mu}\mathsf{S}_{\lambda\nu\kappa}) \\ &= \mathfrak{T}_{\lambda\mu\nu}\end{aligned} \tag{19.26}$$

The density tensor $\mathfrak{T}_{\lambda\mu\nu}$, as the moment of $\Theta_{\mu\nu}$, might be more properly considered to represent the total angular momentum, and correspondingly the moment of $t_{\mu\nu}$

$$\mathbf{S}_{\lambda\mu\nu} = \mathscr{S}_{\lambda\mu\nu} + \partial^{\kappa}(x_{\lambda}\mathsf{S}_{\mu\nu\kappa} - x_{\mu}\mathsf{S}_{\lambda\nu\kappa}) \tag{19.27}$$

as the spin-angular momentum tensor. However, the total integrated angular momentum is given by

$$c\mathscr{J}_{ij} = \int (x_i \Theta_{j4} - x_j \Theta_{i4})\, d\tau \qquad (19.28a)$$
$$= \int \mathscr{J}_{ij4}\, d\tau = \int \mathfrak{T}_{ij4}\, d\tau$$

so that the distinction is not material, and similarly the volume-integrated spin-angular momentum

$$c\mathscr{S}_{ij} = \int \mathbf{S}_{ij4}\, d\tau = \int \mathscr{S}_{ij4}\, d\tau \qquad (19.28b)$$
$$= \int \frac{\partial \mathscr{L}}{\partial \psi^{\cdot 4}} \mathfrak{T}_{ij\ \mathrm{op}}\, \psi\, d\tau$$

may be calculated from either $\mathbf{S}_{\lambda\mu 4}$ or $\mathscr{S}_{\lambda\mu 4}$; the latter being in general much simpler to use.[11]

§ 20. Complex Fields—Gauge Invariance of the First Kind

In order to obtain a description of charge-bearing fields it is necessary to introduce complex field variables such that the Lagrangian density takes the form

$$\mathscr{L} = \mathscr{L}\left(\psi^r(x),\ \psi^r{}_{,\mu}(x),\ \bar{\psi}^r(x),\ \bar{\psi}^r{}_{,\mu}(x)\right) \qquad (20.1)$$

Treating the variations of the ψ^r, $\psi^r{}_{,\mu}$, and their conjugates as independent, the field equations now become

$$\frac{\delta I}{\delta \psi^r} \equiv \frac{\partial \mathscr{L}}{\partial \psi^r} - \partial_\mu \left(\frac{\partial \mathscr{L}}{\partial \psi^r{}_{,\mu}}\right) = 0 \qquad (20.2)$$
$$\frac{\delta I}{\delta \bar{\psi}^r} \equiv \frac{\partial \mathscr{L}}{\partial \bar{\psi}^r} - \partial_\mu \left(\frac{\partial \mathscr{L}}{\partial \bar{\psi}^r{}_{,\mu}}\right) = 0$$

and the canonical energy-momentum and total angular-momentum-density tensors obviously take the forms

$$\mathscr{T}^{\mu\nu} = \frac{\partial \mathscr{L}}{\partial \psi^r{}_{,\nu}} \psi^{r,\mu} + \text{c.c.} - g^{\mu\nu} \mathscr{L} \qquad (20.3a)$$

$$\mathscr{J}_{\lambda\mu\nu} = \frac{\partial \mathscr{L}}{\partial \psi^{\cdot \nu}} \mathfrak{T}_{\lambda\mu\ \mathrm{op}}\, \psi + \text{c.c.} + x_\lambda \mathscr{T}_{\mu\nu} - x_\mu \mathscr{T}_{\lambda\nu} \qquad (20.3b)$$

other changes being made accordingly.

We now make the assumption that $\mathscr{L}$ is also invariant under gauge transformations of the first kind, viz.

$$\psi^r \rightarrow \psi^{r\prime} = \psi^r e^{i\epsilon_r \lambda} \qquad (20.4)$$
$$\bar{\psi}^r \rightarrow \bar{\psi}^{r\prime} = \bar{\psi}^r e^{-i\epsilon_r \lambda}$$

[11] In order not to interrupt the continuity of development we leave to the next chapter the detailed considerations required to establish the fact, mentioned in § 4, that the spin operator $\boldsymbol{S}_{\mathrm{op}} = (S_{23}, S_{31}, S_{12})_{\mathrm{op}}$ is given by $i\hbar\, (\mathfrak{T}_{23}, \mathfrak{T}_{31}, \mathfrak{T}_{12})_{\mathrm{op}}$.

where in the present (e.m.) field-free case both λ and ϵ_r are constants. Clearly, no summation over the index r is implied here due to its appearance in the factors ϵ_r which are included to allow for the possibility that particles of different charges are connected through these fields, the different phases being proportional to the different charges. The physical correlate of the gauge condition is that *in principle* only expressions bilinear in the ψ^r, $\bar{\psi}^r$, etc., i.e. gauge invariants, are associated with physically measurable quantities (as in ordinary Schrödinger theory) while real fields and the electromagnetic field strengths, in particular, are directly measurable.[1] For an infinitesimal gauge transformation such that second- and higher-order terms in λ can be neglected, the gauge condition on $\mathscr{L}$ leads to

$$\begin{aligned}\delta\mathscr{L} = 0 &= \frac{\partial\mathscr{L}}{\partial\psi^r}\,\delta\psi^r + \frac{\partial\mathscr{L}}{\partial\psi^r{}_{,\mu}}\,\delta\psi^r{}_{,\mu} + \text{c.c.} \qquad (20.5)\\ &= i\epsilon_r\lambda\left\{\left(\frac{\partial\mathscr{L}}{\partial\psi^r}\,\psi^r - \frac{\partial\mathscr{L}}{\partial\bar{\psi}^r}\,\bar{\psi}^r\right) + \left(\frac{\partial\mathscr{L}}{\partial\psi^r{}_{,\mu}}\,\psi^r{}_{,\mu} - \frac{\partial\mathscr{L}}{\partial\bar{\psi}^r{}_{,\mu}}\,\bar{\psi}^r{}_{,\mu}\right)\right\}\end{aligned}$$

Or, with the help of the field equations (20.2),

$$\partial_\mu J^\mu = 0 \qquad (20.6a)$$

which is interpreted as the continuity equation for the associated (real) current 4-vector defined by

$$J^\mu = -i\epsilon_r\left(\frac{\partial\mathscr{L}}{\partial\psi^r{}_{,\mu}}\,\psi^r - \frac{\partial\mathscr{L}}{\partial\bar{\psi}^r{}_{,\mu}}\,\bar{\psi}^r\right) \qquad (20.6b)$$

However, the justification for taking J^μ as the current vector must be left to the following sections in which we consider the interaction with an electromagnetic field.

Before considering these questions we may note the interesting fact that if the field variables are expressed in the form

$$\begin{aligned}\psi^r &= \psi^{r(1)} + i\psi^{r(2)} \qquad (20.7a)\\ \bar{\psi}^r &= \psi^{r(1)} - i\psi^{r(2)}\end{aligned}$$

the gauge transformation (20.4) becomes

$$\begin{aligned}\psi^{r(1)} &\to \psi^{r(1)\prime} = \psi^{r(1)}\cos\lambda\epsilon_r - \psi^{r(2)}\sin\lambda\epsilon_r \qquad (20.7b)\\ \psi^{r(2)} &\to \psi^{r(2)\prime} = \psi^{r(1)}\sin\lambda\epsilon_r + \psi^{r(2)}\cos\lambda\epsilon_r\end{aligned}$$

that is, a rotation in the two-dimensional so-called isotopic-spin space. With this observation the infinitesimal gauge transformation yields (cf. 20.5, 20.6*a*, *b*)

$$\partial_\mu\left\{\frac{\partial\mathscr{L}}{\partial\psi^{r(1)}{}_{,\mu}}\,\psi^{r(2)} - \frac{\partial\mathscr{L}}{\partial\psi^{r(2)}{}_{,\mu}}\,\psi^{r(1)}\right\}\epsilon_r\lambda = 0 \qquad (20.7c)$$

in which the term $\left\{\frac{\partial\mathscr{L}}{\partial\psi^{r(1)}{}_{,4}}\,\psi^{r(2)} - \frac{\partial\mathscr{L}}{\partial\psi^{r(2)}{}_{,4}}\,\psi^{r(1)}\right\}$ is of the form of an angular momentum component in the symbolic isotopic-spin space (cf. § 24*e*).

[1] The really precise statement is in terms of quantized fields. Cf. Pauli (41); Wentzel (43).

§ 21. General Gauge Invariance—Fundamental Consequences

We consider now the general case of matter (m) fields in interaction with the electromagnetic (e.m.) field, the former defined as in the preceding section and the latter by the real (e.m.) potentials

$$\phi_\mu = (\boldsymbol{A},\ -\phi) \tag{21.1a}$$

and their skew-derivatives

$$F_{\mu\nu} = \phi_{\nu},_{\mu} - \phi_{\mu},_{\nu} \tag{21.1b}$$

where, in accord with the usual specification,

$$\boldsymbol{H} = \text{curl}\ \boldsymbol{A} \tag{21.2a}$$

$$\boldsymbol{E} = -\text{grad}\ \phi - \frac{1}{c}\frac{\partial \boldsymbol{A}}{\partial t}$$

$$F_{\mu\nu} = \begin{pmatrix} 0 & H_z & -H_y & E_x \\ -H_z & 0 & H_x & E_y \\ H_y & -H_x & 0 & E_z \\ -E_x & -E_y & -E_z & 0 \end{pmatrix} \tag{21.2b}$$

Maxwell's equations (in free space)

$$\text{div}\ \boldsymbol{E} = \rho \tag{21.3a}$$

$$\text{curl}\ \boldsymbol{H} = \frac{1}{c}\left(\frac{\partial \boldsymbol{E}}{\partial t} + \boldsymbol{I}\right)$$

$$\text{div}\ \boldsymbol{H} = 0 \tag{21.3b}$$

$$\text{curl}\ \boldsymbol{E} = -\frac{1}{c}\frac{\partial \boldsymbol{H}}{\partial t}$$

being subsumed in the corresponding tensor equations

$$F^{\mu\nu},_{\nu} = J^\mu \qquad J^\mu = \left(\frac{\boldsymbol{I}}{c},\ \rho\right) \tag{21.4a}$$

$$F_{\mu\nu},_{\sigma} + F_{\nu\sigma},_{\mu} + F_{\sigma\mu},_{\nu} = 0 \tag{21.4b}$$ [1]

The Lagrangian density is now assumed to be of the form

$$\mathscr{L}(\psi^r;\ \psi^r,_{\mu};\ \bar{\psi}^r;\ \bar{\psi}^r,_{\mu};\ \phi_\mu;\ \phi_{\mu,\nu}) \tag{21.5}$$

and is postulated to be invariant under gauge transformations of the first and second kind, viz.

$$\psi^r \rightarrow \psi^{r\prime} = \psi^r e^{i\epsilon_r\chi} \quad \text{and c.c.} \tag{21.6a}$$

$$\phi^\mu \rightarrow \phi^{\mu\prime} = \phi^\mu + \chi^{\cdot\mu} \tag{21.6b}$$

where χ is now an arbitrary function of position.

[1] This is the necessary and sufficient condition that $F_{\mu\nu}$ be expressible in the form (21.1*b*), and is equivalent to $\check{F}^{\mu\nu},_{\nu} = 0$. Noting this we may write the equations (21.4*a*, *b*) in the simple form $G^{\mu\nu},_{\nu} = J^\mu$ in terms of the self-dual skew-tensor defined by $G_{\sigma\tau} = F_{\sigma\tau} + \check{F}_{\sigma\tau} = \frac{1}{2}\eta^{\mu\nu}{}_{\sigma\tau}F_{\mu\nu}$ (cf. 3.1*a*, 13.14).

We can now show that the condition of general gauge invariance determines three fundamental properties: the existence of a charge-current 4-vector as before, the allowed functional form of $\mathscr{L}$ with respect to the derivatives of the e.m. potentials, and the functional form of $\mathscr{L}$ with respect to the potentials themselves, the last mentioned involving an additional assumption.

According to (21.6a) the gauge transformations of the matter field derivatives $\psi^r{}_{,\mu}$ take the form

$$\psi^r{}_{,\mu} \to \psi^r{}_{,\mu}{}' = \psi^r{}_{,\mu} e^{i\epsilon_r\chi} + i\epsilon_r \chi_{,\mu} \psi^r e^{i\epsilon_r\chi} \tag{21.7}$$

and we may again confine our considerations to infinitesimal gauge transformations such that all higher-order terms in χ and its derivatives can be neglected. The gauge condition on $\mathscr{L}$ takes the form

$$\delta\mathscr{L} = 0 = \frac{\partial\mathscr{L}}{\partial\psi^r}\,\delta\psi^r + \frac{\partial\mathscr{L}}{\partial\psi^r{}_{,\mu}}\,\delta\psi^r{}_{,\mu} + \text{c.c.} + \frac{\partial\mathscr{L}}{\partial\phi_\mu}\,\delta\phi_\mu + \frac{\partial\mathscr{L}}{\partial\phi_{\mu,\nu}}\,\delta\phi_{\mu,\nu} \tag{21.8a}$$

and substituting for the gauge variations as obtained from (21.6, 21.7)

$$\begin{aligned} i\epsilon_r\chi\left\{\left(\frac{\partial\mathscr{L}}{\partial\psi^r}\,\psi^r - \frac{\partial\mathscr{L}}{\partial\bar\psi^r}\,\bar\psi^r\right) + \left(\frac{\partial\mathscr{L}}{\partial\psi^r{}_{,\mu}}\,\psi^r{}_{,\mu} - \frac{\partial\mathscr{L}}{\partial\bar\psi^r{}_{,\mu}}\,\bar\psi^r{}_{,\mu}\right)\right\} \\ + \chi_{,\mu}\left\{\frac{\partial\mathscr{L}}{\partial\phi_\mu} + i\epsilon_r\left(\frac{\partial\mathscr{L}}{\partial\psi^r{}_{,\mu}}\,\psi^r - \frac{\partial\mathscr{L}}{\partial\bar\psi^r{}_{,\mu}}\,\bar\psi^r\right)\right\} \\ + \chi_{,\mu\nu}\frac{\partial\mathscr{L}}{\partial\phi_{\mu,\nu}} = 0 \end{aligned} \tag{21.8b}$$

Hence the conditions

$$\chi_{,\mu\nu}\frac{\partial\mathscr{L}}{\partial\phi_{\mu,\nu}} = 0 \tag{21.9a}$$

$$\chi_{,\mu}\left\{\frac{\partial\mathscr{L}}{\partial\phi_\mu} + i\epsilon_r\left(\frac{\partial\mathscr{L}}{\partial\psi^r{}_{,\mu}}\,\psi^r - \frac{\partial\mathscr{L}}{\partial\bar\psi^r{}_{,\mu}}\,\bar\psi^r\right)\right\} = 0 \tag{21.9b}$$

$$\chi\epsilon^r\left\{\left(\frac{\partial\mathscr{L}}{\partial\psi^r}\,\psi^r - \frac{\partial\mathscr{L}}{\partial\bar\psi^r}\,\bar\psi^r\right) + \left(\frac{\partial\mathscr{L}}{\partial\psi^r{}_{,\mu}}\,\psi^r{}_{,\mu} - \frac{\partial\mathscr{L}}{\partial\bar\psi^r{}_{,\mu}}\,\bar\psi^r{}_{,\mu}\right)\right\} = 0 \tag{21.9c}$$

since χ, $\chi_{,\mu}$, and $\chi_{,\mu\nu}$ may be chosen arbitrarily, so that the terms in (21.8b) must vanish separately. From (21.9a) it follows that

$$\left(\frac{\partial\mathscr{L}}{\partial\phi_{\mu,\nu}} + \frac{\partial\mathscr{L}}{\partial\phi_{\nu,\mu}}\right) = 0 \tag{21.10}$$

or: *the Lagrangian density must be, inter alia, a function of the e.m. field strengths* $F_{\mu\nu}$. Equations (21.9b, c) merely restate the previous definition of the current vector and its continuity equation, *but* clearly this current vector *is not invariant* under the transformations (21.6) for arbitrary $\mathscr{L}$. However, if we take $\mathscr{L}$ of the form

$$\mathscr{L} = \mathscr{L}_{\text{mi}}(\psi^r,\ \psi^r{}_{,\mu},\ \bar\psi^r,\ \bar\psi^r{}_{,\mu},\ \phi_\mu) + \mathscr{L}_{\text{em}}(F_{\mu\nu}) \tag{21.11a}$$

i.e. the sum of two terms describing, respectively, the (m) fields and their interaction with the (e.m.) field *plus* the Lagrangian of the latter in the form allowed by the preceding result, and more particularly

$$\mathscr{L} = \mathscr{L}_{\mathrm{mi}}(\psi^r, \bar{\psi}^r, D_\mu\psi^r, \bar{D}_\mu\bar{\psi}^r) + \mathscr{L}_{\mathrm{em}}(F_{\mu\nu}) \tag{21.11b}$$

with $D_\mu = \partial_\mu - i\epsilon_r\phi_\mu$; then $\left(\dfrac{\partial\mathscr{L}}{\partial\psi^r{}_{,\mu}}\right) = \left(\dfrac{\partial\mathscr{L}}{\partial D_\mu\psi^r}\right)$, equation (21.9*b*) is identically satisfied, and

$$J^\mu = -i\epsilon_r\left\{\frac{\partial\mathscr{L}}{\partial D_\mu\psi^r}\psi^r - \frac{\partial\mathscr{L}}{\partial\bar{D}_\mu\bar{\psi}^r}\bar{\psi}^r\right\} = \frac{\partial\mathscr{L}}{\partial\phi_\mu} \tag{21.12}$$

is gauge-invariant as may be readily verified. Thus, the e.m. potentials must appear in $\mathscr{L}$ in the form

$$\psi^r{}_{,\mu} - i\epsilon_r\phi_\mu\psi^r \quad \text{and} \quad \bar{\psi}^r{}_{,\mu} + i\epsilon_r\phi_\mu\bar{\psi}^r \qquad (21.13)^2$$

This important result may be stated in the following more explicit form. Namely, if we have originally only a set of matter fields (which we refer to as the field-free case) specified by the Lagrangian

$$\mathscr{L}_{\mathrm{m}}(\psi^r, \bar{\psi}^r, \bar{\psi}^r{}_{,\mu}, \bar{\psi}^r{}_{,\mu}) \tag{21.14a}$$

from which obtain all field equations

$$\frac{\partial\mathscr{L}_{\mathrm{m}}}{\partial\psi^r} - \partial_\nu\left(\frac{\partial\mathscr{L}_{\mathrm{m}}}{\partial\psi^r{}_{,\nu}}\right) = 0 \quad \text{and c.c.} \tag{21.14b}$$

(without supplementary conditions) then the presence of an external electromagnetic field simply modifies these according to

$$\mathscr{L}_{\mathrm{mi}}(\psi^r, \bar{\psi}^r, D_\mu\psi^r, \bar{D}_\mu\bar{\psi}^r) \tag{21.15a}$$

and

$$\frac{\partial\mathscr{L}_{\mathrm{mi}}}{\partial\psi^r} - \bar{D}_\nu\left(\frac{\partial\mathscr{L}_{\mathrm{mi}}}{\partial D_\nu\psi^r}\right) = 0 \quad \text{and c.c.} \tag{21.15b}$$

All other quantities previously defined by the ψ^r, $\psi^r{}_{,\mu}\ldots$ now involve the field quantities ψ^r, $D_\mu\psi^r\ldots$

In this connection it is important to our future work to note that

$$\partial_\nu\mathscr{L}_{\mathrm{mi}} = \frac{\partial\mathscr{L}_{\mathrm{mi}}}{\partial\psi^r}(D_\nu\psi^r) + \frac{\partial\mathscr{L}_{\mathrm{mi}}}{\partial D_\mu\psi^r}(D_\nu D_\mu\psi^r) + \text{c.c.} \tag{21.16}$$

which is readily verified with the use of (21.9*c*); and further

$$\partial_\nu(AB) = (\bar{D}_\nu A)B + A(D_\nu B) \tag{21.17}$$

for any two functions of the ψ^r, $D_\nu\psi^r$, etc., which undergo the gauge transformations $A \to Ae^{-i\epsilon_r\chi}$ and $B \to Be^{i\epsilon_r\chi}$.

[2] In quantum mechanics this takes the form $\partial_\mu \to \partial_\mu - \dfrac{ie}{\hbar c}\phi_\mu$ (which incidentally agrees with the usual transcription according to our definition of ϕ_μ), so that $\epsilon_r = e_r/\hbar c$.

§ 22. Interaction of Electromagnetic and Matter Fields

(a) Modified Conservation Laws

The modified forms of the conservation laws in the case of (m) fields interacting with an external (e.m.) field follow immediately if we generalize the (m) field energy-momentum density tensor to gauge-invariant form according to the preceding prescription. On the other hand we can proceed directly from the fundamental invariance property in the form (cf. 19.7)

$$\int_{\Omega} d\omega \left(\delta^* \mathscr{L}_{\text{mi}} + \partial_\nu(\mathscr{L}_{\text{mi}}\, \delta x^\nu)\right) = 0 \tag{22.1a}$$

with

$$\mathscr{L}_{\text{mi}} = \mathscr{L}_{\text{mi}}(\psi^r,\ \bar{\psi}^r,\ D^\mu \psi^r,\ \bar{D}^\mu \bar{\psi}^r)$$

and

$$\delta^* \mathscr{L}_{\text{mi}} = \frac{\partial \mathscr{L}_{\text{mi}}}{\partial \psi^r}\, \delta^* \psi^r + \frac{\partial \mathscr{L}_{\text{mi}}}{\partial D_\nu \psi^r}\, \delta^* D_\nu \psi^r + \text{c.c.} \tag{22.1b}$$

Again considering an infinitesimal translation, we note that (cf. 19.14)

$$\begin{aligned} \delta^* \psi^r &= -\psi^{r.\mu}\, \delta x_\mu \\ &= -D^\mu \psi^r\, \delta x_\mu - i\epsilon_r\, \phi^\mu\, \psi^r\, \delta x_\mu \end{aligned} \tag{22.2a}$$

and

$$\begin{aligned} \delta^*(D_\nu \psi^r) &= -\partial_\mu(D_\nu \psi^r)\, \delta x^\mu \\ &= D_\nu\, \delta^* \psi^r + i\epsilon_r\, \phi_{\nu,\,\mu}\, \psi^r\, \delta x^\mu \end{aligned} \tag{22.2b}$$

with which $\delta^*\mathscr{L}_{\text{mi}}$ takes the form

$$\delta^* \mathscr{L}_{\text{mi}} = \frac{\partial \mathscr{L}_{\text{mi}}}{\partial \psi^r}\, \delta^* \psi^r + \frac{\partial \mathscr{L}_{\text{mi}}}{\partial D_\nu \psi^r}\, (D_\nu\, \delta^* \psi^r + i\epsilon_r\, \phi_{\nu,\,\mu}\, \psi^r\, \delta x^\mu) + \text{c.c.} \tag{22.3a}$$

or, with the help of the field equations (21.15*b*) and the relation (21.17),

$$\delta^* \mathscr{L}_{\text{mi}} = \partial_\nu \left\{ \frac{\partial \mathscr{L}_{\text{mi}}}{\partial D_\nu \psi^r}\, \delta^* \psi^r \right\} + i\epsilon_r\, \phi_{\nu,\,\mu}\, \psi^r\, \frac{\partial \mathscr{L}_{\text{mi}}}{\partial D_\nu \psi^r}\, \delta x^\mu + \text{c.c.} \tag{22.3b}$$

Substituting the value of $\delta^*\psi^r$, it follows from the invariance condition (22.1*a*), by the same argument as before, that

$$\partial_\nu \mathscr{T}^{\mu\nu}_{\text{mi}} = -i\epsilon_r \partial_\nu \left(\phi^\mu\, \psi^r\, \frac{\partial \mathscr{L}_{\text{mi}}}{\partial D_\nu \psi^r} \right) + i\epsilon_r\, \phi_{\nu,\,\mu}\, \psi^r\, \frac{\partial \mathscr{L}_{\text{mi}}}{\partial D_\nu \psi^r} + \text{c.c.} \tag{22.4}$$

where $\mathscr{T}^{\mu\nu}_{\text{mi}}$ is the *gauge-invariant* canonical energy-momentum tensor defined by

$$\mathscr{T}^{\mu\nu}_{\text{mi}} = \frac{\partial \mathscr{L}_{\text{mi}}}{\partial D_\nu \psi^r}\, D^\mu \psi^r + \text{c.c.} - g^{\mu\nu}\, \mathscr{L}_{\text{mi}} \tag{22.5}$$

With the definition of the *gauge-invariant* current vector (21.12), equation (22.4) reduces to

$$\begin{aligned}\partial_\nu \mathscr{T}^{\mu\nu}_{\text{mi}} &= \partial_\nu \Theta^{\mu\nu}_{\text{mi}} \\ &= -i\epsilon_r \left\{ \phi^\mu{}_{,\nu} \psi^r \frac{\partial \mathscr{L}_{\text{mi}}}{\partial D_\nu \psi^r} - \phi_\nu{}^{,\mu} \psi^r \frac{\partial \mathscr{L}_{\text{mi}}}{\partial D_\nu \psi^r} \right\} + \text{c.c.} \\ &= -F^{\mu\nu} J_\nu = f^\mu \end{aligned} \tag{22.6}$$

which expresses the existence of the Lorentz force, and provides the justification for the definition of J^μ as the current 4-vector.

Moreover, if $\Theta^{\mu\nu}_{\text{em}}$ denotes the symmetrical gauge-invariant energy-momentum tensor of the e.m. field, it follows from the above result and the requirement of conservation for the total field system that

$$\Theta^{\mu\nu}_{\text{em}}{}_{,\nu} = F^{\mu\nu} J_\nu \tag{22.7a}$$

the solution of which is given by the familiar expression

$$\Theta^{\mu\nu}_{\text{em}} = -F^{\mu\alpha} F^\nu{}_\alpha + \tfrac{1}{4} g^{\mu\nu} F^{\alpha\beta} F_{\alpha\beta} \tag{22.7b}$$

as follows with the use of equations (21.4*a*, *b*). Conversely, with $\mathscr{L}_{\text{em}}$ given by $-\frac{1}{4}F^{\alpha\beta}F_{\alpha\beta}$, and for arbitrary variations in the e.m. field variables—vanishing on the integration boundaries—we find (cf. 21.11*b*, 21.12)

$$\begin{aligned} 0 = \delta \int \mathscr{L}\, d\omega &= \int \{\delta \mathscr{L}_{\text{mi}} + \delta \mathscr{L}_{\text{em}}\}\, d\omega \\ &= \int \{J^\mu \delta\phi_\mu + F^{\mu\nu} \delta \phi_{\mu,\nu}\}\, d\omega \\ &= \int \{(J^\mu - F^{\mu\nu}{}_{,\nu}) \delta\phi_\mu + \partial_\nu (F^{\mu\nu} \delta \phi_\mu)\}\, d\omega \end{aligned} \tag{22.8}$$

so that the Maxwell equations $F^{\mu\nu}{}_{,\nu} = J^\mu$ obtain, as assumed in passing from (22.7*b*) to (22.7*a*).

Similarly, by consideration of an infinitesimal Lorentz transformation, the invariance property (22.1*a*) together with equations (22.3*b*) and (19.12) lead simply to the result that the divergence of the gauge-invariant (m) angular-momentum-density tensor now equals the moment of the Lorentz 4-force:

$$\begin{aligned} \partial^\nu \mathscr{J}_{(\text{mi})\,\lambda\mu\nu} &= -\{x_\lambda F_{\mu\nu} J^\nu - x_\mu F_{\lambda\nu} J^\nu\} \\ &= x_\lambda f_\mu - x_\mu f_\lambda \end{aligned} \tag{22.9a}$$

with

$$\mathscr{J}_{(\text{mi})\,\lambda\mu\nu} = \frac{\partial \mathscr{L}_{(\text{mi})}}{\partial D^\nu \psi} \mathfrak{T}_{\lambda\mu\ \text{op}}\, \psi + \text{c.c.} + x_\lambda \mathscr{T}^{(\text{mi})}_{\mu\nu} - x_\mu \mathscr{T}^{(\text{mi})}_{\lambda\nu} \tag{22.9b}$$

and $\mathscr{T}^{(\text{mi})}_{\mu\nu}$ given by equation (22.5). Again, from the requirement of angular-momentum conservation for the complete field system, it follows that

$$\partial^\nu \mathscr{J}^{(\text{em})}_{\lambda\mu\nu} = \{x_\lambda F_{\mu\nu} J^\nu - x_\mu F_{\lambda\nu} J^\nu\} \tag{22.10a}$$

where the gauge-invariant e.m. field (total) angular-momentum-density tensor is given by

$$\begin{aligned}\mathscr{J}^{(\text{em})}_{\lambda\mu\nu} &= x_\lambda \Theta^{(\text{em})}_{\mu\nu} - x_\mu \Theta^{(\text{em})}_{\lambda\nu} \\ &= -x_\lambda F_\mu{}^\alpha F_{\nu\alpha} + x_\mu F_\lambda{}^\alpha F_{\nu\alpha} + \tfrac{1}{4}(x_\lambda g_{\mu\nu} - x_\mu g_{\lambda\nu}) F^{\alpha\beta} F_{\alpha\beta}\end{aligned} \tag{22.10b}$$

From the fact that $\Theta^{(\text{em})}_{\mu\nu}$ is symmetric, or by simple direct computation, it follows that $\mathscr{J}^{(\text{em})}_{\lambda\mu\nu}$ satisfies (22.10*a*).

(b) Constitutive Tensors of the Electromagnetic Field

For the sake of completeness, it is desirable to construct the constitutive tensors of the electromagnetic field directly from first principles. It is clear that with the definition of $F_{\mu\nu}$ which comprises one set of the Maxwell equations, the variation principle with Lagrangian $\mathscr{L} = -\frac{1}{4}F^{\alpha\beta}F_{\alpha\beta} = \frac{1}{2}(\boldsymbol{E}^2 - \boldsymbol{H}^2)$ leads directly to the other set:

$$\frac{\partial \mathscr{L}}{\partial \phi_\mu} - \partial_\nu\left(\frac{\partial \mathscr{L}}{\partial \phi_{\mu,\nu}}\right) = F^{\mu\nu}{}_{,\nu} = 0 \tag{22.11}$$

The *canonical* energy-momentum tensor is then

$$\begin{aligned}\mathscr{T}^{\mu\nu} &= \frac{\partial \mathscr{L}}{\partial \phi^\kappa{}_{,\nu}} \phi^{\kappa,\mu} - g^{\mu\nu}\mathscr{L} \\ &= F_\kappa{}^\nu \phi^{\kappa,\mu} + \tfrac{1}{4} g^{\mu\nu} F^{\alpha\beta} F_{\alpha\beta}\end{aligned} \tag{22.12}$$

which is, of course, divergenceless by virtue of the field equations, but is neither gauge-invariant nor symmetric. Correspondingly, the (nominally) orbital-angular momentum tensor

$$\begin{aligned}\mathscr{M}_{\lambda\mu\nu} &= x_\lambda \mathscr{T}_{\mu\nu} - x_\mu \mathscr{T}_{\lambda\nu} \\ &= -x_\lambda F_{\nu\alpha}\phi^\alpha{}_{,\mu} + x_\mu F_{\nu\alpha}\phi^\alpha{}_{,\lambda} + \tfrac{1}{4}(x_\lambda g_{\mu\nu} - x_\mu g_{\lambda\nu}) F^{\alpha\beta} F_{\alpha\beta}\end{aligned} \tag{22.13}$$

is not divergenceless.

In order to construct the symmetric energy-momentum tensor we note that the electromagnetic field variables, the potentials, undergo the infinitesimal Lorentz transformation

$$\begin{aligned}\delta\phi^\alpha &= \omega^{\alpha\nu}\phi_\nu = \omega^{\mu\nu}\delta^\alpha_\mu \delta^\beta_\nu \phi_\beta \\ &= \tfrac{1}{2}\omega^{\mu\nu}\delta^{\alpha\beta}_{\mu\nu}\phi_\beta = \tfrac{1}{2}\omega^{\mu\nu}\mathfrak{T}^{\alpha\beta}_{\mu\nu}\phi_\beta\end{aligned} \tag{22.14}$$

whence (cf. 19.23*a*) the spin density tensor is

$$\begin{aligned}\mathscr{S}_{\lambda\mu\nu} &= \frac{\partial \mathscr{L}}{\partial \phi^{\alpha,\nu}} \mathfrak{T}^{\alpha\beta}_{\lambda\mu}\phi_\beta = F_{\alpha\nu}\delta^{\alpha\beta}_{\lambda\mu}\phi_\beta \\ &= F_{\lambda\nu}\phi_\mu - F_{\mu\nu}\phi_\lambda\end{aligned} \tag{22.15a}$$

and (cf. 19.24*c*)

$$\begin{aligned}\mathsf{S}_{\mu\nu\kappa} &= \tfrac{1}{2}(\mathscr{S}_{\nu\mu\kappa} + \mathscr{S}_{\mu\kappa\nu} + \mathscr{S}_{\nu\kappa\mu}) \\ &= F_{\nu\kappa}\phi_\mu\end{aligned} \tag{22.15b}$$

This gives the spin energy-momentum tensor

$$\mathfrak{t}_{\mu\nu} = \mathfrak{S}_{\mu\nu\kappa}{}^{,\kappa} = F_{\nu\kappa}\,\phi_{\mu}{}^{,\kappa} \tag{22.16}$$

and the symmetric gauge-invariant total tensor

$$\Theta^{\mu\nu} = \mathscr{T}^{\mu\nu} + \mathfrak{t}^{\mu\nu} = -F^{\mu\kappa}F^{\nu}{}_{\kappa} + \tfrac{1}{4}g^{\mu\nu}F^{\alpha\beta}F_{\alpha\beta} \qquad (22.17)^1$$

It is of some interest to examine further the form of the total- and spin-angular momenta, particularly in vector notation. Noting equations (22.13, 22.15a, b), we find

$$\begin{aligned}\mathscr{J}_{\lambda\mu\nu} &= \mathscr{S}_{\lambda\mu\nu} + \mathscr{M}_{\lambda\mu\nu} \qquad (22.18a)\\ &= F_{\lambda\nu}\,\phi_{\mu} - F_{\mu\nu}\,\phi_{\lambda} - F_{\nu\alpha}(x_{\lambda}\,\phi^{\alpha}{}_{,\mu} - x_{\mu}\,\phi^{\alpha}{}_{,\lambda}) + \tfrac{1}{4}(x_{\lambda}g_{\mu\nu} - x_{\mu}g_{\lambda\nu})F^{\alpha\beta}F_{\alpha\beta}\end{aligned}$$

and

$$\begin{aligned}\mathfrak{T}_{\lambda\mu\nu} &= \mathscr{J}_{\lambda\mu\nu} + \partial^{\kappa}(x_{\lambda}F_{\nu\kappa}\,\phi_{\mu} - x_{\mu}F_{\nu\kappa}\,\phi_{\lambda}) \qquad (22.18b)\\ &= x_{\lambda}F_{\nu\kappa}F^{\kappa}{}_{\mu} - x_{\mu}F_{\nu\kappa}F^{\kappa}{}_{\lambda} + \tfrac{1}{4}(x_{\lambda}g_{\mu\nu} - x_{\mu}g_{\lambda\nu})F^{\alpha\beta}F_{\alpha\beta}\end{aligned}$$

With the usual transcription to 3-vector notation, $\mathfrak{J} = (\mathscr{J}_{23}, \mathscr{J}_{31}, \mathscr{J}_{12})$, the first of these gives

$$\mathfrak{J} = \frac{1}{c}\int\{[\boldsymbol{E} \wedge \boldsymbol{A}] + [\mathbf{x} \wedge \nabla](\boldsymbol{A} \,.\, \boldsymbol{E})_E\}d\tau \tag{22.19a}$$

where $[\boldsymbol{E}\wedge\boldsymbol{A}]$ represents the vector spin-density and the subscript in the second term indicates that $\boldsymbol{E}$ is constant in the differentiation. This may be rewritten in the form

$$\begin{aligned}\mathfrak{J} &= \frac{1}{c}\int\{\mathbf{x} \wedge [\boldsymbol{E} \wedge \boldsymbol{H}] - [\mathbf{x} \wedge \boldsymbol{A}]\,\mathrm{div}\,\boldsymbol{E}\}d\tau \qquad (22.19b)\\ &= \frac{1}{c}\int \mathbf{x} \wedge [\boldsymbol{E} \wedge \boldsymbol{H}]d\tau\end{aligned}$$

the latter by virtue of the condition div $\boldsymbol{E} = 0$. Thus, the total angular momentum derives from the moment of the familiar Poynting vector. As it must be, of course, this is just the result directly obtaining from the equivalent integral over $\mathfrak{T}_{ij4}$ as is seen from (22.18b).

As has been discussed, for example, by Pauli,[2] the spin of light quanta is manifested in the *quantized* theory through the fact that the lowest (total) angular momentum state corresponds to the quantum number $j = 1$ rather than $j = 0$. In the work of §§ 29, 40 we shall show that this result may also be quite easily established within the framework of the present theory. More generally, we shall find that all fields associated with particles of zero rest mass admit a special degeneracy which amounts to an extended type of gauge transformation, of which that for the e.m. field is a particular case. The consequence of this is that for such fields there exist only 2 rather than $2s + 1$ independent states of polarization corresponding to spin s (except zero), and these correspond to the *axial* quantum numbers s and $-s$.

[1] Note that with the self-dual tensor $G_{\mu\nu}$ mentioned in footnote 1, § 21, this may be written as $\Theta_{\mu\nu} = \frac{1}{2}G_{\mu}{}^{\alpha}\,\overline{G}_{\alpha\nu}$.

[2] Pauli (33). See also Wentzel (43); Fierz (38, 39).

§ 23. Possible (e.m.) Field Modification of the Matter Field Lagrangian

In § 21 we found how the e.m. potentials may enter the Lagrangian $\mathscr{L}_{(\text{mi})}$, or equivalently the canonical tensor $\mathscr{T}^{\mu\nu}_{(\text{mi})}$, *assuming* that the field strengths occur only in the e.m. Lagrangian $\mathscr{L}_{(\text{em})}$. However, this does not exclude the possibility of modifying $\mathscr{L}_{(\text{mi})}$ through the addition, in a gauge-invariant way, of terms explicitly dependent on the field strengths. The point of such a modification is that, in consequence, the magneto-mechanical ratio for a particle described by a given field no longer takes on the value intrinsic to that field, but rather becomes adjustable at will.[1] By the same token, an arbitrary magnetic moment can be associated even with a neutral particle, although this case requires special consideration. This resultant freedom carries with it, in a sense, the burden of too much riches, and is so artificial as to seem entirely unnecessary—as indeed recent theoretical developments suggest. Nonetheless, the introduction of such so-called Pauli terms has its usefulness for the more phenomenological investigations, and warrants at least brief consideration.

If now we admit a Lagrangian density of the form $\mathscr{L}_{(\text{mi})}(\psi^r, D^\mu\psi^r, \ldots F^{\mu\nu})$ then instead of equation (22.6), we obtain

$$\mathscr{T}^{\mu\nu}_{(\text{mi}),\nu} = -F^{\mu\nu}J_\nu - \tfrac{1}{2}\frac{\partial \mathscr{L}}{\partial F^{\alpha\beta}}F^{\alpha\beta,\mu} \tag{23.1}$$

This requires that we modify $\mathscr{T}^{\mu\nu}_{(\text{mi})}$, J^ν to some $\mathsf{T}^{\mu\nu}_{(\text{mi})}$, J^ν through additional terms such that

$$\mathsf{T}^{\mu\nu}_{(\text{mi}),\nu} = -F^{\mu\nu}\mathsf{J}_\nu \tag{23.2}$$

as required. The new J_ν is easily obtained from the generalized definition (cf. 22.8)

$$\mathsf{J}^\nu = \frac{\delta\int \mathscr{L}_{(\text{mi})}\,d\omega}{\delta\phi_\nu} = \frac{\partial\mathscr{L}_{(\text{mi})}}{\partial\phi_\nu} - \partial_\mu\left(\frac{\partial\mathscr{L}_{(\text{mi})}}{\partial F_{\nu\mu}}\right) = J^\nu - \partial_\mu\left(\frac{\partial\mathscr{L}_{(\text{mi})}}{\partial F_{\nu\mu}}\right) \tag{23.3}$$

which also satisfies a continuity equation and guarantees that the Maxwell equations continue to obtain from the variation principle. Then equation (23.2) is satisfied if we define $\mathsf{T}^{\mu\nu}_{(\text{mi})}$ by

$$\mathsf{T}^{\mu\nu}_{(\text{mi})} = \mathscr{T}^{\mu\nu}_{(\text{mi})} - F^{\mu}{}_{\lambda}\frac{\partial\mathscr{L}_{(\text{mi})}}{\partial F_{\lambda\nu}} \tag{23.4}$$

as follows from equations (23.1, 23.3) with the help of the Maxwell equations

$$F^{\mu\nu,\lambda} + F^{\lambda\mu,\nu} + F^{\nu\lambda,\mu} = 0 \tag{23.5}$$

[1] The question of the value of the magnetic moment involves, in general, the detailed examination of the non-relativistic limit, since in this limit alone does the question as to spin and magnetic moment as individual entities become meaningful in the strict sense (cf., for example, Pauli (33); Proca (36)). More generally, it can be easily shown that simply by the transformation properties a particle of spin s can be thus endowed with an additional (not intrinsic) magnetic dipole, electric quadrupole, . . . up to 2^{2s}-pole.

§ 24. Field Theory—Canonical Hamiltonian Formulation

We wish to show now that by suitable generalization of the Hamiltonian theory of point mechanics the theory of wave fields can also be developed in canonical form. Indeed, we shall find that all the familiar concepts of Hamiltonian mechanics, including the theory of Poisson Brackets, contact transformations, etc., can be taken over with but formal changes. The primary objective of this programme is, of course, simply that the passage to quantized wave fields can, almost entirely, be contained in the familiar transition from classical to quantal P.B.s, i.e., commutators.[1] Moreover, consideration of the infinitesimal contact transformations, as generators of the inhomogeneous Lorentz group, provides far simpler proof of the conservation theorems and the Lorentz covariance of the whole formalism than does direct examination of the field variables under various transformations.[2]

(a) *Canonically Conjugate Field Variables—Hamiltonian Equations of Motion*

Introducing explicitly the total (integrated) Lagrangian

$$L = \int \mathscr{L}\, d\tau \tag{24.1}$$

we can cast the field equations (cf. 19.3)

$$\frac{\partial}{\partial t}\left(\frac{\partial \mathscr{L}}{\partial \dot{\psi}^r}\right) = \frac{\partial \mathscr{L}}{\partial \psi^r} - \partial^k\left(\frac{\partial \mathscr{L}}{\partial \psi^{r,k}}\right) \tag{24.2a}$$

in the form

$$\frac{\partial}{\partial t}\frac{\delta L}{\delta \dot{\psi}^r} = \frac{\delta L}{\delta \psi^r} \tag{24.2b}$$

where by definition the *functional* derivatives are

$$\frac{\delta L}{\delta \psi^r} = \frac{\partial \mathscr{L}}{\partial \psi^r} - \partial^k\left(\frac{\partial \mathscr{L}}{\partial \psi^{r,k}}\right) \qquad (24.3)^3$$

$$\frac{\delta L}{\delta \dot{\psi}^r} = \frac{\partial \mathscr{L}}{\partial \dot{\psi}^r}$$

[1] However, it must be pointed out that this canonical quantization procedure only gives sensible results for integral spin fields, the Dirac theory for spin $\frac{1}{2}$, and in general all fields associated with half-integral spin, requiring the introduction of anti-commutators. This difference in procedure, which is imposed by physical considerations, leads to the connection between spin and statistics (cf. §§ 32*b*).

[2] For detailed treatment of this latter aspect of the proof of invariance also, see Heisenberg-Pauli (29, 30); cf. Rosenfeld (30*a*, *b*); Weiss (36, 38); Fuchs (39); de Wet (48). Most of these papers give some aspects of both approaches as well as various analyses of the corresponding q-number theory, and can be profitably consulted.

[3] The symbol $(\delta L/\delta\psi^r)_x$, which is called the functional derivative of the functional L with respect to $\psi^r(x)$ at the point x, is defined by

$$\delta L = \int d\tau \left(\frac{\delta L}{\delta \psi^r}\right)_x \delta\psi^r(x) = \int d\tau \left\{\frac{\partial L}{\partial \psi^r} - \partial^k\left(\frac{\partial L}{\partial \psi^{r,k}}\right)\right\}\delta\psi^r$$

the suffix x in $(\delta L/\delta\psi^r)_x$ usually being omitted. Similarly, if $\mathscr{L}(x)$ is considered as a functional of the functions $\psi^r(x)$, $\psi^{r,k}(x)$, . . . (for a given x^4), the functional derivative of $\mathscr{L}(x')$ with respect to ψ^r at a point x is given by

$$\left(\frac{\delta \mathscr{L}(x')}{\delta \psi^r}\right)_x = \left(\frac{\partial \mathscr{L}}{\partial \psi^r}\right)\delta(x - x') - \partial^k\left\{\left(\frac{\partial \mathscr{L}}{\partial \psi^{r,k}}\right)\delta(x - x')\right\}$$

The functional derivative of $L = \int \mathscr{L}(x')\, d\tau'$ then follows by direct integration.

In analogy with ordinary mechanics we define generalized momenta conjugate to the field variables according to

$$\pi_{k\mu} = \frac{\partial \mathscr{L}}{\partial \psi^{r,\mu}} \tag{24.4a}$$

in which notation the *ordinary* momentum conjugate to ψ^r is

$$\pi_r = \frac{\partial \mathscr{L}}{\partial \dot{\psi}^r} = \frac{1}{c}\frac{\partial \mathscr{L}}{\partial \psi^{r,4}} \tag{24.4b}$$

The Hamiltonian (density) defined in the usual way is then (cf. 19.17a)

$$\mathscr{H}(\pi_r, \psi^r, \psi^{r,k}) = \pi_r \dot{\psi}^r - \mathscr{L} \tag{24.5}$$

and from

$$\begin{aligned} \delta\mathscr{H} &= \frac{\partial \mathscr{H}}{\partial \pi_r}\,\delta\pi_r + \frac{\partial \mathscr{H}}{\partial \psi^r}\,\delta\psi^r + \frac{\partial \mathscr{H}}{\partial \psi^{r,k}}\,\delta\psi^{r,k} \\ &= \dot{\psi}^r\,\delta\pi_r - \left(\frac{\partial \mathscr{L}}{\partial \psi^r}\,\delta\psi^r + \frac{\partial \mathscr{L}}{\partial \psi^{r,k}}\,\delta\psi^{r,k}\right) \end{aligned} \tag{24.6}$$

we find

$$\begin{aligned} \frac{\partial \mathscr{H}}{\partial \pi_r} &= \dot{\psi}^r \\ -\frac{\partial \mathscr{H}}{\partial \psi^r} &= \frac{\partial \mathscr{L}}{\partial \psi^r} \\ -\frac{\partial \mathscr{H}}{\partial \psi^{r,k}} &= \frac{\partial \mathscr{L}}{\partial \psi^{r,k}} \end{aligned} \tag{24.7}$$

For the corresponding total Hamiltonian $H = \int \mathscr{H}\,d\tau$ we have

$$\delta H = \int \left\{\frac{\delta H}{\delta \pi_r}\,\delta\pi_r + \frac{\delta H}{\delta \psi^r}\,\delta\psi^r\right\} d\tau \tag{24.8a}$$

where by definition

$$\begin{aligned} \frac{\delta H}{\delta \psi^r} &= \frac{\partial \mathscr{H}}{\partial \psi^r} - \partial^k\left(\frac{\partial \mathscr{H}}{\partial \psi^{r,k}}\right) \\ \frac{\delta H}{\delta \pi_r} &= \frac{\partial \mathscr{H}}{\partial \pi_r} - \partial_k\left(\frac{\partial \mathscr{H}}{\partial \pi_{r;k}}\right) \end{aligned} \tag{24.8b}$$

which with equations (24.2b, 24.7) result in the generalized Hamiltonian equations of motion

$$\begin{aligned} \frac{\delta H}{\delta \psi^r} &= -\dot{\pi}_r = -\pi_{r4}{}^{,4} \\ \frac{\delta H}{\delta \pi_r} &= \dot{\psi}^r = c\psi^{r,4} \end{aligned} \tag{24.9}$$

In keeping with its role as the total energy of the system, we find after a partial integration

$$\begin{aligned} \frac{dH}{dt} &= \int \left[\left\{\frac{\partial \mathscr{H}}{\partial \psi^r} - \partial^k\left(\frac{\partial \mathscr{H}}{\partial \psi^{r,k}}\right)\right\}\dot{\psi}^r + \left\{\frac{\partial \mathscr{H}}{\partial \pi_r} - \partial_k\left(\frac{\partial \mathscr{H}}{\partial \pi_{r;k}}\right)\right\}\dot{\pi}_r\right] d\tau \\ &= \int \left[\frac{\delta H}{\delta \psi^r}\,\dot{\psi}^r + \frac{\delta H}{\delta \pi_r}\,\dot{\pi}_r\right] d\tau = 0 \end{aligned} \tag{25.10}$$

(*b*) *Canonical Poisson Brackets*

If $F = \int \mathscr{F}\, d\tau$ denotes an arbitrary functional of the field variables ψ^r, $\psi^{r,k}$, π_r, $\pi_{r,k}$, we find in the general case

$$\frac{dF}{dt} = \frac{\partial F}{\partial t} + \int d\tau \left\{ \frac{\delta F}{\delta \psi^r}\, \dot{\psi}^r + \frac{\delta F}{\delta \pi_r}\, \dot{\pi}_r \right\} \tag{24.11a}$$

or, with equations (24.9),

$$\frac{dF}{dt} = \frac{\partial F}{\partial t} + \int d\tau \left\{ \frac{\delta F}{\delta \psi^r} \frac{\delta H}{\delta \pi_r} - \frac{\delta F}{\delta \pi_r} \frac{\delta H}{\delta \psi^r} \right\} \tag{24.11b}$$

$$= \frac{\partial F}{\partial t} + [F, H]$$

which defines the generalized P.B. expression. Thus, for any two such functionals F, G:

$$[F, G] = \int d\tau \left\{ \frac{\delta F}{\delta \psi^r} \frac{\delta G}{\delta \pi_r} - \frac{\delta F}{\delta \pi_r} \frac{\delta G}{\delta \psi^r} \right\} \tag{24.12}$$

and it is easily shown that the usual P.B. relations are also satisfied, viz.

$$[F, G] = -[G. F] \tag{24.13a}$$

$$[F_1 + F_2, G] = [F_1, G] + [F_2, G] \tag{24.13b}$$

$$[F_1 F_2, G] = F_1[F_2, G] + [F_1, G]F_2 \tag{24.13c}$$

$$\big[F_1[F_2, F_3]\big] + \big[F_2, [F_3, F_1]\big] + \big[F_3, [F_1, F_2]\big] = 0 \tag{24.13d}$$

The ψ's and π's can be expressed in functional form in the sense of

$$\psi^s(x, t) = \int \psi^s(x', t)\, \delta(x - x')\, d\tau' \tag{24.14a}$$

$$\pi_s(x, t) = \int \pi_s(x', t)\, \delta(x - x')\, d\tau' \tag{24.14b}$$

so that, for example,

$$\frac{\delta \psi^r(x, t)}{\delta \psi^s(x', t)} = \delta_s{}^r\, \delta(x - x') \tag{24.14c}$$

Then, the basic P.B.s are given by

$$[\psi^r(x), \psi^s(x')] = 0 \tag{24.15a}$$

$$[\pi_r(x), \pi_s(x')] = 0 \tag{24.15b}$$

$$[\psi^r(x), \pi_s(x')] = \int d\tau'' \left\{ \frac{\delta \psi^r(x)}{\delta \psi^t(x'')} \frac{\delta \pi_s(x')}{\delta \pi_t(x'')} \right\} \tag{24.15c}$$

$$= \int d\tau''\, \delta_t{}^r\, \delta(x - x'')\, \delta_s{}^t\, \delta(x' - x'')$$

$$= \delta_s{}^r\, \delta(x - x')$$

$$[\psi^r(x'), \pi_{s,k}(x)] = \delta_s{}^r\, \partial_k\, \delta(x - x') \tag{24.15d}$$

etc., and as usual all further P.B.s are essentially obtainable from the basic P.B.s by algebraic processes. In particular, the easily verified relations

$$[\psi^r, F] = \frac{\delta F}{\delta \pi_r} \tag{24.16a}$$

$$[\pi_r, F] = -\frac{\delta F}{\delta \psi^r} \tag{24.16b}$$

give, as a particular case, the equations of motion

$$\begin{aligned} [\psi^r, H] &= \dot{\psi}^r \\ [\pi_r, H] &= \dot{\pi}_r \end{aligned} \tag{24.17}$$

Finally, from the expression for the integrated total momentum (cf. 19.17b)

$$P^k = \int \pi_r \psi^{r,k} \, d\tau \tag{24.18}$$

we obtain

$$[\psi^r, P^k] = \psi^{r,k} \tag{24.19a}$$

$$[\pi_r, P^k] = \pi_r{}^{,k} \tag{24.19b}$$

and by induction

$$[F, P^k] = F^{,k} \tag{24.20}$$

for an arbitrary functional F.

(c) Contact Transformations

It is well known that the conditions for a contact-transformation in point mechanics, i.e. a transformation from one set of canonically conjugate coordinates and momenta to another, may be expressed in a variety of equivalent forms;[4] and the same holds true in field mechanics by virtue of the similarity of variational and partial differentiation and their parallel relation to functionals and functions, respectively. For our present purposes it is most convenient to express the conditions for a contact transformation in terms of P.B.s as follows:

If $\{\psi^{1\prime}, \psi^{2\prime}, \ldots \pi_n'\}$ denote $2n$ functions of the $2n$ variables $\{\psi^1, \psi^2, \ldots \pi_n\}$,

$$\begin{aligned} \psi^{r\prime} &= \psi^r\{\psi^1, \psi^2, \ldots \pi_n\} \\ \pi_r' &= \pi_r\{\psi^1, \psi^2, \ldots \pi_n\} \end{aligned} \tag{24.21a}$$

then the conditions which must be satisfied in order that the indicated transformations be a contact transformation may be expressed as

$$\begin{aligned} [\psi^{r\prime}(x), \psi^{s\prime}(x')] &= 0 \\ [\pi_r'(x), \pi_s'(x')] &= 0 \\ [\psi^{r\prime}(x), \pi_s'(x')] &= \delta_s{}^r \, \delta(x - x') \end{aligned} \tag{24.21b}$$

[4] e.g. in terms of bilinear covariants, Lagrange Brackets, Poisson Brackets, etc., cf., for example, Whittaker (37); Joos (51).

In other words, the new variables are canonically conjugate so that the basic P.B.s are invariant under the transformation; the proof being essentially the same as for the analogous theorem in point mechanics (cf. Whittaker, *loc. cit.*, Chapter XI). More generally, for any two functionals of the field variables we have, in virtue of (24.21*b*),

$$[F, G]_{\psi,\pi} = [F, G]_{\psi',\pi'} \tag{24.22a}$$

which is a special case of the easily verified relation

$$[F, G]_{\psi,\pi} = [F, G]_{f_r,f_s}[f_r, f_s]_{\psi,\pi} \tag{24.22b}$$

obtaining for two functionals of the variables $\{f_1, f_2, \ldots\}$ which in turn are functions of the ψ's and π's.

From the definition it is clear that the set of all contact transformations on $2n$ variables constitutes an infinite group; every contact transformation has a unique inverse, and the product of two group elements—defined by mere sequential performance of contact transformations—again constitutes a contact transformation. The primary importance of the invariance of P.B.s under contact transformations is, for our present purposes, due to the fact that we wish to prove the Lorentz invariance of the Hamiltonian formalism, and of the P.B.s in particular. This will be established when we show that to every Lorentz transformation of the field variables there corresponds a suitably chosen contact transformation. Further, since all finite contact transformations can be built up by infinitesimal contact transformations, we may confine our attention to the latter for the desired proof.

(*d*) *Generators of Infinitesimal Contact Transformations*

The fundamental principle underlying the theory of contact transformations is the fact that the variation principle will lead to the canonical Hamiltonian equations in the new variables provided that the variation integrand expressed in terms of the new field variables is of the same form as in the original variables, viz.

$$\int d\tau\, \pi_r \dot{\psi}^r - H = \int d\tau\, \pi_r' \dot{\psi}^{r'} - H' + \frac{dU}{dt} \tag{24.23}$$

The time derivative is added for generality, but obviously does not affect the equations of motion since it adds only a constant to the final variational integral. The functional U may be chosen in a variety of ways, e.g. we take $U[\psi, \psi', t]$, regarding the $2n$ variables $\{\psi', \psi\}$ as the independent variables. Then identifying coefficients in (24.23) we find

$$\pi_r = \frac{\delta U}{\delta \psi^r} \qquad \pi_r' = -\frac{\delta U}{\delta \psi^{r'}} \qquad H' = H + \frac{\partial U}{\partial t} \tag{24.24}$$

We now take (24.23) in the form

$$\int d\tau \{\pi_r'\, d\psi^{r'} - \pi_r\, d\psi^r\} + (H - H')\, dt + dU = 0 \tag{24.25a}$$

and confining our attention to infinitesimal contact transformations

$$\begin{aligned} \delta\psi^r &= \psi^{r\prime} - \psi^r = \lambda\Phi^r(\psi^1, \ldots \pi_n) \\ \delta\pi_r &= \pi_r{}' - \pi_r = \lambda\chi_r(\psi^1, \ldots \pi_n) \end{aligned} \tag{24.25b}$$

the fundamental equation for contact transformations (24.25*a*) becomes

$$\int d\tau\{\pi_r d\,\delta\psi^r + \delta\pi_r\, d\psi^r\} + (H - H')\,dt + dU = 0 \tag{24.25c}$$

Finally, we choose U of the form

$$U = -\int d\tau\,\pi_r\,\delta\psi^r + \lambda S \tag{24.26}$$

where S is the so-called *generating functional*, and assuming $S[\psi, \pi, t]$, we find

$$\int d\tau(\delta\pi_r\, d\psi^r - d\pi_r\,\delta\psi^r) + (H - H')\,dt + \lambda\, dS = 0 \tag{24.27a}$$

from which

$$\delta\psi^r = \lambda\frac{\delta S}{\delta\pi_r} \qquad \delta\pi_r = -\lambda\frac{\delta S}{\delta\psi^r} \qquad H' - H = \lambda\frac{\partial S}{\partial t} \tag{24.27b}$$

Alternatively, noting equations (24.16), we have

$$\begin{aligned} \delta\psi^r &= \lambda[\psi^r, S] \\ \delta\pi_r &= \lambda[\pi_r, S] \end{aligned} \tag{24.28a}$$

and the *numerical* change in any function $F\{\psi, \pi\}$ is

$$\delta F = F\{\psi', \pi'\} - F\{\psi, \pi\} = \lambda[F, S] \tag{24.28b}$$

In particular, we shall see that if S is an integral or constant of the motion, then the infinitesimal contact transformation generated by S is just the change in the variables of the system due to its natural development.

(*e*) *Conservation Theorems and Lorentz Invariance*

The functional form change induced in F by the contact transformation may be obtained from the numerical change δF, since from $F'[\psi', \pi'] = F[\psi, \pi]$, which defines the new functional F' in order that the physical quantity represented by F be the same at the given point in the two systems, it follows that

$$\begin{aligned} F\{\psi, \pi\} &= F'\{\psi, \pi\} + F'\{\psi', \pi'\} - F'\{\psi, \pi\} \\ &= F'\{\psi, \pi\} + F\{\psi', \pi'\} - F\{\psi, \pi\} \end{aligned} \tag{24.29a}$$

to terms of order λ^2 (cf. 24.28*b*). Hence

$$\left\{F'\{\psi, \pi\} - F\{\psi, \pi\}\right\} = -\left\{F\{\psi', \pi'\} - F\{\psi, \pi\}\right\} = -\delta F \tag{24.29b}$$

from which it follows that if the quantity F is *form-invariant* under the transformation generated by S, then $[F, S] = 0$. Applied to the Hamiltonian functional H, we obtain the important result that if H is form-invariant under S

$$\delta H = \lambda[H, S] = -\lambda \frac{\partial S}{\partial t} = 0 \tag{24.30}$$

or, that S is a constant of the motion.

The conservation theorems and proof of Lorentz invariance now follow very simply by suitably choosing the generator S, e.g.

I. $$S = H = \int \mathscr{H}\, d\tau \quad \text{and} \quad \lambda = \delta t$$

$$\begin{aligned} \delta\psi^r &= \delta t[\psi^r, H] \\ \delta\pi_r &= \delta t[\pi_r, H] \end{aligned} \tag{24.31}$$

that is, the natural development of the system in time as given by the Hamiltonian equations of motion. Further, if H is form-invariant under translation in time, then $\dot{H} = 0$.

II. $$S = P^k = \int d\tau\, \pi_r \psi^{r,k} \quad \text{and} \quad \lambda = -\delta x_k$$

$$\begin{aligned} \delta\psi^r &= -\delta x_k[\psi^r, P^k] \\ &= -\delta x_k \psi^{r,k} \end{aligned} \tag{24.32}$$

according to (24.19*a*), so that the contact transformation generated by P^k describes the substantial variation induced by the translation δx_k (cf. 19.14). If H is form-invariant under this translation, then again P^k, which is the total linear momentum, is a constant of the motion.

III. $$\begin{aligned} S = \mathscr{J}_{ij} &= \frac{1}{c}\int d\tau \mathscr{J}_{ij4} \\ &= \int d\tau \left(x_i \mathfrak{G}_j - x_j \mathfrak{G}_i + \frac{1}{c}\frac{\partial \mathscr{L}}{\partial \psi^{r,4}} \mathfrak{T}_{ij}{}^r{}_s \psi^s \right) \\ &= \int d\tau\, \pi_r(x_i \psi^r{}_{,j} - x_j \psi^r{}_{,i} + \mathfrak{T}_{ij}{}^r{}_s \psi^s) \end{aligned} \quad \text{(cf. 19.21)}$$

$$\begin{aligned} \delta\psi^t &= \lambda[\psi^t, \mathscr{J}_{ij}] \\ &= \lambda(x_i \psi^t{}_{,j} - x_j \psi^t{}_{,i} + \mathfrak{T}_{ij}{}^t{}_s \psi^s) \end{aligned} \tag{24.33}$$

which, on comparison with (19.12*b*), viz.

$$\delta^* \psi^t = \tfrac{1}{2}\omega^{\mu\nu}(x_\mu \psi^t{}_{,\nu} - x_\nu \psi^t{}_{,\mu} + \mathfrak{T}_{\mu\nu}{}^t{}_s \psi^s)$$

is seen to be a rotation of the system through the angle $\lambda\ (=\omega^{ij})$ about the perpendicular k-axis (i, j, $k = 1, 2, 3$ cyclically); and $\mathscr{J}_{ij}\ (=\mathscr{J}_k)$, which is a component of the total angular momentum, is a constant of the motion if H is form-invariant under this rotation.

IV. $$S = Q = \frac{e}{\hbar}\int d\tau(\psi^{r(1)}\pi_r{}^{(2)} - \psi^{r(2)}\pi_r{}^{(1)})$$

for a charge field described by the variables (cf. § 20, equation 20.7c) $\psi^{r(1)}$, $\psi^{r(2)}$, $\pi_r^{(1)}$, $\pi_r^{(2)}$. Then,

$$\delta\psi^{r(1)} = -\frac{\lambda e}{\hbar}\psi^{r(2)} \tag{24.34}$$

$$\delta\psi^{r(2)} = \frac{\lambda e}{\hbar}\psi^{r(1)}$$

$$\delta\pi_r^{(1)} = -\frac{\lambda e}{\hbar}\pi_r^{(2)}$$

$$\delta\pi_r^{(2)} = \frac{\lambda e}{\hbar}\pi_r^{(1)}$$

i.e. a rotation through $\lambda e/\hbar$ in the isotopic spin space; and if H is invariant under this rotation—which is just the gauge invariance of the first kind—then Q which is just the total charge is a constant[5] of the motion.

V. To prove the Lorentz invariance we take

$$\begin{aligned} S = \mathscr{I}_{k4} &= \frac{1}{c}\int \mathscr{I}_{k44}\, d\tau \\ &= \frac{1}{c}\int d\tau\left(x_k T_{44} - x_4 T_{k4} + \frac{\partial\mathscr{L}}{\partial\psi^{s,4}}\mathfrak{T}_{k4}{}^s{}_t\psi^t\right) \\ &= \frac{1}{c}\int d\tau(x_k\mathfrak{H} - x_4 c\mathfrak{G}_k + c\pi_s\mathfrak{T}_{k4}{}^s{}_t\psi^t) \end{aligned} \qquad \text{(cf. 19.20}b\text{)}$$

Then,

$$\delta\psi^r = \lambda[\psi^r, \mathscr{I}_{k4}] = \lambda\delta\mathscr{I}_{k4}/\delta\pi_r$$
$$\delta\pi_r = \lambda[\pi_r, \mathscr{I}_{k4}] = -\lambda\delta\mathscr{I}_{k4}/\delta\psi^r$$

can be evaluated immediately on noting that from equations (24.2a) and (24.6)

$$\delta\mathfrak{H} = \dot\psi^r\,\delta\pi_r - \dot\pi_r\,\delta\psi^r - \partial^j\left(\frac{\partial\mathscr{L}}{\partial\psi^{r,j}}\,\delta\psi^r\right)$$

We find

$$\begin{aligned} \delta\psi^r &= \frac{\lambda}{c}\left(x_k\frac{\partial\psi^r}{\partial t} - x_4 c\psi^r{}_{,k} + c\mathfrak{T}_{k4}{}^r{}_t\psi^t\right) \\ &= \lambda(x_k\psi^r{}_{,4} - x_4\psi^r{}_{,k} + \mathfrak{T}_{k4}{}^r{}_t\psi^t) \end{aligned} \tag{24.35a}$$

which is seen to be an infinitesimal Lorentz rotation corresponding to $x^\mu \to x^\mu + \omega^{\mu\nu}x_\nu$ with $\omega^{k4} = \lambda$. Similarly,

$$\begin{aligned} \delta\pi_r &= -\frac{\lambda}{c}\left(-x_k\frac{\partial\pi_r}{\partial t} + x_4 c\pi_{r,k} + c\pi_s\mathfrak{T}_{k4}{}^s{}_r + \frac{\partial\mathscr{L}}{\partial\psi^{r,k}}\right) \\ &= \lambda\left(x_k\pi_{r,4} - x_4\pi_{r,k} - \pi_s\mathfrak{T}_{k4}{}^s{}_r - \frac{1}{c}\pi_{rk}\right) \end{aligned} \tag{24.35b}$$

which is just the corresponding specialized form of the readily obtained infinitesimal Lorentz rotation of π_r associated with (19.12b), viz.

$$\delta^*\pi_{r4} = c\,\delta^*\pi_r = \tfrac{1}{2}\omega^{\mu\nu}(x_\mu\pi_{r4,\nu} - x_\nu\pi_{r4,\mu} - \mathfrak{T}_{\mu\nu r}{}^s\pi_{s4}) + \omega_4{}^\nu\pi_{r\nu}$$

[5] It is seen that the integrand is of the form of a component of angular momentum as noted in § 20, and on quantization this naturally leads to $Q = \pm ne$.

The Lorentz invariance of the basic P.B.s follows from their invariance under contact transformations and the preceding realization of the infinitesimal Lorentz rotations. More explicitly,

$$[\psi^{r\prime}, \pi_s{}'] - [\psi^r, \pi_s] = \delta[\psi^r, \pi_s] \tag{24.36}$$
$$= \lambda\big[[\psi^r, \pi_s], \mathscr{J}_{k4}\big] = \lambda[\delta_s{}^r\,\delta(x - x'), \mathscr{J}_{k4}] = 0$$

so that the form of the P.B.s is the same in both space-time frames. Finally, we wish to verify a previous remark to the effect that the total linear momentum and Hamiltonian constitute a 4-vector

$$\mathscr{P}_\mu = \left(P_k, \frac{H}{c}\right) = \frac{1}{c}\int d\tau(\pi_{r4}\,\psi^r{}_{,\mu} - g_{\mu 4}\mathscr{L}) \tag{24.37}$$

We readily find

$$\begin{aligned} \delta\mathscr{P}_i &= \lambda[\mathscr{P}_i, \mathscr{J}_{k4}] \\ &= -\lambda[\mathscr{J}_{k4}, P_i] \\ &= -\lambda\mathscr{J}_{k4,i} \qquad \text{(by 24.20)} \\ &= -\frac{\lambda}{c}\,\delta_{ik}H \end{aligned} \tag{24.38a}$$

and

$$\begin{aligned} \delta H &= \lambda[H, \mathscr{J}_{k4}] \\ &= \lambda\left(\frac{\partial\mathscr{J}_{k4}}{\partial t} - \frac{d\mathscr{J}_{k4}}{dt}\right) \\ &= -\lambda c\mathscr{P}_k \end{aligned} \tag{24.38b}$$

so that the components transform correctly under the rotation in the (x^4, x^k)-plane.

By essentially direct, though lengthy, computation we may establish the P.B.s

$$[\mathscr{J}_{\kappa\lambda}, \mathscr{J}_{\mu\nu}] = g_{\lambda\mu}\mathscr{J}_{\kappa\nu} - g_{\kappa\mu}\mathscr{J}_{\lambda\nu} + g_{\kappa\nu}\mathscr{J}_{\lambda\mu} - g_{\lambda\nu}\mathscr{J}_{\kappa\mu} \tag{24.39}$$

etc. However, the results obtained in cases II, III, and V show that the ten quantities $\mathscr{P}_\kappa$, $\mathscr{J}_{\kappa\lambda}$ are related to the infinitesimal operators of the inhomogeneous Lorentz group[6] in such a way that their Poisson Brackets with any dynamical variable give the change in that variable due to a translation or rotation, respectively, in space-time. Hence, when the dynamical variable is itself a component of $\mathscr{P}_\kappa$ or $\mathscr{J}_{\kappa\lambda}$, whose behaviour under coordinate transformations is already known (cf. § 5), it follows that the Poisson Bracket of any pair of components of $\mathscr{P}_\kappa$, $\mathscr{J}_{\kappa\lambda}$ is determined by the structure of the Lorentz group—independently of the nature of the particular system.

[6] We note that $\mathscr{J}_{\kappa\lambda}$ can be written as

$$\mathscr{J}_{\kappa\lambda} = \int \pi\mathscr{J}_{\kappa\lambda\,\mathrm{op}}\,\psi\,d\tau$$

where the $\mathscr{J}_{\kappa\lambda\,\mathrm{op}}$ $(= x_\kappa\partial_\lambda - x_\lambda\partial_\kappa + \mathfrak{T}_{\kappa\lambda\,\mathrm{op}})$ are, apart from a possible numerical factor, the infinitesimal operators of the homogeneous Lorentz group and satisfy the familiar C.R. of the form of equation (4.12). Similarly,

$$\mathscr{P}_\mu = \int \pi\partial_\mu\psi\,d\tau; \text{ the } \partial_\mu, \mathscr{J}_{\kappa\lambda\,\mathrm{op}} \text{ satisfying the C.R.. } [\mathscr{J}_{\kappa\lambda\,\mathrm{op}}, \partial_\mu] = (g_{\lambda\mu}\partial_\kappa - g_{\kappa\mu}\partial_\lambda).$$

CHAPTER IV

Relativistic Wave-Equations—Field Aspect

§ 25. Introduction

The theory of relativistic wave-equations suitable for the description of particles of arbitrary spin has proceeded on several different main lines of development. Apart from the requirements imposed by relativistic invariance and the quantum mechanical principle of superposition (that is, linearity in the wave functions) which are common to all, these may be divided into three basic classes of theories according to which of several general postulates are assumed to define what is termed an *elementary* particle. Broadly speaking, the various definitions are equivalent to different combinations of the following essentially physical postulates: (A) the requirement of a unique rest-mass; (B) the requirement of a unique spin; (C) the requirement that either the total energy or the total charge of the field associated with the particle is *positive definite.*

To the extent that a discussion of these postulates is feasible at this stage of the development, the first follows from the direct transcription of the classical relativistic form $(E/c)^2 = p^2 + m^2c^2$ and is reflected in the mathematical formulation in the equivalent postulate that each field component shall satisfy the second-order wave equation $(\square + \kappa^2)\psi = 0$ where $\kappa = mc/\hbar$; the second asserts itself mathematically in the further conditions to be imposed on the wave function both in respect to transformation character, which is to say the classification under the Lorentz group, and the conditions imposed in order that only one spin value be realized, that is, the field equations; the third comes from the q-number theory and appears to be a necessary condition if the field is to be quantized. It is related to the connection between spin and statistics and, as we shall see, the postulates (B) and (C) are intimately related—if not equivalent.

The Dirac-Fierz-Pauli[1] theory satisfies all three postulates and its basic mathematical form is the second-order equation, together with the necessary supplementary field equations (which are of the form of divergence conditions) on ψ's of appropriate transformation and symmetry character such that a given field manifests but one spin under all conditions. However, we shall find that for spin > 1 this theory cannot be cast in the simple Lagrangian form (without supplementary conditions) of Chapter III—except through rather involved and not entirely satisfactory artifices. With the same reservation, this means that the transition from the

[1] For brevity and definiteness we refer to this and other theories by the names of the principal first investigators. Needless to say, the development of this and the other theories has proceeded on the work of many writers, most of whom are referred to either in the text or the bibliography.

field-free to the (e.m.) interaction case by the usual transcription $\partial_\mu \rightarrow \partial_\mu - i\epsilon\phi_\mu$ does not lead to consistent field equations in the general case.

The Chandra theory satisfies postulates (A) and (C), but the basic (and complete) mathematical form in this theory is the first-order equation of the Dirac type, say, $(\alpha^\mu\partial_\mu - i\kappa)\psi = 0$; the entire character of the equation and its consequences devolves on the α's which are elements of the various matrix algebras specified by the different possible commutation rules on the α's such that the second-order equation obtains by iteration on this basic form. Here the introduction of the electromagnetic field leads to no evident difficulties since there are no other conditions or field equations with which to conflict, but for spin > 1 this theory leads to what we may term compound spin character, that is, the spin is not unique and higher spin states appear for a single particle in the ultra-relativistic region. The assumption that the elementary particles are described by equations of the first order, as originally argued by Dirac, is entirely predicated on the transformation theory of quantum mechanics and the one-particle interpretation. The subsequent development and interpretation of the many-particle theory does not lend special support to this view, but since the prototype form—the Dirac equation—is known to be essentially correct, there is no cogent ground on which to rule out this formulation on the basis of presently available information; except if we arbitrarily choose to insist on postulate (B) which, after all, lies in the realm of speculation on things yet unknown.

The Bhabha theory adopts the above linear form as basic, though primarily to avoid the (e.m.) interaction difficulty mentioned in connection with the D.F.P. theory, and rejects the first postulate (A)—which it is found is equivalent to rejecting all three postulates for the general case of spin > 1. The commutation rules for the α's are now determined by the assumption that the infinitesimal transformations $\mathfrak{T}_{\mu\nu}$ (corresponding to ψ) are given by $\mathfrak{T}_{\mu\nu} = c(\alpha_\mu\alpha_\nu - \alpha_\nu\alpha_\mu)$ with c a constant and the α's irreducible; that is, by abstraction from the electron and meson cases[2] (which correspond to the results of § 16 for first- and second-rank 4-spinors). In consequence it is found that for spin > 1, iteration leads to an equation of higher than second order, which corresponds to the fact that elementary particles so defined have compound spin and rest-mass character, that is, a single particle is capable of several states of spin and rest mass in the ultra-relativistic region. Moreover, as noted above, postulate (C) is not satisfied in the general case, so that quantization is not possible—except through the introduction of new procedures whose physical meaning is obscure.[3] There are therefore perhaps more grounds for doubt in this theory which, so to speak, disavows its humbler origins by the forfeiture of the physical correspondence implicit in (A).

The striking fact is that (for good reasons!) all three theories agree in the cases corresponding to spins 0, 1, and $\frac{1}{2}$, which we refer to as the special fields, and these are even more sharply distinguished by the definiteness of the energy density in the former and the charge density in the latter—characteristics not found, in general,

[2] The γ's and β's of § 16 correspond to the electron and meson equations in their linear or particle-like forms which we consider later in this chapter. However, as noted in § 16, the representation of the β's obtained was not irreducible, which corresponds physically to the fact that the unrestricted second-rank 4-spinors give a combined theory of spin 0 and 1 particles, cf. §§ 30*b*, *c*.

[3] That is, the so-called quantization on *indefinite* metric in the system Hilbert space. Cf. Dirac (42, 43); Pauli (43).

in the higher spin cases. These general considerations cannot perhaps be construed as meaning that particles of higher spin do not occur in nature but, by contrast, neither can they be considered as favourable to this possibility. In any case the special fields are likely to be of greater interest to the general reader since they constitute, almost exclusively, the media for most current research in the meson theories of nuclear forces and related topics.

The work of the present chapter will be largely in terms of the D.F.P. theory though, as noted above, the threads of the three theories become entirely interwoven in so far as the study of the special fields is concerned.

Before considering the field equations corresponding to spins 0, 1, and $\frac{1}{2}$, which are rather special in many respects, we wish to examine, though briefly and in a general way, some of the general considerations which delineate fields suitable for the description of particles of arbitrary spin. However, we do not propose to follow immediately upon this with the corresponding field equations since this would present the reader with a maze of fairly complex equations whose meaning is not entirely transparent. Rather, in establishing the underlying concepts—which devolve principally on the precise definition and analysis of spin for all cases—the subsequent examination of the special cases becomes more meaningful; the mathematical character of the entities used for the descriptions of particles of any specified spin follow from well-defined principles rather than seemingly *ad hoc* choice. Moreover, in this way we obviate the need to give detailed consideration to the special cases, which are comparatively familiar, except in so far as we wish to emphasize certain aspects of interest such as the interrelation of various formalisms.

The totality of field variables subsumed in the wave-function symbol ψ must, in any case, be expressible as the direct sum of quantities irreducible under the Lorentz group, so that it suffices to consider the prototype primitive form, that is, the irreducible wave equation

$$(\Box + \kappa^2)\psi(k, l) = 0 \tag{25.1}$$ [4]

Under restriction to the subgroup of spatial rotations $\psi(k, l)$ behaves as the direct product of two irreducible quantities which in turn reduces into several irreducible $\psi(j)$ where $j = k + l, k + l - 1, \ldots | k - l |$; the $\psi(j)$ with integral (half-integral) j transforming under this group according to single-(double-)valued representations $\mathfrak{D}(j)$.

In the absence of further restrictions or conditions on the original field quantities represented by $\psi(k, l)$ it is found (cf. *infra*) that the given field describes particles of various spins—which here corresponds to the fact that the reduction leads to *several* $\psi(j)$ irreducible under the rotation group. However, it is not true, as might be casually inferred, that the $2(k + l) + 1$-component quantity $\psi(k + l)$, for example, describes particles of spin $(k + l)$; since there would then be no correlation between the spin value and the number of independent plane-wave solutions or states of polarization obtaining for given values of the wave-number vector components $k_\mu(k_\mu k^\mu = \kappa^2)$ in the plane-wave factor $\exp(ik_\mu x^\mu)$.

Such an inference regarding the spin value would only be correct if the orbital

[4] N.B. It is clear that for the tensor representations, $k = l$, this single equation is invariant under the full Lorentz group, but in the contrary case we require a pair of spinors to maintain reflection invariance. cf. § 17 and Table I, p. 46.

momentum were known to be zero, i.e. in the rest system, $k_\mu = (0, 0, 0, \pm\kappa)$, of the particle. Hence, to find the spin value (or values) associated with a given field we must transform to the rest system where all the spatial components k_i are zero and the wave function depends only on the time. In this reference frame we reduce those field components which do not vanish in consequence of the field equations (that is, equations *in addition* to (25.1)) into parts irreducible under spatial rotations,[5] and then *each* such set of $r = 2s + 1$ independent components lying in the subspace $\mathfrak{R}(s)$ corresponds to a particle of spin s. In consequence of the Lorentz invariance, it follows that for an arbitrary reference frame r (or Σr) momentum eigenfunctions always belong to a given k_μ, though in general ψ will then have components lying outside the subspace $\mathfrak{R}(s)$. In an arbitrary reference frame all we can definitely state without further analysis is that the quantities $\psi(j)$ obtaining by reduction of $\psi(k, l)$ describe only particles of integral (half-integral) spins according as they transform under single-(double-)valued representations $\mathfrak{D}(j)$ of the rotation group, i.e. integral or half-integral $(k + l)$ respectively (cf. § 17).

If the field is to describe particles of only one spin value, there must be only one such irreducible set of components—and this prescribes the fundamental conditions on the choice of field quantities and conditions thereon for any chosen spin.

The case of zero rest mass is only slightly different. There is, of course, no rest system but we may examine the field equations in the special reference frame in which, for example, the momentum is directed along the x^3-axis,[6] i.e. $k_1 = k_2 = 0$. Then, since the orbital angular momentum of a system has zero component along the momentum axis, any angular-momentum components on this axis must correspond to spin (cf. §§ 27*c*, 29, 34*d*). As previously mentioned, we find in this case a special gauge degeneracy in consequence of which the zero rest-mass field obtaining from the corresponding non-zero rest-mass field of spin s, say, in the limit $\kappa \to 0$, has but two independent components rather than $2s + 1$. These correspond to the spin projections $\pm s$ on the x^3-axis; or, what amounts to right- and left-circular polarization.[7]

From an operational viewpoint the foregoing discussion in the case of *non-zero rest mass* is essentially contained in the observation that the two invariant operators

$$\begin{aligned} P &= p_\nu p^\nu \\ W &= \{\tfrac{1}{2}\mathscr{J}^{\lambda\mu}\mathscr{J}_{\lambda\mu}p_\nu p^\nu - \mathscr{J}^{\lambda\mu}\mathscr{J}_{\lambda\nu}p_\mu p^\nu\} \end{aligned} \tag{25.2}$$

[5] The subgroup of the homogeneous Lorentz transformations which leave a chosen momentum 4-vector invariant is referred to as the "little group". In the present case the chosen vector is $p_\mu{}^0 = (0, 0, 0 \pm mc)$ so that the little group is the subgroup of rotations whose infinitesimal operators

$$\mathscr{J}_{kl} = i\hbar\left(p_k\frac{\partial}{\partial p_l} - p_l\frac{\partial}{\partial p_k}\right) + \mathscr{S}_{kl}$$

commute with $p^0{}_\mu$. See Wigner (39); Bargmann-Wigner (48).

[6] The infinitesimal operators of the "little group" are then $\mathscr{J}_{12}$, $\mathscr{J}_{13} - \mathscr{J}_{14}$, $\mathscr{J}_{23} - \mathscr{J}_{24}$ which leave this momentum 4-vector invariant, whatever the values of k_3 and k_4 $(= \pm k_3)$, cf. § 4. The chosen momentum 4-vector is usually taken as $p_\mu{}^0 = (0, 0, 1, 1)$. See the references of the preceding footnote.

[7] As we shall see, this is equivalent to the statement that the two independent components transform according to the irreducible (one-dimensional) representations $D(\pm s)$ of the Abelian subgroup (of the little group of the preceding footnote) comprising the corresponding axial rotations. From the fact that the "little groups" in the two cases $(\kappa \neq 0, \kappa = 0)$ have an isomorphic subgroup (the rotations about an axis), and that therefore the representations of this subgroup will be equivalent, we can infer that the *total* spin of the zero rest-mass field is also s, as noted above.

commute with the ten infinitesimal operators $\mathscr{J}_{\kappa\lambda}$, p_κ in virtue of the C.R. satisfied by the latter (cf. § 4), and are therefore represented by scalar matrices in any irreducible representation of the Lorentz group.[8] Moreover, since

$$\mathscr{J}^{\lambda\mu} = x^\lambda p^\mu - x^\mu p^\lambda + i\hbar\mathfrak{T}^{\lambda\mu}_{\text{op}} = \mathscr{M}^{\lambda\mu} + \mathscr{S}^{\lambda\mu} \tag{25.3}$$

all terms in W involving $\mathscr{M}^{\lambda\mu}$ cancel and we may write

$$W = -\hbar^2\{\tfrac{1}{2}\mathfrak{T}^{\lambda\mu}\mathfrak{T}_{\lambda\mu}p_\nu p^\nu - \mathfrak{T}^{\lambda\mu}\mathfrak{T}_{\lambda\nu}p_\mu p^\nu\} \tag{25.4}$$

Hence, in the rest system of a plane wave, i.e. $\psi(p)$ at the special value $p_\mu{}^0 = (0, 0, 0, \pm mc)$,

$$\begin{aligned} W\psi &= -\hbar^2\{(\mathfrak{T}_{12})^2 + (\mathfrak{T}_{23})^2 + (\mathfrak{T}_{31})^2\}(mc)^2\psi \\ &= (\hbar mc)^2\, s(s+1)\psi \end{aligned} \tag{25.5}$$

if in this reference frame ψ reduces to the irreducible quantity $\psi(s)$. Therefore,

$$\boldsymbol{S}^2 = WP^{-1} = \{\tfrac{1}{2}\mathscr{J}^{\lambda\mu}\mathscr{J}_{\lambda\mu}p_\nu p^\nu - \mathscr{J}^{\lambda\mu}\mathscr{J}_{\lambda\nu}p_\mu p^\nu\}/p_\nu p^\nu \tag{25.6}$$

is the operator representing the spin magnitude, and in virtue of the Lorentz invariance

$$\boldsymbol{S}^2\psi = s(s+1)\hbar^2\psi \tag{25.7}$$

in *any* reference frame.

§ 26. Special Tensor Equations

(a) Spin 0: Scalar and Pseudoscalar Fields

The simplest example of a relativistically invariant wave equation is the Klein-Gordon equation

$$(\Box + \kappa^2)\phi = 0 \tag{26.1}$$

in which ϕ is a real scalar and therefore describes an uncharged field. This equation may be obtained from the Lagrangian density

$$\mathscr{L} = \tfrac{1}{2}(\phi,_\mu\phi^{,\mu} - \kappa^2\phi^2) \tag{26.2}$$

and the corresponding canonical energy-momentum tensor (cf. 19.16*a*)

$$\mathscr{T}_{\mu\nu} = \phi,_\mu\phi,_\nu - g_{\mu\nu}\mathscr{L} \tag{26.3}$$

is symmetric so that there is no spin contribution to the angular momentum density.

With the momentum conjugate to ϕ given by $\pi = \dfrac{\partial\mathscr{L}}{\partial\dot\phi} = \dfrac{1}{c^2}\dot\phi$, the Hamiltonian takes the form

$$\begin{aligned} H = \int \mathscr{H}\, d\tau &= \int \{(\phi,_4)^2 - \tfrac{1}{2}\phi,_\mu\phi^{,\mu} + \tfrac{1}{2}\kappa^2\phi^2\}\, d\tau \\ &= \tfrac{1}{2}\int \{c^2\pi^2 + (\nabla\phi)^2 + \kappa^2\phi^2\}\, d\tau \end{aligned} \tag{26.4}$$

[8] The operator W was first introduced, in the present connection, by W. Pauli; cf. Lubanski (42*b*, *c*).

in which, as previously noted, the energy density $\mathscr{H} = \mathscr{T}_{41}$ is positive definite. The Hamiltonian equations of motion (24.9, 24.17) are then given by

$$\dot{\phi} = \frac{\delta H}{\delta \pi} = \frac{\partial \mathscr{H}}{\partial \pi} = c^2 \pi \tag{26.5}$$

$$-\dot{\pi} = \frac{\delta H}{\delta \phi} = \frac{\partial \mathscr{H}}{\partial \phi} - \partial^k \left(\frac{\partial \mathscr{H}}{\partial \phi_{\cdot k}} \right) = (\kappa^2 - \nabla^2)\phi$$

from which (26.1) follows as required.

The volume-integrated linear and angular momenta are readily found to be (cf. 19.17*a*, 19.21)

$$\boldsymbol{P} = (P_1, P_2, P_3) = \int \pi \nabla \phi \, d\tau \tag{26.6}$$

$$\boldsymbol{\mathcal{J}} = (\mathscr{J}_{23}, \mathscr{J}_{31}, \mathscr{J}_{12}) = -\int \pi (\boldsymbol{x} \wedge \boldsymbol{\nabla}) \phi \, d\tau$$

The most general solution of (26.1) can be expressed as a sum of plane waves, that is, an expansion in momentum eigenstates, and it is of some interest to note that the plane waves serve as normal coordinates for the total energy and momentum —which appear as sums of the contributions from the various modes. This procedure, which conveys the essential physical content of the equations, is exactly *parallel* to that in the q-number theory as given by Wentzel (43)[1] so that we need not reproduce it here.

For the corresponding charged scalar field we have equation (26.1) and its conjugate which now follow from the Lagrangian

$$\mathscr{L} = \bar{\phi}_{,\mu} \phi^{,\mu} - \kappa^2 \bar{\phi} \phi \tag{26.7}$$

The canonical energy-momentum tensor becomes

$$\mathscr{T}_{\mu\nu} = \bar{\phi}_{,\nu} \phi_{,\mu} + \bar{\phi}_{,\mu} \phi_{,\mu} - g_{\mu\nu} \mathscr{L} \tag{26.8}$$

and with the momenta $\pi = \dfrac{1}{c^2} \dot{\bar{\phi}}$, $\bar{\pi} = \dfrac{1}{c^2} \dot{\phi}$, conjugate to ϕ, $\bar{\phi}$, the Hamiltonian is

$$H = \int \{ c^2 \bar{\pi} \pi + (\nabla \bar{\phi} \cdot \nabla \phi) + \kappa^2 \bar{\phi} \phi \} \, d\tau \tag{26.9}$$

from which the field equations follow in usual fashion. Here

$$\boldsymbol{P} = \int (\pi \nabla \phi + \bar{\pi} \nabla \bar{\phi}) \, d\tau \tag{26.10}$$

and from the charge-current density vector (cf. 20.6*b*)

$$J_\mu = -i\epsilon \{ \bar{\phi}_{,\mu} \phi - \phi_{,\mu} \bar{\phi} \} \tag{26.11a}$$

[1] See also Pauli-Weisskopf (34); Pauli (41).

we obtain the total charge

$$Q = \int \rho \, d\tau = -i\epsilon c \int (\pi\phi - \bar{\pi}\bar{\phi}) \, d\tau \tag{26.11b}$$

We also note that in this simple case the so-called charge-conjugate solution of the wave equation (26.1), that is the solution corresponding to charge $-e$ if ϕ corresponds to charge e, is simply the complex conjugate $\bar{\phi}$. Thus, in the presence of an electromagnetic field, $(\partial_\mu \partial^\mu + \kappa^2)\phi = 0$ goes into $(D_\mu D^\mu + \kappa^2)\phi = 0$ which remains correct if we make the transformation $\phi \to \bar{\phi}$, $e \to -e$. This relation is more complicated in the case of half-integral spin fields, as illustrated by the case of spin $\frac{1}{2}$ (cf. § 30).

From the point of view of invariance it is clear that the field variable ϕ may also be taken to be a pseudoscalar. This introduces no essential change in the structure of the equations in the absence of interaction with other fields, but in the contrary case it leads to entirely different types of interaction terms (cf. Kemmer (38)). The structural difference between the scalar and pseudoscalar fields is nicely illustrated in the alternative formulation of (26.1) in terms of the equivalent first-order system

$$\begin{aligned} \phi_{,\sigma} &= \kappa \chi_\sigma \\ \chi^\sigma{}_{,\sigma} &= -\kappa\phi \end{aligned} \tag{26.12}$$ [2]

In the *charged* case, for example, these equations (and conjugates) derive from a Lagrangian of the form

$$\mathscr{L} = \bar{\phi}_{,\mu}\chi^\mu + \phi_{,\mu}\bar{\chi}^\mu - \kappa\bar{\phi}\phi - \kappa\bar{\chi}_\mu\chi^\mu \tag{26.13}$$

ϕ, χ^μ and their conjugates being varied independently.

The *dual* system to (26.12), or so-called pseudoscalar case, is then given by

$$\begin{aligned} \check{\phi}_{,\sigma} &= \kappa \check{\chi}_\sigma \\ \check{\chi}^\sigma{}_{,\sigma} &= -\kappa\check{\phi} \end{aligned} \tag{26.14a}$$ [3]

or

$$\begin{aligned} \phi^{\lambda\mu\nu\sigma}{}_{,\sigma} &= -\kappa\chi^{\lambda\mu\nu} \\ \chi_{\sigma\lambda\nu,\mu} + \chi_{\lambda\nu\mu,\sigma} + \chi_{\nu\mu\sigma,\lambda} + \chi_{\mu\sigma\lambda,\nu} &= -\kappa\phi_{\mu\sigma\lambda\nu} \end{aligned} \tag{26.14b}$$ [3]

by the transformation (cf. § 3)

$$\begin{aligned} \check{\chi}^\mu &= \left(\frac{1}{3!}\right) g^{-1/2} \epsilon^{\mu\nu\sigma\tau} \chi_{\nu\sigma\tau} \qquad & \chi_{\nu\sigma\tau} &= g^{1/2} \epsilon_{\mu\nu\sigma\tau} \check{\chi}^\mu \\ \check{\phi} &= \left(\frac{1}{4!}\right) g^{-1/2} \epsilon^{\mu\nu\sigma\tau} \phi_{\mu\nu\sigma\tau} \qquad & \phi_{\mu\nu\sigma\tau} &= g^{1/2} \epsilon_{\mu\nu\sigma\tau} \check{\phi} \end{aligned} \tag{26.15}$$

[2] This formulation was first introduced in this tensor form by Kemmer (38) and is also contained in the more general formulation of Dirac (36). It is well known that any system of linear differential equations can always be replaced by a system of first-order equations by suitably defining new functions which are given by derivatives of the original functions, cf. for example Courant-Hilbert (37). In the special connection of field theory the above is the simplest example of a more general result obtaining in the case of higher spin fields, cf. §§ 28*a*, *b*.

[3] This dual system as well as that in the vector case (§ 26*b*) was first investigated by Kemmer (38), cf. Bhabha (38).

(b) Spin 1: Vector and Pseudovector Fields

The simplest tensor generalization of the scalar equation is given by

$$(\Box + \kappa^2) U^\mu = 0 \tag{26.16}$$

in which the components of the vector field U^μ are treated as independent scalars. However, as it stands this theory would have to be interpreted as describing the combined system of a spin 1 *and* a spin 0 particle since the wave function given by the column matrix $\| U^\mu \|$, classified as $\psi(\frac{1}{2}, \frac{1}{2})$ under the proper Lorentz group, reduces to $\psi(1) \oplus \psi(0)$, that is a 3-vector and a scalar, under the rotation group. From the transformation properties of U^μ (§ 5) we obtain:

$$\mathfrak{T}_{12}{}^{\alpha}{}_{\beta} = \begin{Vmatrix} 0 & -1 & 0 & 0 \\ 1 & 0 & 0 & 0 \\ 0 & 0 & 0 & 0 \\ 0 & 0 & 0 & 0 \end{Vmatrix} \quad \mathfrak{T}_{23} = \begin{Vmatrix} 0 & 0 & 0 & 0 \\ 0 & 0 & -1 & 0 \\ 0 & 1 & 0 & 0 \\ 0 & 0 & 0 & 0 \end{Vmatrix} \quad \mathfrak{T}_{31} = \begin{Vmatrix} 0 & 0 & 1 & 0 \\ 0 & 0 & 0 & 0 \\ -1 & 0 & 0 & 0 \\ 0 & 0 & 0 & 0 \end{Vmatrix}$$

$$\mathfrak{T}_{14} = \begin{Vmatrix} 0 & 0 & 0 & 1 \\ 0 & 0 & 0 & 0 \\ 0 & 0 & 0 & 0 \\ 1 & 0 & 0 & 0 \end{Vmatrix} \quad \mathfrak{T}_{24} = \begin{Vmatrix} 0 & 0 & 0 & 0 \\ 0 & 0 & 0 & 1 \\ 0 & 0 & 0 & 0 \\ 0 & 1 & 0 & 0 \end{Vmatrix} \quad \mathfrak{T}_{34} = \begin{Vmatrix} 0 & 0 & 0 & 0 \\ 0 & 0 & 0 & 0 \\ 0 & 0 & 0 & 1 \\ 0 & 0 & 1 & 0 \end{Vmatrix} \tag{26.17}$$

and (cf. § 25)

$$-\{(\mathfrak{T}_{12})^2 + (\mathfrak{T}_{23})^2 + (\mathfrak{T}_{31})^2\} = \begin{Vmatrix} 2 & 0 & 0 & 0 \\ 0 & 2 & 0 & 0 \\ 0 & 0 & 2 & 0 \\ 0 & 0 & 0 & 0 \end{Vmatrix} \tag{26.18}$$

Furthermore, in the charged case for example, equation (26.16) and its conjugate obtain from the Lagrangian function

$$\mathscr{L} = -\bar{U}^{\alpha,\beta} U_{\alpha,\beta} + \kappa^2 U_\alpha \bar{U}^\alpha \tag{26.19a}$$

from which the energy density is given by

$$\mathscr{H} = -\bar{U}_{\mu,4} U^\mu{}_{,4} - U_{\mu,4} \bar{U}^\mu{}_{,4} + \bar{U}^{\alpha,\beta} U_{\alpha,\beta} - \kappa^2 U_\alpha \bar{U}^\alpha \tag{26.19b}$$

Thus, with the sign of (26.19*a*) chosen so that the space-like components of U^μ are associated with positive energy terms, the components in U_4 contribute negative terms. This difficulty is eliminated by imposing the supplementary so-called Lorentz condition

$$U^\mu{}_{,\mu} = 0 \tag{26.20}$$

which, *in the rest system*, requires $U_4 = 0$—so that the total integrated energy is positive, and hence in general by virtue of the Lorentz invariance of the theory. By the same token $\psi(\frac{1}{2}, \frac{1}{2})$ reduces to just the one irreducible quantity $\psi(1)$ in the rest system which means that the last row and column of (26.18) are struck out.

The equations (26.16 and 20) do not follow from a simple variation principle, but the alternative formulation in terms of a first-order system (due to Proca (36)) can be so obtained. We write

$$U_{\nu,\mu} - U_{\mu,\nu} = \kappa G_{\mu\nu} \qquad (26.21)^{4}$$
$$G_{\mu\nu},^{\nu} = \kappa U_{\mu}$$

from which all quantities satisfy the second-order wave equation, and the Lorentz condition follows from the antisymmetry of $G_{\mu\nu}$. Treating the U_α, $G_{\alpha\beta}$ ($= -G_{\beta\alpha}$) as independent in the variation, these equations follow from the Lagrangian

$$\mathscr{L} = \tfrac{1}{2}\{\tfrac{1}{2}\kappa G^{\alpha\beta} G_{\alpha\beta} + G^{\alpha\beta}(U_{\alpha,\beta} - U_{\beta,\alpha}) + \kappa U^{\alpha} U_{\alpha}\} \qquad (26.22)$$

and the canonical energy momentum tensor is given by

$$\mathscr{T}_{\lambda\mu} = \frac{\partial \mathscr{L}}{\partial U^{\nu,\mu}} U^{\nu},_{\lambda} - g_{\lambda\mu}\mathscr{L} \qquad (26.3a)$$
$$= G_{\nu\mu} U^{\nu},_{\lambda} - g_{\lambda\mu}\mathscr{L}$$

With the help of the field equations (26.21) this may be written as

$$\mathscr{T}_{\lambda\mu} = G_{\nu\mu}(-\kappa G^{\nu}{}_{\lambda} + U_{\lambda},^{\nu}) - g_{\lambda\mu}\mathscr{L} \qquad (26.23b)$$
$$= -\kappa G_{\nu\mu} G^{\nu}{}_{\lambda} + \kappa U_{\mu} U_{\lambda} - g_{\lambda\mu}\mathscr{L} - \partial^{\nu}(G_{\mu\nu} U_{\lambda})$$

or

$$\Theta_{\lambda\mu} = -\kappa G_{\nu\mu} G^{\nu}{}_{\lambda} + \kappa U_{\mu} U_{\lambda} - g_{\lambda\mu}\mathscr{L} \qquad (26.24)$$
$$= \mathscr{T}_{\lambda\mu} + \partial^{\nu}(G_{\mu\nu} U_{\lambda})$$

the divergenceless term $\partial^{\nu}(G_{\mu\nu} U_{\lambda})$ being just the spin term $\mathsf{S}_{\lambda\mu\nu},^{\nu}$ which we would obtain by direct application of the general prescription of Chapter III, § 19(*c*).

The conjugate momenta are given by $\pi_\mu = \dfrac{1}{c}\dfrac{\partial \mathscr{L}}{\partial U^{\mu,4}} = \dfrac{1}{c} G_{\mu 4}$, so that $\pi_4 \equiv 0$, and the energy density

$$\mathscr{H} = \Theta_{44} = \kappa\{U_4 U_4 - G_{\nu 4} G^{\nu}{}_4 - \tfrac{1}{2}(U^{\alpha} U_{\alpha} - \tfrac{1}{2} G^{\alpha\beta} G_{\alpha\beta})\} \qquad (26.25)$$
$$= \tfrac{1}{2}\kappa\{c^2(\boldsymbol{\pi})^2 + \frac{1}{\kappa^2}(\nabla \wedge \boldsymbol{U})^2 + (\boldsymbol{U})^2 + (U_4)^2\}$$

is positive definite. Since $\pi_4 \equiv 0$, we would like to eliminate U_4 entirely and for this purpose we transcribe the field equations to vector form. By analogy with the

[4] These equations admit solutions of the form

$$\text{(i)}\ U_\mu = U_\mu(p) e^{i p_\sigma x^\sigma/\hbar}$$
$$\text{(ii)}\ G_{\mu\nu} = G_{\mu\nu}(p) e^{i p_\sigma x^\sigma/\hbar}$$

where $U_\mu(p)$, $G_{\mu\nu}(p)$ are constant, if and only if $p_\sigma p^\sigma = (mc)^2$. The Lorentz condition $U^{\mu},_{\mu} = 0$ becomes $p_\mu U^{\mu}(p) = 0$, so that we may choose $U^{\mu}(p)$ arbitrarily and then equations (26.21) determine $G_{\mu\nu}(p)$, viz.

$$G_{\mu\nu}(p) = \frac{i}{mc}\left\{p_\mu U_\nu(p) - p_\nu \dot{U}_\mu(p)\right\}$$

There are clearly *three* possible linearly independent choices for $U_\mu(p)$, so that for each set of p_σ satisfying $p_\sigma p^\sigma = (mc)^2$ there exist three linearly independent solutions which correspond to the three different spin orientations.

electromagnetic case we write

$$G_{\mu\nu} = \begin{Vmatrix} 0 & G_z & -G_y & c\pi_x \\ -G_z & 0 & G_x & c\pi_y \\ G_y & -G_x & 0 & c\pi_z \\ -c\pi_x & -c\pi_y & -c\pi_z & 0 \end{Vmatrix} \tag{26.26}$$

$$U_\mu = (\boldsymbol{U}, -\phi)$$

with which equations (26.21) take the form

$$\kappa \boldsymbol{G} = \nabla \wedge \boldsymbol{U} \tag{26.27a}$$

$$\kappa c \boldsymbol{\pi} = -\nabla\phi - \frac{1}{c}\dot{\boldsymbol{U}} \tag{26.27b}$$

$$c(\nabla \cdot \boldsymbol{\pi}) = -\kappa\phi \tag{26.27c}$$

$$\nabla \wedge \boldsymbol{G} - \dot{\boldsymbol{\pi}} = -\kappa \boldsymbol{U} \tag{26.27d}$$

the first and third being considered as the defining equations for $\boldsymbol{G}$ and ϕ. Then the Hamiltonian takes the form

$$H = \tfrac{1}{2}\kappa \int \left\{ c^2(\boldsymbol{\pi})^2 + \frac{1}{\kappa^2}(\nabla \wedge \boldsymbol{U})^2 + \left(\frac{c}{\kappa}\right)^2 (\nabla \cdot \boldsymbol{\pi})^2 + \boldsymbol{U}^2 \right\} d\tau \tag{26.28}$$

in which $\boldsymbol{U}$ and $\boldsymbol{\pi}$ now play the roles of canonically conjugate variables, and we find the equations of motion

$$\dot{\boldsymbol{U}} = -\frac{\delta H}{\delta \boldsymbol{\pi}} = -\kappa c^2 \boldsymbol{\pi} + \frac{c^2}{\kappa}\nabla(\nabla \cdot \boldsymbol{\pi}) \qquad (26.29)^5$$

$$\dot{\boldsymbol{\pi}} = \frac{\delta H}{\delta \boldsymbol{U}} = \kappa \boldsymbol{U} + \frac{1}{\kappa}\nabla \wedge (\nabla \wedge \boldsymbol{U})$$

In view of (26.27*c*) and (26.27*a*) these are just (26.27*b*) and (26.27*d*), respectively.

Here the integrated linear momentum appears as

$$\boldsymbol{P} = -\int \nabla (\boldsymbol{\pi} \cdot \boldsymbol{U})_\pi \, d\tau \tag{26.30a}$$

from the canonical tensor (26.23*b*), or equivalently

$$\boldsymbol{P} = \int \{\boldsymbol{U}(\nabla \cdot \boldsymbol{\pi}) - \kappa(\boldsymbol{\pi} \wedge \boldsymbol{G})\} \, d\tau \tag{26.30b}$$

from the symmetrical tensor (26.24). The integrated total angular momentum is therefore

$$\mathcal{J} = -\int \boldsymbol{x} \wedge \{\boldsymbol{U}(\nabla \cdot \boldsymbol{\pi}) - \kappa(\boldsymbol{\pi} \wedge \boldsymbol{G})\} \, d\tau \tag{26.31a}$$

in which the integrand can be written as the moment of the canonical momentum density (26.30*a*) plus a spin contribution, viz.

$$\begin{aligned} \boldsymbol{U}(\nabla \cdot \boldsymbol{\pi}) - \kappa(\boldsymbol{\pi} \wedge \boldsymbol{G}) &= \boldsymbol{U}(\nabla \cdot \boldsymbol{\pi}) - \boldsymbol{\pi} \wedge (\nabla \wedge \boldsymbol{U}) \\ &= -\nabla(\boldsymbol{\pi} \cdot \boldsymbol{U})_\pi + (\boldsymbol{\pi} \cdot \nabla)\boldsymbol{U} + \boldsymbol{U}(\nabla \cdot \boldsymbol{\pi}) \end{aligned}$$

[5] N.B. The reversal of sign in these two equations is due to the definition of $\boldsymbol{U}$, $\boldsymbol{\pi}$ and the signature of the metric.

and after an integration by parts we find

$$\mathcal{J} = \int \boldsymbol{\pi} \cdot (\boldsymbol{x} \wedge \nabla) \boldsymbol{U} \, d\tau + \int (\boldsymbol{\pi} \wedge \boldsymbol{U}) \, d\tau \tag{26.31b}$$

The spin term is, of course, just that which follows directly from the spin density tensor (cf. § 19c) $\mathscr{S}_{\lambda\mu\nu} = G_{\lambda\nu} U_\mu - G_{\mu\nu} U_\lambda$, corresponding to the $S_{\lambda\mu\nu}$ found in equation (26.24).

For the corresponding charged theory we merely note that, as the reader may verify, we find

$$\mathscr{L} = \tfrac{1}{2}\kappa \bar{G}^{\alpha\beta} G_{\alpha\beta} + \tfrac{1}{2}\bar{G}^{\alpha\beta}(U_{\alpha,\beta} - U_{\beta,\alpha}) + \tfrac{1}{2}G^{\alpha\beta}(\bar{U}_{\alpha,\beta} - \bar{U}_{\beta,\alpha}) + \kappa \bar{U}^\alpha U_\alpha \tag{26.32}$$

$$\Theta^\lambda{}_\mu = \kappa\{\bar{U}^\lambda U_\mu + U^\lambda \bar{U}_\mu - \bar{G}^{\lambda\nu} G_{\mu\nu} - G^{\lambda\nu}\bar{G}_{\mu\nu} - g^\lambda{}_\mu(\bar{U}^\alpha U_\alpha - \tfrac{1}{2}\bar{G}^{\alpha\beta} G_{\alpha\beta})\} \tag{26.33}$$

$$\pi_k = \frac{1}{c}\bar{G}_{k4} \qquad \pi_4 = \bar{\pi}_4 \equiv 0$$

and

$$\Theta_{44} = \mathfrak{H} = \kappa\left\{c^2 \bar{\pi} \cdot \pi + \left(\frac{c}{\kappa}\right)^2 (\nabla \cdot \bar{\pi})(\nabla \cdot \pi) + \bar{U} \cdot U + \frac{1}{\kappa^2}(\nabla \wedge \bar{U}) \cdot (\nabla \wedge U)\right\} \tag{26.34}$$

The charge-current density vector

$$\begin{aligned} J_\mu &= -i\epsilon(\bar{G}_{\alpha\mu} U^\alpha - G_{\alpha\mu}\bar{U}^\alpha) \\ J_4 &= i\epsilon c(\pi \cdot U - \bar{\pi} \cdot \bar{U}) \end{aligned} \tag{26.35a}$$

can be written in the form

$$J_\mu = -\frac{i\epsilon}{\kappa}\{\bar{U}^\alpha U_{\alpha,\mu} - U^\alpha \bar{U}_{\alpha,\mu} + \partial_\alpha(\bar{U}_\mu U^\alpha - U_\mu \bar{U}^\alpha)\} \tag{26.35b}$$

where it is seen that the two parts satisfy the continuity equation separately. The first term, which is similar in structure to that of the scalar theory (26.11a), we interpret as the usual convection current while the second represents the magnetization current characteristic of the vector field.[6]

[6] To exhibit the magnetic moment of the vector particle as a separate entity, we must consider the non-relativistic limit, for which the usual conditions are: $pc \ll E$, $v/c \ll 1$, $U_4 \ll 1$; the last stating that the non-relativistic approximation to U_μ is a 3-vector. Then, the current vector $J_\mu = \left(\frac{\boldsymbol{I}}{c}, \rho\right)$ becomes

$$\begin{aligned} \boldsymbol{I} &\approx \frac{i\epsilon c}{\kappa}\{\nabla(\bar{U} \cdot U)_{\bar{u}} - \nabla(U \cdot \bar{U})_u + \bar{U}(\nabla \cdot U) - U(\nabla \cdot \bar{U}) + (U \cdot \nabla)\bar{U} - (\bar{U} \cdot \nabla)U\} \\ &\approx \frac{i\epsilon c}{\kappa}\{\nabla(\bar{U} \cdot U)_{\bar{u}} - \nabla(U \cdot \bar{U})_u + \nabla \wedge (\bar{U} \wedge U)\} \end{aligned}$$

$$\rho = i\epsilon c(\pi \cdot U - \bar{\pi} \cdot \bar{U})$$

From the equations of motion, $\kappa c\pi = -\nabla\bar{\phi} - \frac{1}{c}\dot{\bar{U}}$ and c.c. (cf. 26.27b), this last becomes

$$\rho \approx \frac{-i\epsilon}{\kappa c}(\dot{\bar{U}} \cdot U - \dot{U} \cdot \bar{U}) \approx e\bar{V} \cdot V$$

after substituting $U \sim \exp -iEt/\hbar$, $E \sim mc^2$ and absorbing the factor $\sqrt{2/\hbar c}$ in the definition of V to obtain the usual form of ρ. With this $\boldsymbol{I}$ becomes

$$\boldsymbol{I} \approx \frac{ie\hbar}{2m}\{\bar{V} \cdot \nabla V - V \cdot \nabla\bar{V}\} + \left(\frac{e\hbar}{2mc}\right) c\nabla \wedge \left(\frac{V \wedge \bar{V}}{i}\right)$$

the first term giving just the usual current expression, while the second shows that the magnetization density is $\left(\frac{e\hbar}{2mc}\right)\left\{\frac{V \wedge \bar{V}}{i}\right\}$. The expression in brackets can be shown to represent the expectation value of the spin of the vector particle. See Proca (36, 38).

Finally we note that the system dual to (26.21), the so-called pseudovector case, is simply

$$\check{U}_{\nu,\mu} - \check{U}_{\mu,\nu} = \kappa \check{G}_{\mu\nu} \qquad (26.36a)$$
$$\check{G}^{\mu\nu}{}_{,\nu} = \kappa \check{U}^{\mu}$$

or

$$U^{\lambda\mu\nu}{}_{,\nu} = \kappa G^{\lambda\mu} \qquad (26.36b)$$
$$G_{\mu\nu,\lambda} + G_{\nu\lambda,\mu} + G_{\lambda\mu,\nu} = \kappa U_{\lambda\mu\nu}$$

by the transformations (§ 3)

$$\check{G}^{\mu\nu} = \tfrac{1}{2} g^{-1/2} \epsilon^{\mu\nu\sigma\tau} G_{\sigma\tau} \qquad G_{\mu\nu} = \tfrac{1}{2} g^{1/2} \epsilon_{\mu\nu\sigma\tau} \check{G}^{\sigma\tau} \qquad (26.37)$$
$$\check{U}^{\mu} = \left(\frac{1}{3!}\right) g^{-1/2} \epsilon^{\mu\nu\sigma\tau} U_{\nu\sigma\tau} \qquad U_{\nu\sigma\tau} = g^{1/2} \epsilon_{\mu\nu\sigma\tau} \check{U}^{\mu}$$

§ 27. Special Spinor Equations

(a) Spin $\frac{1}{2}$: The Dirac Field

The simplest wave equation corresponding to spin $\frac{1}{2}$ would *appear* to be

$$(\Box + \kappa^2)\psi_A = 0 \qquad (27.1)$$

which admits solutions of the form

$$\psi_A = \psi_A(p) \exp (i p_\sigma x^\sigma / \hbar) \qquad (27.2)$$

provided that $p_\sigma p^\sigma = (mc)^2$. There are thus two independent solutions corresponding to a given propagation vector, and the spin is determined by the values assigned to the constant spinor $\psi_A(p)$. The general state of spin is given by

$$\| \psi_A \| = \psi_1 a_1 + \psi_2 a_2 \qquad (27.3a)$$

where

$$a_1 = \begin{Vmatrix} 1 \\ 0 \end{Vmatrix} \qquad (27.3b)$$
$$a_2 = \begin{Vmatrix} 0 \\ 1 \end{Vmatrix}$$

are the values taken on by $\| \psi_A \|$ when, in the rest system, the x^3-component of spin is positive or negative, respectively; since by the work of §§ 11, 13*b* (cf. § 17) we have

$$\mathfrak{T}_{12A}{}^{B} = \tfrac{1}{2} \begin{Vmatrix} -i & 0 \\ 0 & i \end{Vmatrix} \qquad \mathfrak{T}_{23A}{}^{B} = \tfrac{1}{2} \begin{Vmatrix} 0 & -i \\ i & 0 \end{Vmatrix} \qquad \mathfrak{T}_{31A}{}^{B} = \tfrac{1}{2} \begin{Vmatrix} 0 & -1 \\ 1 & 0 \end{Vmatrix} \qquad (27.4a)$$

so that from § 25

$$S_{23\,\mathrm{op}} = \tfrac{1}{2}\hbar \begin{Vmatrix} 0 & 1 \\ 1 & 0 \end{Vmatrix} \qquad S_{31\,\mathrm{op}} = \tfrac{1}{2}\hbar \begin{Vmatrix} 0 & -i \\ i & 0 \end{Vmatrix} \qquad S_{12\,\mathrm{op}} = \tfrac{1}{2}\hbar \begin{Vmatrix} 1 & 0 \\ 0 & -1 \end{Vmatrix} \qquad (27.4b)$$
$$\boldsymbol{S}^2 = \tfrac{3}{4}\hbar^2 \begin{Vmatrix} 1 & 0 \\ 0 & 1 \end{Vmatrix}$$

However, to obtain a system which is also invariant under improper Lorentz transformations we must introduce a second spinor, say ϕ^A, and the Dirac field equations

$$g^{\sigma \dot{A} B}\psi_{B,\sigma} = \kappa \phi^{\dot{A}} \qquad (27.5a)^{1}$$
$$g^{\sigma}{}_{\dot{A}D}\phi^{\dot{A}}{}_{,\sigma} = -\kappa \psi_D$$

or

$$\partial^{\dot{A}B}\psi_B = \kappa \phi^{\dot{A}} \qquad (27.5b)$$
$$\partial_{\dot{A}D}\phi^{\dot{A}} = -\kappa \psi_D$$

from which it follows that both spinors satisfy the second-order equation.[2] It is evident that these equations are covariant under arbitrary coordinate and constant spin transformations, considered as unrelated. Let us consider the previously noted special *form* invariance of these equations; namely, the numerical invariance of the g spin-tensor coefficients under simultaneous application of an arbitrary Lorentz transformation and its spin inverse, considered as related geometric transformations (cf. § 10*d*). For this purpose it is convenient to write equations (27.5*a*) in the form

$$g^{\sigma \dot{A}}{}_{B}\psi^{B}{}_{,\sigma} = -\kappa \phi^{\dot{A}} \qquad (27.6a)$$
$$\bar{g}^{\sigma \dot{D}}{}_{A}\phi^{\dot{A}}{}_{,\sigma} = -\kappa \psi^{D}$$

or

$$g^{\sigma}\partial_{\sigma}\psi = -\kappa \bar{\phi} \qquad (27.6b)$$
$$\bar{g}^{\sigma}\partial_{\sigma}\bar{\phi} = -\kappa \psi$$

with

$$g^{\sigma} = \left\| g^{\sigma \dot{A}}{}_{B} \right\| \qquad \psi = \left\| \begin{matrix} \psi^1 \\ \psi^2 \end{matrix} \right\| \qquad \phi = \left\| \begin{matrix} \phi^1 \\ \phi^2 \end{matrix} \right\|$$

Then if we perform the Lorentz transformation $x^{\sigma'} = L^{\sigma}{}_{\tau}x^{\tau}$, equations (27.6*b*) become

$$g^{\sigma}L^{\tau}{}_{\sigma}\partial_{\tau}{}'\psi = -\kappa \bar{\phi} \qquad (27.7)$$
$$\bar{g}^{\sigma}L^{\tau}{}_{\sigma}\partial_{\tau}{}'\bar{\phi} = -\kappa \psi$$

and if we further make the spin transformation ($S = \| S^{A}{}_{B} \|$)

$$\psi' = S\psi \qquad \phi' = S\phi \qquad (27.8a)$$

these equations revert to (27.6) in the transformed quantities.[3] However, if $L^{\tau}{}_{\sigma}$ is

[1] The simplest 2-spinor equation is the two-component form

$$g^{\mu \dot{A} B}\psi_{B,\mu} = 0 \qquad \text{or} \qquad \partial^{\dot{A}B}\psi_B = 0 \qquad \text{(i)}$$

and conjugate

$$g^{\mu \dot{B} A}\psi_{\dot{B},\mu} = 0 \qquad \text{or} \qquad \partial^{\dot{B}A}\psi_{\dot{B}} = 0 \qquad \text{(ii)}$$

proposed by Weyl. These lead to a current vector

$$J^{\mu} = \tfrac{1}{2}\psi_{\dot{A}}\, g^{\mu \dot{A} B}\psi_B \qquad \text{(iii)}$$

which satisfies the continuity equation, but being a null-vector (that is, on the light-cone, cf. (9.2*b*)) represents a zero rest-mass particle. These equations lead to the usual second-order equation by iteration, but are only covariant with respect to the proper Lorentz transformations. For this reason they have not proved useful in physical theory. However, see § 32.

[2] e.g. $\partial_{\dot{A}D}\partial^{\dot{A}B}\psi_B = \Box\, \delta^{B}_{D}\psi_B = \kappa \partial_{\dot{A}D}\phi^{\dot{A}} = -\kappa^2 \psi_D$.

[3] By theorem I′, § 10*d*, $g^{\tau} = \bar{S}g^{\sigma}S^{-1}L^{\tau}{}_{\sigma}$.

an improper Lorentz transformation, we must make the spin transformation ($S = \| S^A{}_B \|$)

$$\psi' = \bar{S}\bar{\phi} \qquad \phi' = \bar{S}\bar{\psi} \tag{27.8b}$$

to regain the *form* (27.6).[4]

The equations (27.5*a*) admit solutions of the form

$$\begin{aligned} \psi_A &= \psi_A(p) \exp (ip_\nu x^\nu/\hbar) \\ \phi^A &= \phi^A(-p) \exp (ip_\nu x^\nu/\hbar) \end{aligned} \tag{27.9a}$$

where $\psi_A(p)$ and $\phi^A(-p)$ are constant spinors, provided that $p_\nu p^\nu = (mc)^2$. Then

$$\begin{aligned} g^{\sigma \dot{A} B} p_\sigma \psi_B(p) &= -imc\, \phi^A(-p) \\ g^\sigma{}_{AD} p_\sigma \phi^A(-p) &= imc\, \psi_D(p) \end{aligned} \tag{27.9b}$$

so that choosing $\psi_B(p)$ arbitrarily, $\phi^A(-p)$ is determined and there are, *for each set of* p_σ, just two linearly independent solutions—corresponding to the two different spin orientations.

For a charged field we have the equations

$$\begin{aligned} g^{\mu \dot{A} B} \psi_{B},_{\mu} &= \kappa \phi^A \\ g^\mu{}_{AB} \phi^A,_\mu &= -\kappa \psi_B \end{aligned} \tag{27.10a}$$

and their conjugates

$$\begin{aligned} g^{\mu \dot{B} A} \psi_{\dot{B}},_{\mu} &= \kappa \phi^A \\ g^\mu{}_{\dot{B} A} \phi^A,_\mu &= -\kappa \psi_{\dot{B}} \end{aligned} \tag{27.10b}$$

These can be obtained from the real Lagrangian

$$\mathscr{L} = -\tfrac{1}{2} i \{ \psi_C g^{\mu \dot{D} C} \psi_{\dot{D}},_\mu - \psi_{\dot{C}} g^{\mu \dot{C} D} \psi_D,_\mu + \phi^{\dot{C}} g^\mu{}_{\dot{C} D} \phi^D,_\mu - \phi^C g^\mu{}_{\dot{D} C} \phi^{\dot{D}},_\mu - 2\kappa (\psi_C \phi^C - \psi_{\dot{C}} \phi^{\dot{C}}) \} \tag{27.11}$$ [5]

which, it should be noted, vanishes if the field equations are satisfied.

The canonical energy-momentum tensor is readily found to be

$$\mathscr{T}_{\mu\nu} = -\tfrac{1}{2} i \{ \psi_C g_\nu{}^{AC} \psi_A,_\mu + \phi^{\dot{C}} g_{\nu \dot{C} A} \phi^A,_\mu - \psi_{\dot{C}} g_\nu{}^{\dot{C} A} \psi_A,_\mu - \phi^C g_{\nu A \dot{C}} \phi^A,_\mu \} \tag{27.12}$$

and in this case the symmetrical tensor turns out to be just

$$\Theta_{\mu\nu} = \tfrac{1}{2} (\mathscr{T}_{\mu\nu} + \mathscr{T}_{\nu\mu}) \tag{27.13}$$ [6]

The charge-current density is given by

$$J^\mu = \epsilon g^{\mu \dot{A} C} (\psi_A \psi_C + \phi_A \phi_C) \tag{27.14}$$

[4] By theorem II′, § 10*d*, $g^\tau = S \bar{g}^\sigma \bar{S}^{-1} L^\tau{}_\sigma$.

[5] e.g.
$$\frac{\partial \mathscr{L}}{\partial \psi_A} - \partial_\nu \left(\frac{\partial \mathscr{L}}{\partial \psi_A,_\nu} \right) = \tfrac{1}{2} g^{\mu \dot{D} A} \psi_{\dot{D}},_\mu + \tfrac{1}{2} g^{\nu \dot{C} A} \psi_{\dot{C}},_\nu - \kappa \phi^A = 0$$
omitting the common multiplier.

[6] The derivation of this result from the general principles of Chapter III is quite lengthy, and is therefore omitted; particularly, since it follows quite simply in the equivalent 4-spinor formulation of the theory; cf. § 30*a*.

so that the charge density is positive-definite while the energy density

$$\Theta_{44} = \mathfrak{H} = -\tfrac{1}{2}i\{\psi_C g_4{}^{AC}\psi_{A,4} + \phi^{\dot{C}} g_4{}^{\dot{C}A}\phi^A{}_{,4} - \psi_{\dot{C}} g_4{}^{\dot{C}A}\psi_{A,4} - \phi^C g_{4AC}\phi^A{}_{,4}\} \quad (27.15)$$

is not. We shall leave the further consideration of these and related matters to the work of § 30, since the 4-spinor formalism is both more convenient and useful.

However, it is of interest to consider here the transformation properties of the tensors determined by the solutions of the Dirac equations, viz.

$$T = \psi^A \phi_A \quad (27.16a)$$
$$A^\rho = \psi^A g^\rho{}_{AB}\psi^B \quad (b)$$
$$B^\rho = \phi^A g^\rho{}_{AB}\phi^B \quad (c)$$
$$C^\rho = \psi^A g^\rho{}_{AB}\phi^B \quad (d)$$
$$P^{\sigma\tau} = S^{\sigma\tau}{}_{AB}\psi^A\phi^B \quad (e)$$
$$M^{\sigma\tau} = S^{\sigma\tau}{}_{AB}\psi^A\psi^B \quad (f)$$
$$R^{\sigma\tau} = S^{\sigma\tau}{}_{AB}\phi^A\phi^B \quad (g)$$

which are evidently scalars under transformations of spin coordinates in P_1, and transform as tensors of the type indicated by the indices under coordinate transformations in R_4.

Under *linear* geometric transformations in spin space, they transform as tensors of the type indicated by the indices, e.g.

$$A^{\rho'} = \psi^{A'} g^\rho{}_{AB}\psi^{B'} = \bar{S}^A{}_C\psi^{\dot{C}} g^\rho{}_{AB} S^B{}_D\psi^D = L^\rho{}_\sigma\psi^A g^\sigma{}_{AB}\psi^B = L^\rho{}_\sigma A^\sigma \quad (27.17)$$

and, similarly, B^ρ, C^ρ, transform as vectors, T as a scalar, and $P^{\sigma\tau}$, $M^{\sigma\tau}$, $R^{\sigma\tau}$, as self-dual skew-tensors. However, under antilinear transformations—$\psi^A \to \psi^{A'} = \bar{S}^A{}_B\bar{\phi}^B$—we find

$$T \to T' = \psi^{A'}\phi_A{}' = -\bar{S}^A{}_B\bar{\phi}^B\bar{S}_A{}^C\bar{\psi}_C = \bar{\phi}^B\bar{\psi}_B = -\bar{T} \quad (27.18a)$$
$$A^\rho \to A^{\rho'} = L^\rho{}_\sigma B^\sigma \quad (b)$$
$$B^\rho \to B^{\rho'} = L^\rho{}_\sigma A^\sigma \quad (c)$$
$$C^\rho \to C^{\rho'} = L^\rho{}_\sigma\bar{C}^\sigma \quad (d)$$
$$P^{\sigma\tau} \to P^{\sigma\tau\prime} = L^\sigma_\mu L^\tau_\nu\bar{P}^{\mu\nu} \quad (e)$$
$$M^{\sigma\tau} \to M^{\sigma\tau\prime} = L^\sigma_\mu L^\tau_\nu\bar{R}^{\mu\nu} \quad (f)$$
$$R^{\sigma\tau} \to R^{\sigma\tau\prime} = L^\sigma_\mu L^\tau_\nu\bar{M}^{\mu\nu} \quad (g)$$

where $L^\sigma{}_\tau$ is now the *improper* Lorentz transformation determined by $g^\sigma = S\bar{g}^\tau\bar{S}^{-1}L^\sigma{}_\tau$. It follows that the quantities $\Delta = T - \bar{T}$, $J^\sigma = A^\sigma + B^\sigma$, $F_{\sigma\tau} = \bar{P}_{\sigma\tau} + P_{\sigma\tau}$, undergo a tensor transformation of the type indicated, whenever ψ and ϕ undergo a spin transformation of either linear or antilinear type. These quantities, as well as the Dirac equation, are also invariant under the gauge transformations

$$\phi_\sigma \to \phi_\sigma + \chi_{,\sigma}$$
$$\psi^A \to e^{i\epsilon\chi}\psi^A$$
$$\phi^A \to e^{-i\epsilon\chi}\phi^A$$

(*b*) *Spin* 1: *The Proca Field*

Let us consider now the transcription of the tensor equations (26.21) of the Proca field into spinor form. This will serve as a further illustration of spinor methods and, more particularly, provide equations for comparison with those proposed by Dirac (36) for higher-spin fields (cf. § 28*b*) in the first non-trivial case of spin 1. The higher-spin cases may well be unrealized in Nature, but we find that even in the present special case—which is known to correspond to a particular genus of particles—the two formulations of the theory are not completely equivalent.

Allowing for electromagnetic interaction, we take the Proca field equations in the form

$$D_\mu U_\nu - D_\nu U_\mu = \kappa G_{\mu\nu} \tag{27.19a}$$

$$D_\nu G^{\mu\nu} = \kappa U^\mu \tag{27.19b}$$

which, with $\check{G}^{\mu\nu} = \frac{1}{2} g^{-1/2} \epsilon^{\mu\nu\sigma\tau} G_{\sigma\tau}$ and $F_{\mu\nu} = \phi_{\nu},_{\mu} - \phi_{\mu},_{\nu}$, lead to

$$D_\nu \check{G}^{\mu\nu} = -i\epsilon U_\tau \check{F}^{\mu\tau} \tag{27.20a}$$

$$\kappa D_\mu U^\mu = -\tfrac{1}{2} i\epsilon G^{\mu\nu} F_{\mu\nu} \tag{27.20b}$$

respectively.

From § 13*b* we have

$$G_{\mu\nu} = S_{\mu\nu AB}\phi_1{}^{AB} + \bar{S}_{\mu\nu AB}\bar{\phi}_2{}^{AB} \tag{27.21}$$

$$\check{G}_{\mu\nu} = S_{\mu\nu AB}\phi_1{}^{AB} - \bar{S}_{\mu\nu AB}\bar{\phi}_2{}^{AB}$$

and inversely

$$\phi_1{}^{AB} = \tfrac{1}{8} G_{\mu\nu} S^{\mu\nu AB} \tag{27.22}$$

$$\bar{\phi}_2{}^{AB} = \tfrac{1}{8} G_{\mu\nu} \bar{S}^{\mu\nu AB}$$

Substituting from (27.21) into the sum of equations (27.19*b*) and (27.20*a*) gives

$$D_\nu S^{\mu\nu}{}_{AB}\phi_1{}^{AB} = \tfrac{1}{2}\kappa U^\mu - \tfrac{1}{2} i\epsilon U_\tau \check{F}^{\mu\tau} \tag{27.23}$$

Multiplying by $g_\mu{}^{\dot{C}D}$, and using the equation (13.4)

$$g_\mu{}^{\dot{C}D} S^{\mu\nu}{}_{AB} = -\delta^D{}_A g^{\nu\dot{C}}{}_B - \delta^D{}_B g^{\nu\dot{C}}{}_A \tag{27.24}$$

(27.23) reduces to

$$g^{\nu\dot{C}}{}_A D_\nu \phi_1{}^{AD} = -\kappa(\tfrac{1}{4}U^\mu) g_\mu{}^{\dot{C}D} + i\epsilon(\tfrac{1}{4}U_\tau)\check{F}^{\mu\tau} g_\mu{}^{\dot{C}D} \tag{27.25}$$

$$= -\kappa\psi^{\dot{C}D} + i\epsilon\chi^{\dot{C}D}$$

and similarly, from the difference of equations (27.19*b*) and (27.20*a*), and multiplication by $\bar{g}_\mu{}^{\dot{C}D}$, we get

$$\bar{g}^{\nu\dot{C}}{}_A D_\nu \bar{\phi}_2{}^{AD} = -\kappa(\tfrac{1}{4}U^\mu) g_\mu{}^{\dot{D}C} - i\epsilon(\tfrac{1}{4}U_\tau)\check{F}^{\mu\tau} g_\mu{}^{\dot{D}C} \tag{27.26}$$

$$= -\kappa\psi^{\dot{D}C} - i\epsilon\chi^{\dot{D}C}$$

Again, post-multiplying (27.19*a*) by $S^{\mu\nu}{}_{CD}$, and noting (27.22),

$$(D_\mu U_\nu - D_\nu U_\mu) S^{\mu\nu}{}_{CD} = \kappa G_{\mu\nu} S^{\mu\nu}{}_{CD} = 8\kappa\phi_{1CD} \tag{27.27a}$$

or

$$(D^\mu g^{\nu AB}\psi_{AB} - D^\nu g^{\mu AB}\psi_{AB}) S_{\mu\nu CD} = 4\kappa\phi_{1CD} \tag{27.27b}$$

and with (27.24) this gives

$$D_\nu(g^{\nu A}{}_D\psi_{AC} + g^{\nu A}{}_C\psi_{AD}) = 2\kappa\,\phi_{1CD} \tag{27.27c}$$

Now $D_\nu(g^{\nu A}{}_C\psi_{AD} - g^{\nu A}{}_D\psi_{AC}) = \rho\epsilon_{CD}$, since the left-hand side is skew in C and D and there is only one second-rank skew-spinor, namely ϵ_{CD}. Multiplying both sides by ϵ^{CD} and summing, we find

$$\rho = D_\nu g^{\nu AD}\psi_{AD} = \tfrac{1}{2}D_\nu U^\nu = -\frac{i\epsilon}{4\kappa}G^{\mu\nu}F_{\mu\nu}$$

Thus, with (27.27*c*)

$$g^{\nu A}{}_C D_\nu\psi_{AD} = \kappa\,\phi_{1CD} - \frac{i\epsilon}{8\kappa}G^{\mu\nu}F_{\mu\nu}\epsilon_{CD} \tag{27.28}$$

Similarly, we find

$$(D_\mu U_\nu - D_\nu U_\mu)\bar{S}^{\mu\nu}{}_{CD} = \kappa G_{\mu\nu}\bar{S}^{\mu\nu}{}_{CD} = 8\kappa\,\bar{\phi}_{2CD} \tag{27.29a}$$

or

$$(D^\mu g^{\nu AB}\psi_{AB} - D^\nu g^{\mu AB}\psi_{AB})\bar{S}_{\mu\nu CD} = 4\kappa\,\bar{\phi}_{2CD} \tag{27.29b}$$

and with (27.24),

$$D_\nu(\bar{g}^{\nu\dot{B}}{}_D\psi_{\dot{C}B} + \bar{g}^{\nu\dot{B}}{}_C\psi_{\dot{D}B}) = 2\kappa\,\bar{\phi}_{2CD} \tag{27.29c}$$

Again,

$$D_\nu(\bar{g}^{\nu\dot{B}}{}_C\psi_{\dot{D}B} - \bar{g}^{\nu\dot{B}}{}_D\psi_{\dot{C}B}) = \tau\epsilon_{\dot{C}\dot{D}} = -\frac{i\epsilon}{4\kappa}G^{\mu\nu}F_{\mu\nu}$$

by the same process as above, and with (27.29*c*),

$$\bar{g}^{\nu\dot{B}}{}_C D_\nu\psi_{\dot{D}B} = \kappa\,\bar{\phi}_{2CD} - \frac{i\epsilon}{8\kappa}G^{\mu\nu}F_{\mu\nu}\bar{\epsilon}_{CD} \tag{27.30}$$

The equations (27.25, 26, 28, 30) are the spinor equivalents of the tensor equations (27.19*a*, *b*). Thus we have:

$$g^{\nu\dot{C}}{}_A D_\nu\phi_1{}^{AD} = -\kappa\psi^{\dot{C}D} + i\epsilon\chi^{\dot{C}D} \tag{27.31i}$$

$$\bar{g}^{\nu\dot{C}}{}_A D_\nu\bar{\phi}_2{}^{AD} = -\kappa\psi^{\dot{D}C} - i\epsilon\chi^{\dot{D}C} \tag{ii}$$

$$g^{\nu A}{}_C D_\nu\psi_{AD} = \kappa\,\phi_{1CD} - \frac{i\epsilon}{8\kappa}G^{\mu\nu}F_{\mu\nu}\epsilon_{CD} \tag{iii}$$

$$\bar{g}^{\nu\dot{B}}{}_C D_\nu\psi_{\dot{D}B} = \kappa\,\bar{\phi}_{2CD} - \frac{i\epsilon}{8\kappa}G^{\mu\nu}F_{\mu\nu}\bar{\epsilon}_{CD} \tag{iv}$$

On the other hand, anticipating the results of § 28(*b*), the Dirac equations for spin 1 are:

$$\partial^{AK}\psi^{\dot{B}}_K = \kappa\,\phi^{A\dot{B}} \qquad (\phi^{A\dot{B}}\text{ symmetric}) \tag{27.32v}$$

$$\partial_{AK}\phi^{A\dot{B}} = -\kappa\,\psi^{\dot{B}}_K \tag{vi}$$

and conjugates. These *do not* lead to the above spinor Proca equations by the substitution $\partial \to D$ in the presence of a field. However, if U_μ, $G_{\mu\nu}$ are *real* and no field is present, then $\phi_1{}^{AB} = \phi_2{}^{AB}$, ψ_{AB} is Hermitian, and the set (i–iv) becomes identical with the set (v, vi) and conjugates. Conversely, if the set (v, vi) and conjugates are transcribed to tensor form (cf. Taub 39*a*, *b*), they do not agree with the Proca equations (26.21)—except for *real* U_μ.

A most important aspect of this difference—in its bearing on the introduction of e.m. field interaction—is brought out by examining the transformation properties of the above sets of field equations; and for this purpose we take the latter in the more convenient form:

$$\left.\begin{aligned} g^{\mu\dot{A}}{}_{C}\,\psi_{\dot{A}D},_{\mu} &= \kappa\,\phi_{CD} \\ g^{\mu\dot{A}}{}_{B}\,\phi^{BC},_{\mu} &= -\kappa\,\psi^{\dot{A}C} \end{aligned}\right\} \qquad \begin{aligned} &(27.33a) \\ &(b) \end{aligned}$$

$$\left.\begin{aligned} g^{\mu\dot{C}}{}_{A}\,\psi^{\dot{D}A},_{\mu} &= -\kappa\,\phi^{\dot{C}\dot{D}} \\ g^{\mu\dot{B}}{}_{A}\,\phi_{\dot{B}\dot{C}},_{\mu} &= \kappa\,\psi_{\dot{C}A} \end{aligned}\right\} \qquad \begin{aligned} &(c) \\ &(d) \end{aligned}$$

The numerical invariance of the coefficients $g^{\mu\dot{A}}{}_{B}$ in equations (27.31) under arbitrary Lorentz transformations, provided ϕ_{1AB}, ϕ_{2AB} and $\psi^{\dot{A}}{}_{B}$ are replaced by appropriate linear combinations, may be proved by the same method as for the Dirac field of the preceding section. However, in order to make comparison between the sets (27.31, 33), we follow a somewhat different procedure.

Since both sets are spinor equations, they are invariant in form under arbitrary (constant) spin coordinate transformations, and similarly with respect to *linear* geometric spin transformations which behave formally as coordinate transformations. When we make a *proper* Lorentz transformation on the x^μ, this induces a transformation on the $g^{\mu\dot{A}}{}_{B}$, but this can be compensated (cf. § 12) by a linear transformation of the above type which will leave the equations in the same form with the same numerical coefficients.

To prove the numerical invariance of the g-coefficients in equations (27.31) under *improper* Lorentz transformations, it suffices to prove it for the particular transformation which reverses the sign of the x^2-coordinate (cf. 10.12, 13). From the values given for the $g^{\mu\dot{A}}{}_{B}$ in § 8*c*, it can readily be verified that

$$g^{\mu\dot{A}}{}_{B}\,D_\mu = g^{\mu\dot{A}}{}_{B}{}'\,D_\mu{}' = \bar{g}^{\mu\dot{A}}{}_{B}\,D_\mu{}'$$

where $x^{\mu\prime}$ and $\phi_\mu{}'$ are obtained from x^μ and ϕ_μ by the transformation $x^2 \to -x^2$. If we perform this transformation on the x^μ, followed by the transformation

$$\begin{aligned} \phi_1{}^{AB} &\to \phi_1{}^{AB\prime} = \bar{\phi}_2{}^{AB} \\ \bar{\phi}_2{}^{AB} &\to \bar{\phi}_2{}^{AB\prime} = \phi_1{}^{AB} \\ \psi^{\dot{A}B} &\to \psi^{\dot{A}B\prime} = \psi^{\dot{B}A} \\ \chi^{\dot{A}B} &\to \chi^{\dot{A}B\prime} = -\chi^{\dot{B}A} \end{aligned} \qquad (27.34)$$

it follows that the equations (27.31) remain numerically invariant, and the spinors $\phi_1{}^{AB\prime}$, $\phi_2{}^{AB\prime}$, $\psi^{\dot{A}B\prime}$ and $\chi^{\dot{A}B\prime}$ are the *same* linear combinations of the $G_{\mu\nu}{}'$, $U_\mu{}'$ and $\check{F}^{\mu\tau\prime}U_\tau{}'$ as the original spinors were of the original $G_{\mu\nu}$, U_μ and $\check{F}^{\mu\tau}U_\tau$.

Similarly, it is seen that if we perform the transformation $x^2 \to -x^2$, then equations (27.33) remain numerically invariant if we perform the transformation

$$\begin{aligned} \phi^{BC} &\to \phi^{BC\prime} = \bar{\phi}^{BC} \\ \psi^{\dot{A}C} &\to \psi^{\dot{A}C\prime} = \bar{\psi}^{\dot{A}C} \end{aligned} \qquad (27.35)$$

under which each equation of (27.33) goes into its complex conjugate ($a \rightleftharpoons c$, $b \rightleftharpoons d$).

The important point to be noted is that if, in the Dirac equations (27.33), we take account of an external field by the usual modification $\partial_\nu \to D_\nu$, then the equations are no longer invariant, since $D_\nu \to \bar{D}_\nu$ on taking the complex conjugate. Thus, the terms in $\chi^{\dot{C}D}$ in (27.31i, ii) and $G^{\mu\nu}F_{\mu\nu}$ in (27.31iii, iv) show that it *is not* possible to obtain the spinor equations of the meson in a field from those of a free meson by the prescription $\partial_\nu \to D_\nu$—if this procedure is used for the tensor equations. This is a particular instance of the previously noted difficulty connected with the introduction of e.m. field interaction (cf. § 31) in the higher-spin cases, since the transformation $\partial_\nu \to D_\nu$ violates the conditions imposed in obtaining the irreducible free-particle wave-field.

If this transformation is to be valid for equations (27.31), the terms in $\chi^{\dot{C}D}$ and $G^{\mu\nu}F_{\mu\nu}$ would have to be eliminated, which would then correspond to the Proca equations (27.19) with the *supplementary* conditions

$$G_{\mu\nu}F^{\mu\nu} = 0 \qquad \check{F}^{\mu\tau}U_\tau = 0$$

The second of these cannot be satisfied for arbitrary e.m. fields, since it requires the vanishing of the determinant of $\check{F}^{\mu\tau}$ if U_τ is to be different from zero, and this implies $F_{\mu\nu}\check{F}^{\mu\nu} = 0$ or $E \cdot H = 0$.

(i) *The Method of Fusion.*—In the absence of an external field, equations (27.31) become

$$g^{\nu A}{}_{B}\,\phi_{1BD},_\nu = \kappa\psi^A_D \tag{27.36a}$$
$$g^\nu{}_{AD}\,\psi^A{}_{C},_\nu = -\kappa\,\phi_{1CD} \tag{b}$$
$$g^{\nu AB}\,\phi_{2BD} = \kappa\bar{\psi}_{\dot{D}}{}^A,_\nu \tag{c}$$
$$g^\nu{}_{AD}\,\bar{\psi}_{\dot{C}}{}^A,_\nu = -\kappa\,\phi_{2CD} \tag{d}$$

If now we consider the solutions of these equations and the specification of the associated spin, we find that the (spin 1) Proca particle can be represented as the composition, or fusion, of two (spin $\frac{1}{2}$) Dirac particles when suitable restrictions are placed on their masses, momenta, and spins. This type of interpretation is at the base of de Broglie's *méthode de fusion*[7] which essentially amounts to building up all higher-spin fields in terms of *fusion* of Dirac (spin $\frac{1}{2}$) fields; in particular, de Broglie's theory of light.[(8)] We shall, therefore, briefly examine this process here, but it must be emphasized that the rather artificial restrictions which must be imposed on the constituent particles indicate the purely formal nature of this interpretation—quite apart from other fundamental difficulties.

The equations (27.36) have solutions of the plane-wave form

$$\phi_{1AB} = \phi_{1AB}(k)\exp(ik_\mu x^\mu) \tag{27.37a}$$
$$\phi_{2AB} = \phi_{2AB}(-k)\exp(-ik_\mu x^\mu) \tag{b}$$
$$\psi^A_B = \psi^A_B(k)\exp(ik_\mu x^\mu) \tag{c}$$

where $\phi_{1AB}(k)$, etc., are constant spinors, provided

$$k_\mu k^\mu = \kappa^2 \tag{27.38}$$

[7] de Broglie (43); cf. Taub (40).

[8] Cf. also de Broglie (40); Geheniau (38); Petiau (41, 46).

We may choose $\phi_{1AB}(k)$, for example, arbitrarily and then equations (27.36a) and (27.36d) determine $\psi^{A}{}_{\dot{B}}(k)$ and $\phi_{2\dot{A}\dot{B}}(-k)$ such that (27.36b, c) are satisfied when (27.38) holds. Since $\phi_{1AB}(k)$ is symmetric, there are three possible linearly independent choices for it, so that for each set of k_μ which satisfy (27.38) there exist three linearly independent solutions—as there must be, corresponding to the three different spin orientations. From (27.36b, d) we obtain, by multiplying by ϵ^{CD} and summing,

$$k_\nu g^{\nu}{}_{A\dot{B}}\psi^{A\dot{B}}(k) = 0 \qquad (27.39)$$

which corresponds to the tensor equation $k_\mu U^\mu(k) = 0$, where $U_\mu = U_\mu(k)\exp(ik_\mu x^\mu)$. From equations (27.36$b$, d), (27.38), and (8.14), we get

$$\begin{aligned}\phi_{1CD}(k)\phi_1{}^{CD}(k) &= \bar{\phi}_{2\dot{C}\dot{D}}(-k)\bar{\phi}_2{}^{\dot{C}\dot{D}}(-k) \qquad (27.40)\\ &= -\psi_{\dot{C}D}(k)\psi^{\dot{C}D}(k)\end{aligned}$$

and from (13.5) it follows that the first of these is equivalent to the statement $G_{\mu\nu}(k)\breve{G}^{\mu\nu}(k) = 0$, which is satisfied for solutions of the type (27.37), since $G_{\mu\nu}(k) = \frac{i}{\kappa}\{k_\mu U_\nu(k) - k_\nu U_\mu(k)\}$. The second of equations (27.40) states that $G_{\mu\nu}(k)G^{\mu\nu}(k) = -2U^\mu(k)U_\mu(k)$.

Now, since ϕ_{1AB} is a symmetric second-order spinor, it may be written in the form

$$\phi_{1AB} = \psi_A\chi_B + \chi_A\psi_B \qquad (27.41)$$

where the spinors χ_A, ψ_A are determined, up to a factor, by the equations

$$\begin{aligned}\phi_{1AB}\psi^B &= (\chi_B\psi^B)\psi_A\\ \phi_{1AB}\chi^B &= -(\chi_B\psi^B)\chi_A\end{aligned}$$

From (27.41) it follows that

$$2\mid\phi_{1AB}\mid = \phi_{1AB}\phi_1{}^{AB} = -2(\psi^A\chi_A)^2$$

and the requirement of linear independence of ψ_A, χ_A implies $\mid\phi_{1AB}\mid \neq 0$. We may now define two other spinors ϕ^A, η^A by the equations

$$\psi^{A}{}_{\dot{B}} = \phi^A\chi_B + \eta^A\psi_B \qquad (27.42)$$

or

$$\begin{aligned}\psi^{A}{}_{\dot{B}}\psi^B &= (\psi^A\chi_A)\phi^A\\ \psi^{A}{}_{\dot{B}}\chi^B &= -(\psi^A\chi_A)\eta^A\end{aligned}$$

and $\phi_{2\dot{A}\dot{B}}$ is then given in terms of ψ^A, χ^A, ϕ^A and η^A by (27.36d).

We can now show that by taking ψ^A, ϕ^A and χ^A, η^A as plane-wave solutions of the Dirac equation for a particle of spin $\frac{1}{2}$ and masses m_1, m_2 (suitably related to m) respectively, we obtain a solution of equations (27.36) of the form (27.37). Let

$$\left.\begin{aligned}\psi^A &= \psi^A(k_1)\exp(ik_{1\mu}x^\mu)\\ \phi^A &= \phi^A(-k_1)\exp(-ik_{1\mu}x^\mu)\end{aligned}\right\} \qquad (27.43)$$

$$\left.\begin{aligned}\chi^A &= \chi^A(k_2)\exp(ik_{2\mu}x^\mu)\\ \eta^A &= \eta^A(-k_2)\exp(-ik_{2\mu}x^\mu)\end{aligned}\right\}$$

which, if (ψ^A, ϕ^A) and (χ^A, η^A) are to represent free particles of spin $\frac{1}{2}$ and masses m_1, m_2, respectively, must satisfy the Dirac equations (cf. 27.5*a*)

$$\left.\begin{aligned} g^{\nu AB} k_{1\nu} \psi_B(k_1) &= -i\kappa_1 \phi^{\dot{A}}(-k_1) \\ g^{\nu}{}_{\dot{A}D} k_{1\nu} \phi^{\dot{A}}(-k_1) &= i\kappa_1 \psi_D(k_1) \end{aligned}\right\} \tag{27.44}$$

$$\left.\begin{aligned} g^{\nu AB} k_{2\nu} \chi_B(k_2) &= -i\kappa_2 \eta^{\dot{A}}(-k_2) \\ g^{\nu}{}_{\dot{A}D} k_{2\nu} \eta^{\dot{A}}(-k_2) &= i\kappa_2 \chi_D(k_2) \end{aligned}\right\}$$

with $k_{1\nu} k_1{}^{\nu} = \kappa_1{}^2$, $k_{2\nu} k_2{}^{\nu} = \kappa_2{}^2$, and $\psi^A(k_1)\chi_A(k_1) \neq 0$, in order that $| \phi_{1AB} | \neq 0$. Equations (27.36*a*, *b*, 41, 42) then require, if $m \neq 0$, that the constitutive particles satisfy the expected relations

$$\begin{aligned} m &= m_1 + m_2 \\ k_\nu &= k_{1\nu} + k_{2\nu} \\ k_{1\nu} &= \left(\frac{m_1}{m}\right) k_\nu \\ k_{2\nu} &= \left(\frac{m_2}{m}\right) k_\nu \end{aligned} \tag{27.45}$$

Finally, we consider the relation between spins of the Proca particle and the two Dirac particles. If we substitute (27.43) into (27.42) and the latter into (27.36*d*), we obtain

$$\phi_{2CD} = \{\phi_C(-k_1)\eta_D(-k_2) + \phi_D(-k_1)\eta_D(-k_2)\} \exp(-ik_\nu x^\nu) \tag{27.46}$$

and (27.36*c*) is satisfied by this value of ϕ_{2CD}. It follows that we may obtain solutions of equations (27.36) of the form (27.37) where

$$\begin{aligned} \phi_{1AB}(k) &= \psi_A(k_1)\chi_B(k_2) + \chi_A(k_2)\psi_B(k_1) && (27.47a) \\ \phi_{2AB}(-k) &= \phi_A(-k_1)\eta_B(-k_2) + \phi_B(-k_1)\eta_A(-k_2) && (b) \\ \psi_{\dot{A}B}(k) &= \phi_{\dot{A}}(-k_1)\chi_B(k_2) + \eta_{\dot{A}}(-k_2)\psi_B(k_1) && (c) \end{aligned}$$

The spin of the vector particle is determined by the values assigned to the constant spinor $\phi_{1AB}(k)$, whereas the spins of the two spinor particles are determined by the values assigned to the two constant spinors $\psi_A(k_1)$ and $\chi_A(k_2)$.[9] The relation between them is given by (27.47*a*), which states that spin of the Proca particle is the vector sum of the spins of the two Dirac particles. To see this, we let

$$a_1 = b_1 = \begin{Vmatrix} 1 \\ 0 \end{Vmatrix} \qquad a_2 = b_2 = \begin{Vmatrix} 0 \\ 1 \end{Vmatrix}$$

denote the values taken on by $\| \psi_A \|$ and $\| \chi_A \|$ when (in the rest system $k_i = k_{1i} = k_{2i} = 0$) the x^3-component of the spins of the Dirac particles are both positive or negative, respectively, and the corresponding general spin states are

$$\begin{aligned} \| \psi_A \| &= \psi_1 a_1 + \psi_2 a_2 \\ \| \chi_A \| &= \chi_1 b_1 + \chi_2 b_2 \end{aligned}$$

[9] Cf. (27.3).

If the two spins are to combine to a resultant spin 1, we must take the symmetric product of the two spin-functions, and letting $A_1 = \| \phi_{1AB} \|$ represent the spin-function of the state with spin 1, we have

$$\begin{aligned} A_1 &= (\psi_1 a_1 + \psi_2 a_2)(\chi_1 b_1 + \chi_2 b_2) + (\chi_1 a_1 + \chi_2 a_2)(\psi_1 b_1 + \psi_2 b_2) \\ &= 2\psi_1\chi_1 V_1 + (\psi_1\chi_2 + \chi_1\psi_2)V_0 + 2\psi_2\chi_2 V_{-1} \\ &= A_{11}\begin{Vmatrix} 1 & 0 \\ 0 & 0 \end{Vmatrix} + A_{12}\begin{Vmatrix} 0 & 1 \\ 1 & 0 \end{Vmatrix} + A_{22}\begin{Vmatrix} 0 & 0 \\ 0 & 1 \end{Vmatrix} \end{aligned}$$

Thus, according as $\| \phi_{1AB} \|$ is

$$\begin{Vmatrix} 1 & 0 \\ 0 & 0 \end{Vmatrix} \qquad \begin{Vmatrix} 0 & 1 \\ 1 & 0 \end{Vmatrix} \qquad \begin{Vmatrix} 0 & 0 \\ 0 & 1 \end{Vmatrix}$$

the x^3-component of the spin of the vector particle in the rest system is $+1$, 0, -1,[10] respectively. The various constitutive tensors and spinors for the Proca field, and their resolution in the preceding way, are given by Taub (*loc. cit.*). The same procedure may, of course, be followed in the synthesis of the scalar, pseudoscalar, and pseudovector fields and, in a special degenerate form, the photon field.

(c) *Spin* 1: *The Maxwell Field*

(i) *In Vacuo.*—An interesting and essentially simple application of spinor analysis is to the transcription of Maxwell's equations into spinor form. Considered on its own merits only, this might appear to be merely a methodological exercise; but, in fact, the resulting equations are very useful in the interpretation of results obtained in field theories corresponding to charged particles.

Taking Heaviside-Lorentz rational units, for the present, the Maxwell equations are (cf. § 21)

$$G^{\mu\nu}{}_{,\nu} = J^{\mu} \tag{27.48a}$$

$$G_{\sigma\tau} = \tfrac{1}{2}\eta^{\mu\nu}_{\sigma\tau} F_{\mu\nu} = \breve{G}_{\sigma\tau} \tag{b}$$

and from our previous work we have

$$\begin{aligned} J^{\mu} &= \tfrac{1}{2} g^{\mu AB} J_{AB} \\ \phi^{\mu} &= \tfrac{1}{2} g^{\mu AB} \phi_{AB} \\ G_{\mu\nu} &= G_{AB} S_{\mu\nu}{}^{AB} \qquad \text{(cf. (13.16))} \end{aligned} \tag{27.49}$$

where the same carrier letter may now be used for both the spinor and tensor forms of a given physical entity, since the index type distinguishes them. On multiplying (27.48*b*) by $S^{\sigma\tau}{}_{CD}$ there obtains (cf. 13.16)

$$\begin{aligned} 8G_{CD} &= \tfrac{1}{2} S^{\sigma\tau}{}_{CD}\, \eta^{\mu\nu}_{\sigma\tau} (\phi_{\nu,\mu} - \phi_{\mu,\nu}) \\ &= \tfrac{1}{4} S^{\sigma\tau}{}_{CD}\, \eta^{\mu\nu}_{\sigma\tau} \{(g_{\nu}{}^{AB}\phi_{AB})_{,\mu} - (g_{\mu}{}^{AB}\phi_{AB})_{,\nu}\} \\ &= S^{\mu\nu}{}_{CD} \{(g_{\nu}{}^{AB}\phi_{AB})_{,\mu} - (g_{\mu}{}^{AB}\phi_{AB})_{,\nu}\} \end{aligned} \tag{27.50a}$$

[10] From the values of the spin tensors $S_{\mu\nu}{}^{AB}$ (cf. § 13*b*) and the relation $\tfrac{1}{2}(G_{\mu\nu} + \breve{G}_{\mu\nu}) = \phi_{1AB} S_{\mu\nu}{}^{AB}$ we have

$$\begin{aligned} \tfrac{1}{2}(G_{14} + \breve{G}_{14}) &= -\tfrac{1}{2}i(G_{23} + \breve{G}_{23}) = \phi_{1AB} S_{14}{}^{AB} = 2(\psi_1\chi_1 - \psi_2\chi_2) = A_{11} - A_{22} \\ \tfrac{1}{2}(G_{24} + \breve{G}_{24}) &= -\tfrac{1}{2}i(G_{13} + \breve{G}_{13}) = \phi_{1AB} S_{24}{}^{AB} = 2i(\psi_1\chi_1 + \psi_2\chi_2) = i(A_{11} + A_{22}) \\ \tfrac{1}{2}(G_{34} + \breve{G}_{34}) &= \tfrac{1}{2}i(G_{12} + \breve{G}_{12}) = \phi_{1AB} S_{34}{}^{AB} = -2(\psi_1\chi_2 + \chi_1\psi_2) = -2A_{12} \end{aligned}$$

and, since the left-hand members transform as vector components under spatial rotations, we have three linear combinations of the spin-functions for the state of spin 1 which transform in the same way.

by virtue of the self-duality of $S^{\mu\nu}{}_{CD}$; and with (13.4)

$$\begin{aligned} G_{CD} &= \tfrac{1}{4}(g^{\nu A}{}_{\dot D}\,\phi_{AC},_\nu + g^{\nu A}{}_{\dot C}\,\phi_{AD},_\nu) \\ &= \tfrac{1}{4}(\partial^A{}_{\dot D}\,\phi_{AC} + \partial^A{}_{\dot C}\,\phi_{AD}) \end{aligned} \tag{27.50b}$$

Similarly, Maxwell's equations (27.48a) take the form

$$S^{\mu\nu AB}\,G_{AB},_\nu = \tfrac{1}{2}g^{\mu\dot A B}\,J_{\dot A B} \tag{27.51a}$$

or

$$\left.\begin{aligned} g^{\nu\dot C}{}_B\,G^{BD},_\nu &= -\tfrac{1}{2}J^{\dot C D} \\ \partial_{\dot C}{}^B\,G_{BD} &= \tfrac{1}{2}J_{\dot C D} \end{aligned}\right\} \tag{27.51b}$$

The wave-equation follows readily from equations (27.50b) and (27.51b), viz.

$$2J_{\dot C D} = \partial_{\dot C}{}^B\,(\partial^A{}_D\,\phi_{AB} + \partial^A{}_B\,\phi_{AD}) \tag{27.52a}$$

$$= \partial_{\dot C}{}^B\partial^A{}_D\,\phi_{AB} + \partial_{\dot C}{}^B\partial_{AB}\,\phi^A{}_D + \partial_{\dot A}{}^B\partial^{\dot A}{}_B\,\phi_{\dot C D} \tag{27.52b}$$

on applying (6.12d) to the undotted indices of the second term in (27.52a);

$$= -\partial_{\dot C B}\partial^{\dot A B}\,\phi_{AD} - \partial_{\dot C D}\partial^A{}_B\,\phi_A{}^B + \partial_{\dot C}{}^B\partial_{AB}\,\phi^A{}_D - \partial_{\dot A B}\partial^{\dot A B}\,\phi_{\dot C D} \tag{27.52c}$$

on applying (6.12d) to the undotted indices of the first term of (27.52b) and (6.12a) to the last. Finally,

$$J_{\dot C D} = -\tfrac{1}{2}\partial_{\dot A B}\partial^{\dot A B}\,\phi_{\dot C D} = -\,\Box\,\phi_{\dot C D} \tag{27.52d}$$

where we have eliminated the second term in (27.52c) by the usual accessory gauge (Lorentz) condition $\phi^\mu{}_{.\mu} = \tfrac{1}{2}\partial_{\dot A B}\phi^{\dot A B} = 0$. This result is, of course, just what we obtain by direct transcription of

$$\Box\,\phi_\mu = -J_\mu \tag{27.52e}$$ [11]

(ii) *In Matter.*—For our later work we shall also need to consider the phenomenological form of Maxwell's equations in matter. In this case the electric displacement and magnetic induction are given by

$$\begin{aligned} \mathcal{D} &= \mathcal{E} + \mathcal{P} \\ \mathcal{B} &= \mathcal{H} + \mathcal{M} \end{aligned} \tag{27.53}$$

where $\mathcal{P}$ and $\mathcal{M}$ represent the electric and magnetic polarization vectors, and we use Temple Script symbols, for the present, to distinguish from our previous equations. The Maxwell equations now take the form

[11] If we introduce a self-dual Hertzian tensor $Z^{\mu\nu}$ such that $\phi^\mu = Z^{\mu\nu},_\nu$, the above divergence condition is automatically satisfied; as is the continuity equation $J^\mu,_\mu = 0$ by introducing a stream potential such that $Q^{\mu\nu},_\nu = J^\mu$ . . . with which $\Box Z^{\mu\nu} = -Q^{\mu\nu}$.
In spinor form, the identity of these conditions is shown, for example, by (cf. (27.51a, b))

$$\partial_{\dot C}{}^B Z_{BD} = \tfrac{1}{2}\phi_{\dot C D} \qquad \text{and} \qquad \partial^{\dot C D}\,\phi_{\dot C D} = 2\partial^{\dot C D}\partial_{\dot C}{}^B Z_{BD} = (\partial^{\dot C D}\partial_{\dot C}{}^B + \partial^{\dot C B}\,\partial_{\dot C}{}^D)Z_{BD} = 0$$

by the symmetry of Z_{BD}, and (6.12a).

$$\left.\begin{aligned} \operatorname{div} \mathcal{E} &= \varrho \\ \operatorname{curl} \mathcal{E} + \frac{1}{c}\frac{\partial \mathcal{B}}{\partial t} &= 0 \end{aligned}\right\} \tag{27.54a}$$

$$\left.\begin{aligned} \operatorname{div} \mathcal{B} &= 0 \\ \operatorname{curl} \mathcal{B} - \frac{1}{c}\frac{\partial \mathcal{E}}{\partial t} &= \frac{\mathcal{J}}{c} \end{aligned}\right\} \tag{27.54b}$$

with

$$\left.\begin{aligned} \varrho &= \rho - \operatorname{div} \mathcal{P} \\ \frac{\mathcal{J}}{c} &= \frac{\boldsymbol{I}}{c} + \operatorname{curl} \mathcal{M} + \frac{1}{c}\frac{\partial \mathcal{P}}{\partial t} \end{aligned}\right\} \tag{27.54c}$$

We introduce self-dual tensors $\mathfrak{G}^{\mu\nu}$, $\mathfrak{R}^{\mu\nu}$ defined by

$$\mathfrak{G}^{\mu\nu} = \left\| \begin{matrix} 0 & \mathcal{B}_3 - i\mathcal{E}_3 & -\mathcal{B}_2 + i\mathcal{E}_2 & -\mathcal{E}_1 - i\mathcal{B}_1 \\ & 0 & \mathcal{B}_1 - i\mathcal{E}_1 & -\mathcal{E}_2 - i\mathcal{B}_2 \\ & & 0 & -\mathcal{E}_3 - i\mathcal{B}_3 \\ & & & 0 \end{matrix} \right\| \tag{27.55}$$

$$\mathfrak{R}^{\mu\nu} = \left\| \begin{matrix} 0 & \mathcal{M}_3 + i\mathcal{P}_3 & -\mathcal{M}_2 - i\mathcal{P}_2 & \mathcal{P}_1 - i\mathcal{M}_1 \\ & 0 & \mathcal{M}_1 + i\mathcal{P}_1 & \mathcal{P}_2 - i\mathcal{M}_2 \\ & & 0 & \mathcal{P}_3 - i\mathcal{M}_3 \\ & & & 0 \end{matrix} \right\| \tag{27.56}$$

and the equations (27.54) take the concise form (cf. 21.3, 4)

$$\mathfrak{G}^{\mu\nu}{}_{,\nu} = \mathscr{J}^{\mu} \tag{27.57}$$

with

$$\mathscr{J}^{\mu} = J^{\mu} + \tfrac{1}{2}(\mathfrak{R}^{\mu\nu}{}_{,\nu} + \bar{\mathfrak{R}}^{\mu\nu}{}_{,\nu})$$

from which transcription to spinor form again follows in simple fashion. The spinor form of (27.57) is

$$\partial_{\dot{C}}{}^{B}\mathfrak{G}_{BD} = \tfrac{1}{2}\mathscr{J}_{\dot{C}D} \tag{27.58}$$

with

$$\mathscr{J}_{\dot{C}D} = J_{\dot{C}D} + \partial_{\dot{C}}{}^{A}\mathfrak{R}_{AD} + \partial_{D}{}^{A}\mathfrak{R}_{A\dot{C}}$$

composed of the conduction and electric-magnetic polarization currents.

(iii) *The Energy-Momentum Spinor.*—The energy-momentum tensor (cf. 22.17)

$$\begin{aligned} \Theta^{\mu\nu} &= -F^{\mu\kappa}F^{\nu}{}_{\kappa} + \tfrac{1}{4}g^{\mu\nu}F^{\alpha\beta}F_{\alpha\beta} \\ &= \tfrac{1}{2}G^{\mu\kappa}\bar{G}_{\kappa}{}^{\nu} \end{aligned} \tag{27.59}$$

will be represented by a fourth-rank (mixed) spinor, with two dotted and two undotted indices, whose divergence must again equal the force spinor corresponding to the 4-force vector. This spinor may be obtained by writing the Maxwell equations (27.51*b*) and their conjugate in the form

$$\left.\begin{aligned} \partial_{\dot{C}B}G^{BD} &= -\tfrac{1}{2}J_{\dot{C}}{}^{D} \\ \partial_{\dot{E}F}G^{\dot{E}\dot{H}} &= -\tfrac{1}{2}J_{F}{}^{\dot{H}} \end{aligned}\right\} \tag{27.60}$$

Multiplying the first by $G^{\dot{C}\dot{H}}$, the second by G^{FD}, and adding, we obtain

$$G^{\dot{C}\dot{H}}\partial_{\dot{C}B}G^{BD}+G^{FD}\partial_{\dot{E}F}G^{\dot{E}\dot{H}}=-\tfrac{1}{2}(G^{\dot{C}\dot{H}}J_{\dot{C}}{}^{D}+G^{FD}J_{F}{}^{\dot{H}}) \tag{27.61}$$

i.e.

$$\partial_{\dot{C}B}\Theta^{\dot{C}\dot{H}BD}=f^{\dot{H}D} \tag{27.62}$$

where

$$f^{\dot{H}D}=-(G^{\dot{C}\dot{H}}J_{\dot{C}}{}^{D}+G^{FD}J_{F}{}^{\dot{H}})$$

represents the force, and

$$\Theta^{\dot{C}\dot{H}BD}=2G^{\dot{C}\dot{H}}G^{BD}$$

the momentum-energy, as is readily found from (27.59) with the help of (13.7) and (13.16*a*).

(iv) *The Photon Spin.*—As remarked in § 25, wave-fields associated with quanta of zero rest-mass are distinguished by the existence of an *extended* gauge-group with respect to which the physically meaningful field tensors are invariant, and which also manifests itself in a special degeneracy in the spin eigenvalues. The simplest instance is the well-known gauge-invariance of the Maxwell or *photon* field which we consider here in regard to the specification of the lowest *total* angular-momentum state.

The free photon wave-equation

$$\Box\,\phi_\mu=0 \tag{27.63a}$$

and the supplementary (Lorentz) condition

$$\partial^\mu\phi_\mu=0 \tag{27.63b}$$

are invariant under the gauge transformations

$$\phi_\mu\to\phi_\mu+\chi_{,\mu}\qquad\Box\,\chi=0$$

We can write the content of equations (27.63) in spinor form as

$$\Box\,\phi^{\dot{B}K}=0 \tag{27.64}$$

$$\left.\begin{aligned}\partial^{\dot{A}}{}_{K}\,\phi^{\dot{B}K}&=\psi^{(+)\dot{A}\dot{B}}\\ \epsilon_{\dot{A}\dot{B}}\psi^{(+)\dot{A}\dot{B}}&=0\end{aligned}\right\} \tag{27.65}$$

where the symmetry of the auxiliary spinor $\psi^{(+)\dot{A}\dot{B}}$ guarantees the Lorentz condition. In a similar way we can define a spinor

$$\psi^{(-)JK}=\partial^{J}{}_{\dot{B}}\phi^{\dot{B}K} \tag{27.66}$$

which, as we shall see, is symmetric as a consequence of equations (27.64, 65).

From the plane-wave solutions

$$\phi^{\dot{B}K}=\phi^{\dot{B}K}(k)\exp(ik_\mu x^\mu)\qquad k_\mu k^\mu=0 \tag{27.67}$$

and

$$\psi^{(+)\dot{A}\dot{B}}(k)=ik_\mu g^{\mu\dot{A}}{}_{K}\phi^{\dot{B}K}(k) \tag{27.68a}$$

$$\psi^{(-)JK}(k)=ik_\mu\bar{g}^{\mu J}{}_{\dot{B}}\phi^{\dot{B}K}(k) \tag{27.68b}$$

it follows that $\phi^{\dot{B}K}(k)$ has only three independent components, since the symmetry of the right-hand side of (27.68a) imposes one condition on the components of $\phi^{\dot{B}K}(k)$. The components of $\psi^{(+)\dot{A}\dot{B}}(k)$ and $\psi^{(-)JK}(k)$ are then expressed in terms of these three independent components. Further, we can show, in a special coordinate system, that one component of $\phi^{\dot{B}K}(k)$ is zero and $\psi^{(+)}(k)$ and $\psi^{(-)}(k)$ have each but one non-zero component.

We choose a space-time frame of reference such that

$$k_1 = k_2 = 0 \qquad k_3{}^2 = k_4{}^2 \tag{27.69}$$

and hence (cf. 8.12)

$$\| ik_\mu g^{\mu A}{}_B \| = \begin{cases} \begin{Vmatrix} 0 & -2ik_4 \\ 0 & 0 \end{Vmatrix} & (k_3 = k_4) \\[2ex] \begin{Vmatrix} 0 & 0 \\ 2ik_4 & 0 \end{Vmatrix} & (k_3 = -k_4) \end{cases} \tag{27.70}$$

Taking the latter for definiteness, it follows from (27.68a) and the symmetry of $\psi^{(+)}(k)$ that

$$\begin{aligned} \psi^{(+)\dot{1}\dot{1}} &= \psi^{(+)\dot{1}\dot{2}} = \psi^{(+)\dot{2}\dot{1}} = 0 \\ \psi^{(+)\dot{2}\dot{2}} &= 2ik_4\,\phi^{\dot{2}1} \end{aligned} \tag{27.71}$$

and, again in this special coordinate system,

$$\phi^{\dot{1}1} = 0 \tag{27.72}$$

so that the number of non-zero components of $\phi^{\dot{B}K}(k)$ is just the number of independent components. With (27.68b), this shows that

$$\psi^{(-)12} = -2ik_4\,\phi^{\dot{1}1} = 0 \tag{27.73a}$$

and similarly,

$$\begin{aligned} \psi^{(-)21} &= \psi^{(-)11} = 0 \\ \psi^{(-)22} &= -2ik_4\,\phi^{\dot{1}2} \end{aligned} \tag{27.73b}$$

from (27 68b) and (27.70). The symmetry of $\psi^{(-)JK}$ in *any* coordinate system follows from this result.

The photon field system is invariant to gauge transformations, or in terms of plane waves, the transformations

$$\phi^{\dot{B}K}(k)e^{ik_\mu x^\mu} \to [\phi^{\dot{B}K}(k) + \phi^{\dot{B}K}(k)'']e^{ik_\mu x^\mu} \tag{27.74a}$$

with

$$\phi^{\dot{B}K}(k)'' = ik_\mu g^{\mu\dot{B}K}\chi \tag{27.74b}$$

Correspondingly,

$$\begin{aligned} \psi^{(+)\dot{A}\dot{B}}(k)'' &= ik_\mu g^{\mu\dot{A}}{}_K\,\phi^{\dot{B}K}(k)'' = -k_\mu k_\nu g^{\mu\dot{A}}{}_K g^{\nu\dot{B}K}\chi = 0 \\ \psi^{(-)JK}(k)'' &= 0 \end{aligned} \tag{27.75}$$

Thus, by suitable choice of χ, one of the three independent components of $\phi^{\dot{B}K}(k)$ can be made zero, but $\psi^{(+)}(k)$ and $\psi^{(-)}(k)$ have each one independent component which cannot be transformed away by gauge transformation. In terms of $\phi^{\dot{B}K}(k)$ these gauge-invariant components are given by

$$\left.\begin{aligned} \psi^{(+)\dot{2}\dot{2}}(k) &= 2ik_4\,\phi^{\dot{2}1} \\ \psi^{(-)22}(k) &= -2ik_4\,\phi^{\dot{1}2} \end{aligned}\right\} \tag{27.76}$$

and the physical meaning of these two components may be seen by considering the effect of a rotation θ about the x^3-axis. This corresponds to the spin transformation (cf. 10.6*b*)

$$\| S^A_{\ B} \| = \pm \begin{Vmatrix} e^{i\theta/2} & 0 \\ 0 & e^{-i\theta/2} \end{Vmatrix}$$

which multiplies $\phi^{\dot{2}1}(k)$ by $e^{i\theta}$ and $\phi^{\dot{1}2}(k)$ by $e^{-i\theta}$ so that these two states[12] correspond to right and left circular polarization, or, as we may say, spin components ± 1 on the x^3-axis. By the special choice of the reference frame, there can be no orbital angular-momentum component on this axis, so that the lowest *total* angular-momentum state corresponds to squared magnitude $\geqslant 2$.

§ 28. General Theory of Fields Associated with Particles of Arbitrary Spin and Non-zero Rest Mass [1]

(*a*) *Tensor Fields*

We have noted the general conditions which must be satisfied by a field system which is to describe particles of only one spin value, and we can now set down the field equations corresponding to any integral spin s. Clearly, these *may* be written in tensor notation, and we assert that one *possible form* is the system[2]

$$(\Box + \kappa^2)U_{\kappa\lambda\ldots\nu} = 0 \tag{28.1a}$$

$$\partial^\kappa U_{\kappa\lambda\ldots\nu} = 0 \tag{28.1b}$$

where $U_{\kappa\lambda\ldots\nu}$ is a symmetric tensor of rank s with *vanishing spurs* (cf. Table I, § 17); the supplementary condition (28.1*b*) again serves to guarantee that the U-field describes particles of spin s, that is, does not also include particles of smaller integral spin, and that there exists an energy-momentum tensor such that the total (volume-integrated) energy is positive. It will be seen that the above field without the supplementary condition describes particles of spin s, $s - 1$, . . . 0; i.e. $s + 1$ species—as illustrated in the vector field case.

Considering the question of spin first, we note that a totally symmetric tensor of rank s has

$$\binom{s+n-1}{n-1}$$

[12] The two states go over into each other by reflections. Cf. § 40*c*.

[1] The general theory was first developed by Dirac (36) in the spinor formalism. Our discussion follows the simpler systematization developed by Fierz (39*a*, *b*) to which the interested reader is referred for the more detailed considerations; in particular as regards quantization.

[2] We give here what appears to be the simplest form. As will be seen from the discussion and the work of the later sections, there are a great number of globally (that is, in the overall, though not in detail) equivalent formulations in all the higher-spin cases.

independent components in an n-dimensional space, and the spur condition states, in virtue of the symmetry, that a tensor of rank $s - 2$ shall vanish, i.e. $U^{\kappa}{}_{\kappa\lambda\ldots\nu} = 0$. In other words, there are as many conditions on the U-field as there are independent components of a similar tensor of rank $s - 2$. In the present case of 4-space it follows that $U_{\kappa\lambda\ldots\nu}$ has $\binom{s+3}{3}$ components subject to $\binom{s+1}{3}$ conditions so that there are $\binom{s+3}{3} - \binom{s+1}{3} = (s+1)^2$ linearly independent components. Further, the divergence condition (28.1b), considered in the rest system, states that all components involving the index 4 are zero; so that for given sign of the energy $U_{\kappa\lambda\ldots\nu}$ has the form

$$U_{kl\ldots n} = U(p)_{kl\ldots n} \exp(i\kappa x^4) \qquad k, l, \ldots = (1, 2, 3) \tag{28.2}$$

where $U(p)_{kl\ldots n}$ is a constant symmetric tensor satisfying the spur condition. Hence, repeating the previous procedure, now in 3-space, we find $\binom{s+2}{2} - \binom{s}{2} = (2s+1)$ linearly independent components in the rest system; or, $(2s + 1)$ independent plane-wave solutions for given propagation vector k_μ satisfying the condition $k_\mu k^\mu = \kappa^2$.

According to the discussion of § 17, $U_{\kappa\lambda\ldots\nu}$ is an irreducible tensor transforming according to the irreducible representation $\mathfrak{D}\left(\frac{s}{2}, \frac{s}{2}\right)$ of the Lorentz group. In the rest system it reduces to a 3-tensor which transforms under the irreducible representation $\mathfrak{D}(s)$ of the Rotation subgroup.[3]

As in the case of the vector field the system (28.1a, b) does not follow from a simple Lagrangian, but here again the second-order wave equation can be replaced by the system of first-order equations

$$\partial_\kappa U_{\lambda\mu\ldots\nu} - \partial_\lambda U_{\kappa\mu\ldots\nu} = \kappa G_{[\kappa\lambda]\mu\ldots\nu} \tag{28.3a}$$

$$\partial^\lambda G_{[\kappa\lambda]\mu\ldots\nu} = \kappa U_{\kappa\mu\ldots} \tag{28.3b}$$

which include the divergence condition on the U-field. It is readily verified that the tensor $G_{[\kappa\lambda]\mu\ldots\nu}$, which is skew in its bracketed indices and symmetric in all the others, satisfies the second-order wave equation

$$(\Box + \kappa^2) G_{[\kappa\lambda]\mu\ldots\nu} = 0 \tag{28.4a}$$

as well as the conditions

$$g^{\lambda\mu} G_{[\kappa\lambda]\mu\ldots\nu} = 0 \tag{28.4b}$$

$$G_{[\kappa\lambda]\mu\ldots} + G_{[\mu\kappa]\lambda\ldots} + G_{[\lambda\mu]\kappa\ldots} = 0$$

$$\partial^\mu G_{[\kappa\lambda]\mu\ldots} = 0 \tag{28.4c}$$

$$\partial_\mu G_{[\kappa\lambda]\ldots} + \partial_\lambda G_{[\mu\kappa]\ldots} + \partial_\kappa G_{[\lambda\mu]\ldots} = 0$$

[3] In the general case the reduction of $\mathfrak{D}\left(\frac{s}{2}, \frac{s}{2}\right)$ leads to $\mathfrak{D}\left(\frac{s}{2}\right) \times \mathfrak{D}\left(\frac{s}{2}\right) = \mathfrak{D}(s) \oplus \mathfrak{D}(s-1) \oplus \ldots \mathfrak{D}(0)$, corresponding to the component count $(s+1)^2 = \sum_{o}^{s}(2i+1)$ of the $(s+1)$ irreducible parts; or, as remarked above, $(s + 1)$ species of particles—in the absence of the supplementary conditions.

It can further be shown that the system (28.4) is also entirely equivalent to the system (28.1)[4] and can equally well be used to describe the given wave-field. Moreover, by continuing the process we obtain a sequence of $(s + 1)$ field tensors

$$U_{\kappa\lambda\,\ldots\,\sigma}, \quad G^{(1)}_{[\kappa\lambda]\,\ldots}, \ldots, \quad G^{(s)}_{[\kappa\lambda][\mu\nu]\,\ldots\,[\rho\tau]}$$

the prototype function $G^{(r)}$ being skew in each of the r pairs of bracketed indices, and symmetric in the remaining $s - r$ indices as well as among the brackets themselves. Each G satisfies a system of the type (28.4) and each successive pair of G's form a first-order system of the type (28.3). We do not give these involved systems of equations since they are of little interest, *per se*, but the important fact is that the non-uniqueness of the field system is reflected in the non-uniqueness of the localization of energy and charge in the field. In the absence of forces we find s equivalent linearly independent energy-momentum tensors and charge-current density vectors —and all lead to the same total (integrated) energy and charge, apart from unessential numerical factors (cf. Fierz, *loc. cit.*). In how far this arbitrariness might be reduced by the introduction of interactions with other fields, or indeed whether it implies a serious physical difficulty, is not known at present.

We give below the Lagrangian and associated tensors for the charged field corresponding to the first derived system (28.3) which are the prototype forms for all the derived systems. As may be easily verified, we have now

$$\begin{aligned}\mathscr{L} &= \tfrac{1}{2}\kappa\bar{G}^{\alpha\beta\,\ldots\,[\rho\sigma]}G_{\alpha\beta\,\ldots\,[\rho\sigma]} + \tfrac{1}{2}\bar{G}^{\alpha\beta\,\ldots\,[\rho\sigma]}(U_{\alpha\beta\,\ldots\,\rho,\sigma} - U_{\alpha\beta\,\ldots\,\sigma,\rho}) \\ &\quad + \tfrac{1}{2}G^{\alpha\beta\,\ldots\,[\rho\sigma]}(\bar{U}_{\alpha\beta\,\ldots\,\rho,\sigma} - \bar{U}_{\alpha\beta\,\ldots\,\sigma,\rho}) + \kappa\bar{U}^{\alpha\beta\,\ldots\,\rho}U_{\alpha\beta\,\ldots\,\rho}\end{aligned} \tag{28.5}$$

$$\mathscr{T}^{\lambda}{}_{\mu} = G_{\alpha\beta\,\ldots\,[\rho\mu]}\bar{U}^{\alpha\beta\,\ldots\,\rho,\lambda} + \text{c.c.} - g^{\lambda}{}_{\mu}\mathscr{L} \tag{28.6a}$$

$$\begin{aligned}\Theta^{\lambda}{}_{\mu} = \kappa\{&\bar{U}^{\alpha\beta\,\ldots\,\lambda}U_{\alpha\beta\,\ldots\,\mu} + U^{\alpha\beta\,\ldots\,\lambda}\bar{U}_{\alpha\beta\,\ldots\,\mu} \\ &- \bar{G}^{\alpha\beta\,\ldots\,[\rho\lambda]}G_{\alpha\beta\,\ldots\,[\rho\mu]} - G^{\alpha\beta\,\ldots\,[\rho\lambda]}\bar{G}_{\alpha\beta\,\ldots\,[\rho\mu]} \\ &- g^{\lambda}{}_{\mu}(\bar{U}^{\alpha\beta\,\ldots\,\sigma}U_{\alpha\beta\,\ldots\,\sigma} - \tfrac{1}{2}\bar{G}^{\alpha\beta\,\ldots\,[\rho\sigma]}G_{\alpha\beta\,\ldots\,[\rho\sigma]})\}\end{aligned} \tag{28.6b}$$

$$J_{\mu} = -i\epsilon\{\bar{G}_{\alpha\beta\,\ldots\,[\rho\mu]}U^{\alpha\beta\,\ldots\,\rho} - G_{\alpha\beta\,\ldots\,[\rho\mu]}\bar{U}^{\alpha\beta\,\ldots\,\rho}\} \tag{28.7}$$

Here it is important to note that, as can be shown by expanding in plane waves, the total charge exhibits both signs but the total energy is definite—though the energy density is not.

As regards introduction of (e.m.) interaction by the usual transcription $\partial_{\mu} \rightarrow D_{\mu}$, the first-order systems of the type (28.3) show no evident contradiction—just as in the vector case, but here again the primitive system (28.1) becomes immediately inconsistent since the D operators do not commute. Therefore, in the tensor formulation of integral-spin fields in terms of the primitive system (28.1) it is necessary to work from an extended Lagrangian formalism (cf. § 31), such as developed by Fierz and Pauli, which gives the supplementary condition (28.1*b*) as one of the variational equations.[5] The situation is more difficult in this regard in the spinor

[4] The counting of the number of independent components of G *and* conditions becomes somewhat involved, and we refer the interested reader to Fierz (39*a*).

[5] It may be remarked that this situation already arises in the vector field theory.

formulation originally developed by Dirac—in which, of course, there is no sharp distinction between the integral and half-integral spin fields. As we have seen (§ 27*b*), the Dirac spinor equations for the integral case are not precisely equivalent to the foregoing tensor equations proposed by Fierz, so that it is not superfluous to consider both formulations in this case.

(*b*) *Spinor Fields*

Let us consider the interpretation of the spinor field equations

$$(\Box + \kappa^2)\psi^{\dot{B}\dot{C}\cdots}_{KL\cdots} = 0 \tag{28.8a}$$

$$\epsilon_{\dot{B}A}\partial^{AK}\psi^{\dot{B}\dot{C}\cdots}_{KL\cdots} = 0 \tag{28.8b}$$

in the general case where $\psi^{\dot{B}\dot{C}\cdots}_{KL\cdots}$ is an irreducible spinor with, say, $2k$ undotted and $2l - 1$ dotted indices, and is symmetric in both sets. This spinor has $2l(2k + 1)$ linearly independent components which, under proper Lorentz transformations, transform among themselves according to the irreducible representation $\mathfrak{D}(k, l - \frac{1}{2})$.[6] The symmetry of $\psi^{\dot{B}\dot{C}\cdots}_{KL\cdots}$ is equivalent to the statement that all its spurs vanish, viz.

$$\epsilon_{\dot{C}\dot{B}}\psi^{\dot{B}\dot{C}\cdots}_{KL\cdots} = 0 \tag{28.9}$$

$$\epsilon^{KL}\psi^{\dot{B}\dot{C}\cdots}_{KL\cdots} = 0$$

and the supplementary condition (28.8*b*) states that the spinors $\partial^{AK}\psi^{\dot{B}\dot{C}\cdots}_{KL\cdots}$ and $\partial_{\dot{B}J}\psi^{\dot{B}\dot{C}\cdots}_{KL\cdots}$ are symmetric in their dotted and undotted indices.

We have now only to show that in consequence of the supplementary condition there are $2s + 1 = 2k + 2l$ linearly independent plane waves for a given propagation vector. In the rest system $\partial^{AK} = g^{\mu AK}\partial_\mu$ has the form $i\kappa g^{4AK} = i\kappa\,\delta^{AK}$, and with respect to spatial rotations δ^{AK} transforms like δ^K_A. Similarly, $\psi^{\dot{B}\dot{C}\cdots}_{KL\cdots}$ transforms like a spinor $\psi'_{KL\ldots,BC\ldots}$ so that (28.8*b*) reads as

$$\epsilon^{BA}\psi'_{AL\ldots,BC\ldots} = 0 \tag{28.10}$$

which states that ψ' is symmetric in the indices A and B, and therefore in all indices. Hence, in the rest system, $\psi^{\dot{B}\dot{C}\cdots}_{KL\cdots}$ is equivalent to a totally symmetric spinor $\psi'_{KL\ldots,BC\ldots}$ of rank $2k + 2l - 1 = 2s$, and has therefore $2k + 2l = 2s + 1$ linearly independent components which, under spatial rotations, transform among themselves according to the irreducible representation $\mathfrak{D}(s)$ of the Rotation group. The spin s is therefore integral (half-integral) if the total number of indices is even (odd).

The equations (28.8) can again be subsumed in a first-order system, the Dirac equations,

$$\partial^{AK}\psi^{\dot{B}\dot{C}\cdots}_{KL\dot{M}\cdots} = \kappa\phi^{A\dot{B}\dot{C}\cdots}_{LM\cdots} \tag{28.11}$$

$$\partial_{AK}\phi^{A\dot{B}\dot{C}\cdots}_{LM\cdots} = -\kappa\psi^{\dot{B}\dot{C}\cdots}_{KL\dot{M}\cdots}$$

[6] The choice of $2k$ undotted and $2l - 1$ dotted indices rather than $2k$ and $2l$ is purely a matter of convenience.

With the use of the spinor-matrix formalism of § 17*b*, ii, these are directly transcribed to the form (cf. especially equations 17.39)

$$\partial^{AK}\psi^{\rho}_{K} = \kappa\phi^{A\sigma} \qquad (28.12)^{7}$$
$$\partial_{\dot{A}K}\phi^{A\sigma} = -\kappa\psi^{\rho}_{K}$$

where we use ρ, σ to denote the matrix index.

§ 29. General Theory of Fields Associated with Particles of Arbitrary Spin and Zero Rest-Mass—Extended Gauge Invariance

Since fields associated with (hypothetical) zero rest-mass quanta of spin > 1 appear to have little likelihood of significance in physical theory,[1] we wish to briefly consider here only such aspects as illustrate the general invariance property—of which the photon field presents the familiar special case.

Again, for integral spin, we may take the system

$$\left.\begin{aligned} \Box U_{\kappa\lambda\mu\ldots} &= 0 \\ U^{\cdot\kappa}{}_{\kappa\lambda\mu\ldots} &= 0 \end{aligned}\right\} \qquad (29.1)$$

or any of the derived systems constructed as in § 28; and similarly for half-integral spin in terms of spinors. The interested reader may profitably consult Fierz (38, 39*a*, *b*) and de Wet (40) for details, including construction of the associated field quantities *and* their quantization. It may be remarked that the treatment of de Wet, which takes spinor forms for both species of quanta, affords the simpler analysis of spin characterization along lines which are a straightforward generalization of the procedure of § 27*c*, ii, for the photon. The essential result is again that for spin s $(>\frac{1}{2})$ there exist but two *really* independent plane-wave solutions rather than the $(2s + 1)$ of the non-zero rest-mass case, and these correspond to spin components $\pm s$ along the momentum vector of the wave.

The particularly interesting feature of the general zero rest-mass case is the appearance of an *extended gauge group*, viz. for integral spin the total energy is invariant to the transformation

$$U_{\kappa\lambda\mu\ldots} \rightarrow U'_{\kappa\lambda\mu\ldots} = U_{\kappa\lambda\mu\ldots} + V_{\kappa\lambda\mu\ldots} \qquad (29.2)$$

provided $V_{\kappa\lambda\mu\ldots}$ is defined by the equation

$$V_{\kappa\lambda\mu\ldots\rho} = W_{\lambda\mu\ldots\rho,\kappa} + W_{\kappa\mu\ldots\rho,\lambda} + W_{\kappa\lambda\ldots\rho,\mu} + \ldots W_{\kappa\lambda\mu\ldots,\rho} \qquad (29.3)$$

and W is fully symmetric, satisfying the conditions

$$\left.\begin{aligned} \Box W_{\kappa\lambda\mu\ldots} &= 0 \\ W^{\cdot\kappa}{}_{\kappa\ldots} &= 0 \\ W^{\kappa}{}_{\kappa\ldots} &= 0 \end{aligned}\right\} \qquad (29.4)$$

A similar gauge transformation appears for half-integral spin, such that the expressions for total energy and charge are unaffected.[2]

[7] Cf. Dirac (36); Fierz (39); especially see § 35*b*, equations (35.17, 18).

[1] The gravitational quanta, spin 2, appear to be the only exceptional case of interest. Cf. Fierz-Pauli (39*a*, *b*); also § 31.

[2] Cf. Fierz and de Wet, *loc. cit.*; also § 40.

§ 30. 4-Spinor Formulation

(a) Spin ½

The 4-spinor form of the Dirac equation follows directly from the work of §§ 14 and 27. We write the equations (27.5*a*) in the form

$$g^{\sigma \dot{A} B} p_\sigma \psi_B = -imc \bar{\phi}^{\dot{A}} \qquad (30.1)$$
$$\bar{g}^{\sigma}{}_{\dot{A} B} p_\sigma \bar{\phi}^{B} = imc \psi_{A}$$

or

$$p_\sigma \left\| \begin{matrix} 0 & -i \, \| \bar{g}^{\sigma}{}_{\dot{A}B} \| \\ i \, \| g^{\sigma \dot{A} B} \| & 0 \end{matrix} \right\| \cdot \left\| \begin{matrix} \psi_B \\ \bar{\phi}^{\dot{B}} \end{matrix} \right\| = mc \left\| \begin{matrix} \psi_A \\ \bar{\phi}^{\dot{A}} \end{matrix} \right\| \qquad (30.2)$$

By the definitions of the 4-spinor Ψ^a and the γ's (cf. § 14), this becomes

$$\gamma_\sigma p_\sigma \Psi = mc\Psi \qquad (30.3)$$

where again, by virtue of the relation $\gamma^\sigma \gamma^\tau + \gamma^\tau \gamma^\sigma = 2g^{\sigma\tau}$, it follows that Ψ satisfies the usual second-order wave-equation. As previously stated, the condition of Lorentz invariance requires that, for each Lorentz transformation

$$x^\mu \to x^{\mu'} = L^\mu{}_\nu x^\nu \qquad (30.4a)$$

there exists a transformation of Ψ

$$\Psi \to T\Psi \qquad (30.4b)$$
$$\gamma^\nu \to T\gamma^\nu T^{-1}$$

such that

$$L^\mu{}_\nu T\gamma^\nu T^{-1} = \gamma^\mu \qquad (30.4c)$$

It is clear that all of these equations are unaltered by a change of basis, or spin frame,

$$\Psi \to S\Psi \qquad \gamma^\mu \to S\gamma^\mu S^{-1} \qquad T \to STS^{-1} \qquad (30.5)$$

In this connection, we may note the relation of the equation

$$(\gamma^\mu \partial_\mu - i\kappa)\Psi = 0 \qquad (30.6)$$

to the more familiar form

$$\left(\frac{\partial}{\partial ct} + \alpha^k \partial_k - i\kappa\beta\right)\Phi = 0 \qquad (30.7)$$

given by Dirac.[1] To go from (30.6) to (30.7), put $\Psi = S^{-1}\Phi$ and pre-multiply (30.6) by $S\gamma^4$, viz.

$$\left(\frac{\partial}{\partial ct} + S(\gamma^4\gamma^k)S^{-1}\partial_k - i\kappa S\gamma^4 S^{-1}\right)\Phi = 0 \qquad (30.8)$$

[1] Dirac, P.A.M.: *Quantum Mechanics* (47).

Then, taking S to be the unitary matrix

$$S = \frac{1}{\sqrt{2}} \left\| \begin{matrix} 1 & 0 & -i & 0 \\ 0 & 1 & 0 & -i \\ -1 & 0 & -i & 0 \\ 0 & -1 & 0 & -i \end{matrix} \right\|$$

we find

$$S\gamma^4 S^{-1} = \left\| \begin{matrix} 1 & 0 & 0 & 0 \\ 0 & 1 & 0 & 0 \\ 0 & 0 & -1 & 0 \\ 0 & 0 & 0 & -1 \end{matrix} \right\| = \beta \qquad S(\gamma^4\gamma^1)S^{-1} = \left\| \begin{matrix} 0 & 0 & 0 & -1 \\ 0 & 0 & -1 & 0 \\ 0 & -1 & 0 & 0 \\ -1 & 0 & 0 & 0 \end{matrix} \right\| = \alpha^1$$

$$S(\gamma^4\gamma^2)S^{-1} = \left\| \begin{matrix} 0 & 0 & 0 & i \\ 0 & 0 & -i & 0 \\ 0 & i & 0 & 0 \\ -i & 0 & 0 & 0 \end{matrix} \right\| = \alpha^2 \qquad S(\gamma^4\gamma^3)S^{-1} = \left\| \begin{matrix} 0 & 0 & -1 & 0 \\ 0 & 0 & 0 & 1 \\ -1 & 0 & 0 & 0 \\ 0 & 1 & 0 & 0 \end{matrix} \right\| = \alpha^3 \tag{30.9}$$

which are the contravariant forms of the matrices given by Dirac (*ibid.*).

The adjoint of (30.6), namely

$$\Psi^*(\gamma^{\mu *}\partial_\mu + i\kappa) = 0 \tag{30.10a}$$

may be written in the form

$$\Psi^+(\gamma^\mu \partial_\mu + i\kappa) = 0 \qquad \Psi^+ = \Psi^* \gamma^4 \tag{30.10b}$$

on post-multiplying (30.10a) by γ^4, and noting that in our spin frame (§ 14b) the γ^k are skew-Hermitian and γ^4 Hermitian. Then, by direct transcription, the Lagrangian, energy-momentum tensor, and current vector of § 27a take the forms:

$$\mathscr{L} = -\frac{i\hbar c}{2}\left\{(\Psi^+{}_{,\mu}\gamma^\mu + i\kappa\Psi^+)\Psi - \Psi^+(\gamma^\mu \Psi_{,\mu} - i\kappa\Psi)\right\} \tag{30.11}$$

$$\Theta_{\mu\nu} = \frac{\hbar c}{4i}\left\{\Psi^+{}_{,\mu}\gamma_\nu \Psi - \Psi^+\gamma_\nu \Psi_{,\mu} + \Psi^+{}_{,\nu}\gamma_\mu \Psi - \Psi^+\gamma_\mu \Psi_{,\nu}\right\} \tag{30.12}$$

$$J_\mu = e\Psi^+\gamma_\mu \Psi \tag{30.13}$$

The symmetric tensor $\Theta_{\mu\nu}$ may, of course, be constructed from the prescription given in § 19c, but we may proceed more directly as follows: We have stated that $\Theta_{\mu\nu} = \frac{1}{2}(\mathscr{T}_{\mu\nu} + \mathscr{T}_{\nu\mu})$, and, since $\Theta_{\mu\nu} = \mathscr{T}_{\mu\nu} + \mathsf{S}_{\mu\nu\lambda}{}^{,\lambda}$ by definition (cf. 19.23), we need only calculate $\mathsf{S}_{\mu\nu\lambda}{}^{,\lambda} = \frac{1}{2}(\mathscr{T}_{\nu\mu} - \mathscr{T}_{\mu\nu})$ and verify that it is skew in the indices ν, λ. Now,

$$\begin{aligned} \partial_\nu \Psi &= \tfrac{1}{2}(\gamma_\lambda \gamma_\nu \partial^\lambda + i\kappa\gamma_\nu)\Psi \\ \partial_\nu \Psi^+ &= \tfrac{1}{2}\Psi^+(\gamma_\nu \gamma_\lambda \partial^\lambda - i\kappa\gamma_\nu) \end{aligned} \tag{30.14}$$

from (30.6, 10b) and the C.R. of the γ's, so that

$$\begin{aligned} \mathscr{T}_{\mu\nu} &= -\tfrac{1}{2}i\hbar c\{\Psi^+{}_{,\mu}\gamma_\nu \Psi - \Psi^+\gamma_\nu \Psi_{,\mu}\} \\ &= \tfrac{1}{4}i\hbar c\{\Psi^+\gamma_\mu(\gamma_\lambda\gamma_\nu\partial^\lambda + i\kappa\gamma_\nu)\Psi - (\Psi^{+,\lambda}\gamma_\nu\gamma_\lambda - i\kappa\Psi^+\gamma_\nu)\gamma_\mu\Psi\} \end{aligned} \tag{30.15}$$

and

$$\tfrac{1}{2}(\mathscr{T}_{\nu\mu} - \mathscr{T}_{\mu\nu}) = \mathsf{S}_{\mu\nu\lambda}{}^{,\lambda} = \tfrac{1}{8}i\hbar c\, \partial^\lambda\{\Psi^+(\gamma_\mu\gamma_\lambda\gamma_\nu - \gamma_\nu\gamma_\lambda\gamma_\mu)\Psi\} \tag{30.16}$$

Further, $\gamma_\lambda \gamma_\nu = g_{\lambda\nu} + \gamma_{\lambda\nu}$ (cf. 14.6a), whence

$$\Theta_{\mu\nu} = \mathscr{T}_{\mu\nu} + \tfrac{1}{8} i\hbar c\, \partial^\lambda \{\Psi^+(\gamma_\mu \gamma_{\lambda\nu} - \gamma_{\nu\lambda}\gamma_\mu)\Psi\} \tag{30.17}$$

and $\Theta_{\mu\nu}{}^{\cdot\nu} = 0$, as required. The integrated spin is then given by (cf. 19.28b)

$$\mathscr{S}_{ij} = \frac{1}{c}\int \mathscr{S}_{ij4}\, d\tau = \int \Psi^* \mathscr{S}_{ij\,\mathrm{op}} \Psi\, d\tau \tag{30.18a}$$

or

$$\mathscr{S}_{ij} = \int \Psi^*(\tfrac{1}{2} i\hbar \gamma_{ij})\Psi\, d\tau = \int \Psi^*(i\hbar \mathfrak{T}_{ij\,\mathrm{op}})\Psi\, d\tau \tag{30.18b}$$

since $\mathscr{S}_{\lambda\mu\nu} = \mathsf{S}_{\mu\lambda\nu} - \mathsf{S}_{\lambda\mu\nu}$ (19.24b), and $\mathfrak{T}_{\mu\nu\,\mathrm{op}} = \frac{1}{2}\gamma_{\mu\nu}$ (16.2b). Thus, as previously noted, $\mathscr{S}_{\mu\nu\,\mathrm{op}} = i\hbar \mathfrak{T}_{\mu\nu\,\mathrm{op}}$, and

$$\mathscr{S}_{12\,\mathrm{op}} = \frac{\hbar}{2}\begin{Vmatrix} 1 & 0 & 0 & 0 \\ 0 & -1 & 0 & 0 \\ 0 & 0 & 1 & 0 \\ 0 & 0 & 0 & -1 \end{Vmatrix} = \frac{\hbar}{2}\begin{Vmatrix} \sigma_3 & 0 \\ 0 & \sigma_3 \end{Vmatrix} \tag{30.18c}$$

$$\mathscr{S}_{23\,\mathrm{op}} = \frac{\hbar}{2}\begin{Vmatrix} 0 & 1 & 0 & 0 \\ 1 & 0 & 0 & 0 \\ 0 & 0 & 0 & 1 \\ 0 & 0 & 1 & 0 \end{Vmatrix} = \frac{\hbar}{2}\begin{Vmatrix} \sigma_1 & 0 \\ 0 & \sigma_1 \end{Vmatrix}$$

$$\mathscr{S}_{31\,\mathrm{op}} = \frac{\hbar}{2}\begin{Vmatrix} 0 & -i & 0 & 0 \\ i & 0 & 0 & 0 \\ 0 & 0 & 0 & -i \\ 0 & 0 & i & 0 \end{Vmatrix} = \frac{\hbar}{2}\begin{Vmatrix} \sigma_2 & 0 \\ 0 & \sigma_2 \end{Vmatrix}$$

which are just the usual Pauli matrices, doubled.

As in previous examples, the current J^μ may be decomposed into a *convection* term and a *polarization* term, each of which satisfies the continuity equations, with the familiar result:

$$J^\mu_{\text{convection}} = \frac{e}{2mc}\{\Psi^+(p^\mu \Psi) - (\bar{p}^\mu \Psi^+)\Psi\} \tag{30.19}$$

$$J^\mu_{\text{polarization}} = \frac{e\hbar}{2mc}\,\partial_\nu (\Psi^+ \mathscr{S}^{\nu\mu}{}_{\mathrm{op}} \Psi)$$

To complete the identification of $\Theta_{\mu\nu}$ and J^μ, we must examine their transformation properties, and here we meet a characteristic difficulty of half-integral spin fields, namely, these quantities *do not* transform correctly under *extended* Lorentz transformations which reverse the time. For example, we find (cf. §§ 14d, e):

$$\left.\begin{aligned} &x^k \to -x^k \qquad T = -i\gamma^4 \\ &J^{k\prime} = -J^k \\ &J^{4\prime} = J^4 \end{aligned}\right\} \tag{30.20a}$$

$$\left.\begin{aligned} &x^4 \to -x^4 \qquad T = \gamma^1\gamma^2\gamma^3 \\ &J^{k\prime} = -J^k \\ &J^{4\prime} = J^4 \end{aligned}\right\} \tag{30.20b}$$

$$\left.\begin{aligned} &x^\mu \to -x^\mu \qquad T = \gamma_5 \\ &J^{\mu\prime} = J^\mu \end{aligned}\right\} \tag{30.20c}$$

Thus, J^μ behaves as an ordinary 4-vector with respect to spatial reflections; but J^4 *does not* change sign when $x^4 \to -x^4$ while the space-like components do experience a change in sign, and a reversal in sign of all coordinates leaves J^μ unchanged. The charge density in this theory is positive-definite and *cannot* change sign. Similarly, the energy-momentum tensor does change sign when all coordinates are reversed.

These are basic difficulties in the classical Dirac field, and are connected with the existence of negative-energy solutions. They can be removed only by quantization in accordance with the exclusion principle and with Heisenberg's rule for symmetrization[2] of operators in the " hole " theory.

(i) *Charge Conjugation.*—From the wave function $\Psi^{(e)}$, which satisfies

$$(\gamma^\mu D_\mu - i\kappa)\Psi = 0 \tag{30.21}$$

corresponding to algebraic charge e, we can derive another wave function $\Psi^{£} = \Psi^{(-e)}$ which satisfies the same equation, but with reversed charge sign:

$$(\gamma^\mu \bar{D}_\mu - i\kappa)\Psi^{£} = 0 \tag{30.22}$$

This so-called charge-conjugate[3] solution is seen to follow from the ordinary conjugate of (30.21), with the definition,

$$\Psi^{£} = £\overline{\Psi} \tag{30.23a}$$

where £ denotes the charge-conjugator and satisfies the relations

$$£\bar{\gamma}^\mu £^{-1} = -\gamma^\mu \tag{30.23b}$$

That a matrix £ with these properties exists, is a simple consequence of the fact that the $\bar{\gamma}^\mu$'s satisfy the same C.R. as the γ^μ's; and, apart from *equivalence*, there is but one irreducible representation of the γ-ring. For the same reasons there exist matrices θ, Λ, satisfying (cf. § 37*a*)

$$\theta\gamma^\mu\theta^{-1} = -\tilde{\gamma}^\mu \qquad \tilde{\theta} = \theta \tag{30.23c}$$
$$\Lambda\gamma^\mu\Lambda^{-1} = \gamma^{\mu *} \qquad \Lambda^* = \Lambda$$

and we may show that $£ = \Lambda^{-1}\bar{\theta}$. We may normalize £ according to $£\bar{£} = 1$, since $\Psi^{££} = \Psi$, and it is readily seen that an £ satisfying these conditions is given by

$$£ = i\gamma^1\gamma^3\gamma^4 = \begin{Vmatrix} 0 & 0 & 0 & 1 \\ 0 & 0 & -1 & 0 \\ 0 & -1 & 0 & 0 \\ 1 & 0 & 0 & 0 \end{Vmatrix} \tag{30.24}$$ [4]

[2] cf. Pauli, W. (41); also § 32.

[3] cf. Kramers (37) Belinfante (39*a*, 40*a*), Pauli (40).

[4] However, if $\Psi^{£}$ is to be a 4-spinor with respect to the same transformations as $\Psi \to \Psi' = T\Psi$, it follows that £ must further satisfy the conditions $£\overline{T} = T£$. With £ defined as above, it follows that $\Psi^{£}$ is a proper 4-spinor only with respect to the Lorentz group. In the general case it is more convenient to introduce a charge-adjoint function, with properties similar to the charge conjugate, but which transforms as a 4-spinor under the full extended group.

All the preceding results of this section may now be rewritten in full covariant 4-spinor form in terms of $\Psi_{\bullet}^{\pounds}$ and $\Psi^{\bullet}$, and we find that the theory *is not* charge-symmetric; that is, the transformation $e \to e^{\pounds} = -e$, $\Psi \to \Psi^{\pounds}$, affects the signs of the various field-density tensors—which should not happen in a proper theory. This difficulty is of the same kind as previously noted regarding the full reflection properties, and again is remedied in the proper transition to the q-number theory. The requirement of charge-symmetry or charge-invariance is intimately related to other requirements leading to the connection between spin and statistics, and we shall come back to it in § 32.

The position of fields connected with *neutral* quanta of half-integral spin is evidently rather different than in the simpler case of integral spin, as illustrated in § 26. It is, therefore, of interest to find that there is a simple formalism for the description of such quanta, involving so-called self-charge-conjugate[5] 4-spinors, $\Psi^{\pounds} = \Psi$, which, in a specially chosen representation, lead to Majorana's (37) theory of *real* Dirac wave-functions.

Thus, if the 4-spinor Ψ is composed of a covariant regular 2-spinor and its contravariant conjugate

$$\Psi = \left\| \begin{matrix} \psi_1 \\ \psi_2 \\ \psi^{\dot{1}} \\ \psi^{\dot{2}} \end{matrix} \right\| \tag{30.25}$$

it follows that $\Psi^{\pounds} = \Psi$. If now we make the change of spin frame $\Psi' = S\Psi$ with S, a unitary matrix, given by

$$S = \tfrac{1}{2}e^{-i\pi/4} \left\| \begin{matrix} 1 & -i & 1 & i \\ i & 1 & -i & 1 \\ 1 & i & -1 & i \\ -i & 1 & -i & -1 \end{matrix} \right\| \tag{30.26}$$

we find

$$\Psi' = \overline{\Psi}' = \Psi'^{\pounds} \tag{30.27a}$$

$$\pounds' = S\pounds\bar{S}^{-1} = I \tag{30.27b}$$

[6]

$$\gamma^{\mu'} = S\gamma^{\mu}S^{-1} \tag{30.27c}$$

$$\gamma^{1'} = \left\| \begin{matrix} 0 & i & 0 & 0 \\ i & 0 & 0 & 0 \\ 0 & 0 & 0 & -i \\ 0 & 0 & -i & 0 \end{matrix} \right\| \qquad \gamma^{2'} = \left\| \begin{matrix} 0 & 0 & -i & 0 \\ 0 & 0 & 0 & -i \\ -i & 0 & 0 & 0 \\ 0 & -i & 0 & 0 \end{matrix} \right\|$$

$$\gamma^{3'} = \left\| \begin{matrix} i & 0 & 0 & 0 \\ 0 & -i & 0 & 0 \\ 0 & 0 & -i & 0 \\ 0 & 0 & 0 & i \end{matrix} \right\| \qquad \gamma^{4'} = \left\| \begin{matrix} 0 & 0 & i & 0 \\ 0 & 0 & 0 & i \\ -i & 0 & 0 & 0 \\ 0 & -i & 0 & 0 \end{matrix} \right\|$$

Therefore,

$$(\gamma^{\mu}\partial_{\mu} - i\kappa)\Psi \to (\gamma^{\mu'}\partial_{\mu} - i\kappa)\Psi' = 0$$

[5] Also termed "neutrettors".

[6] Since $\Psi^{\pounds'} = \Psi'^{\pounds} = \pounds'\overline{\Psi}' = \pounds'\bar{S}\overline{\Psi} = S\Psi^{\pounds} = S\pounds\overline{\Psi}$.

where, in accord with the *reality* of Ψ', all the $\gamma^{\mu\prime}$ have pure-imaginary elements which can be cancelled out—leaving a pure-real Dirac wave-equation. The property $\Psi' = \overline{\Psi}'$ must be Lorentz invariant, and to prove this it suffices to show that the coefficients of the Lorentz transformations of Ψ' are real. This follows immediately, since we had previously (§ 16) $\mathfrak{T}_{\mu\nu\ \mathrm{op}} = \frac{1}{2}\gamma_\mu\gamma_\nu$ for the infinitesimal operators, which are now represented by

$$\mathfrak{T}'_{\mu\nu\ \mathrm{op}} = S\mathfrak{T}_{\mu\nu}S^{-1} = \tfrac{1}{2}\gamma_\mu{}'\gamma_\nu{}' \tag{30.28}$$

and, hence, pure-real.

The preceding is not quite the representation used by Majorana, which is actually realized by the transformation with $S = \dfrac{e^{i\pi/4}}{\sqrt{2}}(1 - i\pounds)$. Here again, we find $\pounds' = I$, $\Psi' = \overline{\Psi}' = \Psi'^{\pounds}$ for $\Psi = \Psi^{\pounds}$ originally, and the $\gamma^{\mu\prime}$'s are again pure-imaginary.

(*b*) *Spin* 0, 1

The meson field equations, in the so-called *particle-like* form, are given by

$$(\beta^\mu\partial_\mu - i\kappa)\Psi = 0 \tag{30.29}$$

with the structure of the β-ring defined by the Duffin (38)-Kemmer (39) commutation rules

$$\beta^\mu\beta^\nu\beta^\rho + \beta^\rho\beta^\nu\beta^\mu = \beta^\mu g^{\nu\rho} + \beta^\rho g^{\nu\mu} \tag{30.30}$$

which characterize the equations as belonging to spin 1, including spin 0.[7]

We shall leave the matrix-algebraic aspects for the considerations of Chapter V, in accordance with our general plan, and examine here the way in which the 4-spinor equation (30.29) provides a synthesis of the pseudoscalar-vector and scalar-pseudovector equations of § 26. Now, Ψ in (30.29) is essentially a 16-component quantity, in accordance with the realization of the β's

$$\beta^\mu = \tfrac{1}{2}[(\gamma^\mu \times I) + (I \times \gamma^\mu)] \tag{30.31}$$

obtained in § 16. However, as noted there, we may express (30.29) in terms of the second-rank 4-spinor Ψ^{ac} as the matrix equation

$$\tfrac{1}{2}\partial_\sigma(\gamma^\sigma\Psi^{\bullet\bullet} + \Psi^{\bullet\bullet}\tilde{\gamma}^\sigma) - i\kappa\Psi^{\bullet\bullet} = 0 \tag{30.32}$$

where

$$\Psi^{\bullet\bullet} = \|\Psi^{ac}\| = \tfrac{1}{4}[\tfrac{1}{2}iG_{\mu\nu}\gamma_5\gamma^{\mu\nu} + U_\mu\gamma_5\gamma^\mu + \check{\chi}_\mu\gamma^\mu + i\check{\varphi} + \rho\gamma_5]\varepsilon^0 \tag{30.33a}$$

according to (15.2*a*). Then,

$$\begin{aligned}\Psi^{\bullet}{}_{\bullet} &= \|\Psi^{ac}\varepsilon_{bc}\| \\ &= \tfrac{1}{4}[\tfrac{1}{2}iG_{\mu\nu}\gamma_5\gamma^{\mu\nu} + U_\mu\gamma_5\gamma^\mu + \check{\chi}_\mu\gamma^\mu + i\check{\varphi} + \rho\gamma_5]\end{aligned} \tag{30.33b}$$

and (30.32) becomes

$$\tfrac{1}{2}\partial_\sigma(\gamma^\sigma\Psi^{\bullet}{}_{\bullet} + \Psi^{\bullet}{}_{\bullet}\gamma^\sigma) - i\kappa\Psi^{\bullet}{}_{\bullet} = 0 \tag{30.34}$$

[7] According to the work of § 16, we have

$$S_{\mu\nu\ \mathrm{op}} = i\hbar\mathfrak{T}_{\mu\nu\ \mathrm{op}} = i\hbar(\beta_\mu\beta_\nu - \beta_\nu\beta_\mu)$$

for the case of a second-rank 4-spinor, and it readily follows that the spin-component eigenvalues are 0, ± 1.

With the equations (cf. §§ 14*b*, *f*)

$$\gamma^{\mu\nu} = \tfrac{1}{2}(\gamma^\mu\gamma^\nu - \gamma^\nu\gamma^\mu) \qquad \widetilde{(\gamma_5\gamma^{\mu\nu}\varepsilon^0)} = \gamma_5\gamma^{\mu\nu}\varepsilon^0 \qquad \widetilde{(\gamma_5\gamma^\mu\varepsilon^0)} = \gamma_5\gamma^\mu\varepsilon^0 \tag{30.35a}$$

$$\gamma^\mu\gamma^\nu + \gamma^\nu\gamma^\mu = 2g^{\mu\nu} \tag{30.35b}$$

$$\gamma_5\gamma_\mu + \gamma_\mu\gamma_5 = 0 \qquad (\gamma_5)^2 = 1 \tag{30.35c}$$

$$\tfrac{1}{2}(\gamma^\sigma\gamma_5\gamma^{\mu\nu} + \gamma_5\gamma^{\mu\nu}\gamma^\sigma) = \tfrac{1}{2}\gamma_5(\gamma^{\mu\nu}\gamma^\sigma - \gamma^\sigma\gamma^{\mu\nu}) = g^{\nu\sigma}\gamma^\mu - g^{\mu\sigma}\gamma^\nu \tag{30.35d}$$

(30.33*b*) and (30.34) give

$$\begin{aligned} &i\partial_\sigma G^{\mu\sigma}\gamma_5\gamma_\mu + \partial_\sigma U_\tau\gamma_5\gamma^{\tau\sigma} + \partial_\sigma\breve{\chi}_\tau g^{\tau\sigma} + i\partial_\sigma\breve{\varphi}\gamma^\sigma \\ &\quad = i\kappa(\tfrac{1}{2}iG_{\tau\sigma}\gamma_5\gamma^{\tau\sigma} + U^\mu\gamma_5\gamma_\mu + \breve{\chi}_\sigma\gamma^\sigma + i\breve{\varphi} + \rho) \end{aligned} \tag{30.36}$$

and from the linear independence of the 16 matrices constituting the γ-ring, it follows that $G^{\mu\sigma}$ and U^μ describe a particle satisfying the vector field equations (26.21); $\breve{\chi}_\tau$ and $\breve{\varphi}$ describe a particle satisfying the pseudoscalar field equations (26.14*a*); and $\rho = 0$. The explicit forms of $G_{\mu\nu}$, U_μ, $\breve{\chi}_\mu$ and $\breve{\varphi}$ in terms of $\Psi^{\cdot\cdot}$ are given by (15.5*a*, *b*, *c*, *d*), respectively—and these determine Ψ^{ab} uniquely. Further, since $\tilde{\Psi}^{\cdot\cdot}$ satisfies (30.32) when $\Psi^{\cdot\cdot}$ does, it follows that the restrictions $\Psi^{ab} = \pm\Psi^{ba}$ are consistent with (30.32). As shown in § 15*b*, the symmetric case corresponds to $\breve{\varphi} = \breve{\chi}_\tau = 0$, and the antisymmetric case to $U_\mu = G_{\mu\nu} = 0$.

(*c*) *Spin* 0, 1—*Dual Cases*

The scalar-pseudovector equations differ from the pseudoscalar-vector equations only in that U_μ, $G_{\mu\nu}$, $\breve{\varphi}$ and $\breve{\chi}_\mu$ are replaced by their duals, so that we are led naturally to consider the previously defined 4-spinor (15.6*a*)

$$\Phi^{\cdot\cdot} = \gamma_5\breve{\Psi}^{\cdot\cdot} = \tfrac{1}{4}[\tfrac{1}{2}i\breve{G}_{\mu\nu}\gamma^{\mu\nu} + \breve{U}_\mu\gamma^\mu + \chi_\mu\gamma_5\gamma^\mu + i\phi\gamma_5 + \breve{\rho}]\varepsilon^0$$

which has the same transformation properties as $\Psi^{\cdot\cdot}$. The relation between (30.32) and the scalar-pseudovector equations can be given either in terms of the spinor $\Phi^{\cdot\cdot}$ or the pseudospinor $\breve{\Psi}^{\cdot\cdot}$. The latter is identical with that given above for the spinor $\Psi^{\cdot\cdot}$ when we replaced $\Psi^{\cdot\cdot}$ by $\breve{\Psi}^{\cdot\cdot}$, and all tensors by pseudotensors and conversely. In terms of the spinor $\Phi^{\cdot\cdot}$, (30.32) becomes

$$\tfrac{1}{2}\partial_\sigma(\gamma^\sigma\gamma_5\Phi^{\cdot\cdot} + \gamma_5\Phi^{\cdot\cdot}\tilde{\gamma}^\sigma) - i\kappa\gamma_5\Phi^{\cdot\cdot} = 0 \tag{30.37}$$

and pre-multiplying by γ_5 we get

$$\tfrac{1}{2}\partial_\sigma(-\gamma^\sigma\Phi^{\cdot\cdot} + \Phi^{\cdot\cdot}\tilde{\gamma}^\sigma) - i\kappa\Phi^{\cdot\cdot} = 0 \tag{30.38}$$

or, in 16-spinor notation,

$$\breve{\beta}^\mu\partial_\mu\Phi = \tfrac{1}{2}\partial_\sigma[(-\gamma^\sigma \times I) + (I \times \gamma^\sigma)]\Phi = i\kappa\Phi \tag{30.39}$$

$$\breve{\beta}^\mu = (\gamma_5 \times I)\beta^\mu(\gamma_5 \times I) \tag{30.40}$$

Thus, if Φ^{ab} is a second-order 4-spinor satisfying (30.38), then $\mathscr{S}p(\varepsilon\Phi^{\cdot\cdot}) = 0$, and the (skew) pseudotensor $\breve{G}_{\mu\nu}$ and the pseudovector $\breve{U}_\mu$ given in terms of $\Phi^{\cdot\cdot}$ by (15.7*a*, *b*) satisfy the pseudovector field equations (26.36*a*), and the vector χ_μ and scalar ϕ defined by (15.7*c*, *d*) satisfy the scalar field equations (26.12). However, since $\tilde{\Phi}^{\cdot\cdot}$

does not satisfy (30.38) as a consequence of $\Phi^{\cdot\cdot}$ satisfying it, the restrictions to the individual pseudovector or scalar cases are no longer obtained by the requirements of symmetry or antisymmetry on Φ. The appropriate restrictions are obtained by setting $\check{\Psi}^{\cdot\cdot} = \pm\tilde{\check{\Psi}}^{\cdot\cdot}$, whence

$$\Phi^{\cdot\cdot} = \gamma_5\tilde{\Phi}^{\cdot\cdot}\tilde{\gamma}_5 \tag{30.41a}$$

and

$$\Phi^{\cdot\cdot} = -\gamma_5\tilde{\Phi}^{\cdot\cdot}\tilde{\gamma}_5 \tag{30.41b}$$

correspond to the pseudovector and scalar cases, respectively.

(*d*) *Spin* 1:—*Degenerate Case*

As remarked in § 27*b*, the spinor formalism developed for the non-zero rest-mass field for spin 1 can be applied, in a special degenerate form, to the description of the photon field. The development of § 27*c* gave such a description in 2-spinor form in terms of the 4-vector potential, and we consider now the equivalent 4-spinor description in terms of the field strengths. As is well known, this is a classic instance of *global*, but not *local*, equivalence of description, the choice of one or other of several such equivalent systems involving assumptions as to possible interactions, the character of the field densities, etc.

By contrast with the work of § 27*c*, the photon field equations in 4-spinor form do not arise from (30.29) in the limit $\kappa \to 0$, which, in view of the constitution of a second-rank 4-spinor (30.33*a*), would lead to trivial results. Instead we modify (30.29) to the form

$$(\beta^\mu\partial_\mu - i\gamma)\Psi = 0 \tag{30.42}$$

where the mass term κ is replaced by a singular idempotent matrix γ such that

$$\gamma^2 = \gamma \qquad (30.43)^8$$
$$\gamma\beta^\mu + \beta^\mu\gamma = \beta^\mu$$

Pre-multiplying (30.42) by $(1 - \gamma)$, we obtain

$$\beta^\mu\partial_\mu(\gamma\Psi) = 0 \tag{30.44a}$$

so that $(\gamma\Psi)$ satisfies (30.29) with $\kappa = 0$; and pre-multiplying (30.42) by $\partial_\kappa\beta^\kappa\beta^\lambda$ we find

$$\partial_\kappa\beta^\kappa\beta^\lambda(\gamma\Psi) = \partial^\lambda(\gamma\Psi) \tag{30.44b}$$

which, with the preceding equation, leads to the second-order wave-equation

$$\partial_\kappa\partial^\kappa(\gamma\Psi) = 0 \tag{30.45}$$

Further, if Φ is a solution of

$$\beta^\kappa\partial_\kappa\Phi = 0 \qquad \text{(cf. 30.44}a\text{)} \tag{30.46a}$$

then, also

$$\gamma\beta^\kappa\partial_\kappa\Phi = \beta^\kappa\partial_\kappa(1-\gamma)\Phi = 0 \tag{30.46b}$$

[8] As previously noted, there exists a Hermitian matrix Λ such that $\Lambda\beta^\mu\Lambda^{-1} = \beta^{\mu *}$, and similarly γ must satisfy $\Lambda\gamma\Lambda^{-1} = \gamma^*$. For the γ-ring (§ 39*c*) γ^4 takes the role of Λ so that from the mode of construction of the β's (cf. § 16) and the properties of the Kronecker product, it follows that in the present case $\Lambda = \gamma^4 \times \gamma^4$, as is readily verified.

and since $\gamma(1 - \gamma)\Phi = 0$, $(1 - \gamma)\Phi$ satisfies (30.42). Thus, $\Psi + (1 - \gamma)\Phi$ is a solution of (30.42), and the transformation

$$\Psi \to \Psi + (1 - \gamma)\Phi \tag{30.46c}$$

corresponds to a *gauge* transformation of the usual theory.

From the relation of the first-rank 16-spinor form of Ψ to the second-rank 4-spinor form, and the tensor constitution of the latter, it is then essentially obvious what form γ must take if (30.45) is to represent the second-order equation in the field strengths—with (30.46c) corresponding to a gauge transformation. The choice of γ also follows from the formation of $\Psi^{\cdot\cdot}$ in terms of 2-spinor constituents (cf equations 15.1, 16.4), and the separation of the β-space into the 10-, 5-, and 1-dimensional parts. Actually, there are two quantities $(1 - \beta)$ and β, say, which satisfy (30.43). They will, of course, be diagonal with elements 1 and 0 only; the elements of $(1 - \beta)$, for example will be

$$\text{diag}\ (1, 1, 0, 0, 1, 1, 0, 0, 0, 0, 1, 1, 0, 0, 1, 1)$$

Corresponding to the choice $\gamma = 1 - \beta$, the 4-spinor form $\hat{\Psi}^{\cdot\cdot}$ of $\gamma\Psi^{(16)}$ will be

$$\hat{\psi}^{\cdot\cdot} \sim \left\| \begin{matrix} \psi^1 & \psi^5 & 0 & 0 \\ \psi^2 & \psi^6 & 0 & 0 \\ 0 & 0 & \psi^{11} & \psi^{15} \\ 0 & 0 & \psi^{12} & \psi^{16} \end{matrix} \right\| \tag{30.47a}$$

and similarly, for $\gamma = \beta$,

$$\hat{\psi}^{\cdot\cdot} \sim \left\| \begin{matrix} 0 & 0 & \psi^9 & \psi^{13} \\ 0 & 0 & \psi^{10} & \psi^{14} \\ \psi^3 & \psi^7 & 0 & 0 \\ \psi^4 & \psi^8 & 0 & 0 \end{matrix} \right\| \tag{30.47b}$$

that is, a form involving only the antisymmetric tensor constituent $G_{\mu\nu}$ *or* the vector U_u, respectively (cf. 30.33*a*).

The procedure in equations (30.34, 36) now shows that the choice $\gamma = 1 - \beta$, with restriction to symmetry, gives the Maxwell equations

$$\partial_\sigma G^{\mu\sigma} = 0$$
$$\partial_\tau U_\sigma - \partial_\sigma U_\tau = G_{\tau\sigma}$$

while the choice $\gamma = \beta$ gives

$$\partial_\sigma G^{\mu\sigma} = U^\mu$$
$$\partial_\sigma U_\tau - \partial_\tau U_\sigma = 0$$

which are *formally* equivalent to the equations of the scalar theory, since it follows that there exists a U such that $U_\sigma = \partial_\sigma U$ and $\partial^\sigma \partial_\sigma U = \partial^\sigma U_\sigma = 0$.

As in the case $\kappa \neq 0$, of the two preceding sections, the equation (30.42) can be derived from the Lagrangian

$$\mathscr{L} = \frac{1}{2i}\{\partial_\kappa \Psi^+ \beta^\kappa \Psi - \Psi^+ \beta^\kappa \partial_\kappa \Psi\} + \Psi^+ \gamma \Psi \tag{30.48}$$

where, as in the spin $\frac{1}{2}$ case, $\Psi^{+} = \Psi^{*}\Lambda$ (cf. footnote 8). However, this Lagrangian is not gauge-invariant so that we must take the form

$$\mathscr{L} = \frac{1}{i}\{\partial_\kappa \Psi^{+}\beta^\kappa\gamma\Psi - \Psi^{+}\gamma\beta^\kappa\partial_\kappa\Psi\} + \Psi^{+}\gamma\Psi \tag{30.49}$$

which is invariant to the transformation $\Psi \to \Psi + (1 - \gamma)\Phi$. The energy-momentum tensor deriving from (30.49) is

$$\mathscr{T}_{\kappa\lambda} = \frac{1}{i}\{\partial_\kappa \Psi^{+}\beta_\lambda\gamma\Psi - \Psi^{+}\gamma\beta_\lambda\partial_\kappa\Psi\} + g_{\kappa\lambda}\Psi^{+}\gamma\Psi \tag{30.50a}$$

which can be written in the form

$$\mathscr{T}_{\kappa\lambda} = \frac{1}{2i}\{\partial_\kappa \Psi^{+}\beta_\lambda\Psi - \Psi^{+}\beta_\lambda\partial_\kappa\Psi\} \tag{30.50b}$$

$$- \frac{1}{2i}\partial^\mu\{\Psi^{+}[(g_{\kappa\lambda}\beta_\mu - g_{\kappa\mu}\beta_\lambda)\gamma - \gamma(g_{\kappa\lambda}\beta_\mu - g_{\kappa\mu}\beta_\lambda)]\Psi\}$$

of which the first term is in the usual form of the meson theory. Again this first term can be expressed as

$$\Psi^{+}(g_{\kappa\lambda} - \beta_\kappa\beta_\lambda - \beta_\lambda\beta_\kappa)\gamma\Psi - \frac{1}{2i}\partial^\mu[\Psi^{+}(\beta_\mu\beta_\kappa\beta_\lambda - \beta_\lambda\beta_\kappa\beta_\mu)\Psi]$$

and the first part may be taken to be the symmetric energy-momentum tensor $\Theta_{\kappa\lambda}$. In consequence of the factor γ, which effectively plays a role similar to κ, this tensor is gauge-invariant. The expression for the total angular-momentum-density tensor is readily obtained in the usual way and is found to be essentially the same in form as for the non-zero rest-mass case, but, as is to be expected, the spin term is *not* gauge-invariant.

It is particularly interesting that we can eliminate the longitudinal waves by imposing the condition $\partial_\mu\beta^\mu\beta^\lambda\Psi = \partial^\lambda\Psi$ on Ψ, which can be accomplished by suitable gauge transformation—and may be shown to correspond to imposing the usual Lorentz condition on U_μ. In this form of the photon field theory, the interaction with charged particles is introduced in a way no different than that for any other field (cf. Harish-Chandra, 46*b*). This brings out in striking fashion the complete particle-like form, or wave-particle correspondence, but we leave the detailed consideration of this aspect to the work of Chapter V.

§ 31. Difficulties in Connection with Electromagnetic Field—Extended Lagrangian Formalism for Spin > 1

Apart from the incidental difficulties already encountered on introducing interactions with an external electromagnetic field, there is the previously noted difficulty of *principle* in that for spin > 1 the field equations become immediately inconsistent on introducing $\partial_\mu \to D_\mu$; and, as is to be expected concomitantly, no longer follow from the simple Lagrangian formalism—without supplementary conditions.[1]

[1] A parallel, and as yet insufficiently investigated, difficulty appears in the quantized theory in the non-commutation of charge densities at points separated by space-like intervals. Cf. Fierz (39*a*).

To see this we may consider the Dirac equations for spin $\frac{3}{2}$, viz. (cf. § 28*b*)

$$\left.\begin{aligned} \kappa\phi_L^{\dot{A}\dot{B}} &= \partial^{\dot{A}K}\psi^{\dot{B}}_{KL} = \partial^{\dot{B}K}\psi^{\dot{A}}{}_{KL} \\ -\kappa\psi^{\dot{B}}{}_{KL} &= \partial_{\dot{A}K}\phi_L^{\dot{A}\dot{B}} = \partial_{\dot{A}L}\phi_K^{\dot{A}\dot{B}} \end{aligned}\right\} \tag{31.1}$$

Dirac (36) proposed to introduce the external field through the form:

$$\left.\begin{aligned} D^{\dot{A}K}\psi^{\dot{B}}_{KL} &= \kappa\phi_L^{\dot{A}\dot{B}} \\ D_{\dot{A}K}\phi_L^{\dot{A}\dot{B}} &= -\kappa\psi^{\dot{B}}_{KL} \end{aligned}\right\} \tag{31.2}$$

where $\partial_{\dot{A}B} \to D_{\dot{A}B}$ corresponds to $\partial_\mu \to D_\mu$. However, from the work of §§ 13*a*, *b* it follows that

$$D_{\dot{A}B}D^{\dot{C}D} - D^{\dot{C}D}D_{\dot{A}B} = 4i\epsilon(F_B{}^D\delta_{\dot{A}}{}^{\dot{C}} + F_{\dot{A}}{}^{\dot{C}}\delta_B{}^D) \tag{31.3a}$$

and

$$D_{\dot{A}B}D^{\dot{A}D} = -D^2\delta_B{}^D + 4i\epsilon F_B{}^D \tag{31.3b}$$

so that with equations (31.2) we find:

$$D_{\dot{A}B}D^{\dot{A}L}\psi^{\dot{D}}_{KL} = \kappa D_{\dot{A}B}\phi_K^{\dot{A}\dot{D}} = -\kappa^2\psi^{\dot{D}}_{BK} \tag{31.4}$$

and

$$-D^2\psi^{\dot{D}}_{KB} + 4i\epsilon F_B{}^L\psi^{\dot{D}}_{KL} = -\kappa^2\psi^{\dot{D}}_{KB} \tag{31.5}$$

Since the right-hand side of equation (31.5) is symmetric in K, B it follows that ψ must satisfy the supplementary or subsidiary condition

$$F_B{}^L\psi^{\dot{D}}_{KL} = F_K{}^L\psi^{\dot{D}}_{BL} \tag{31.6a}$$

or

$$F^{AB}\psi^{\dot{D}}_{AB} = 0 \tag{31.6b}$$

which cannot, in general, be satisfied simultaneously with the other equations.

As remarked by Pauli and Fierz (39*a*, *b*), we might try to circumvent this difficulty by tentatively replacing equations (31.2) by weaker conditions, e.g.

$$\begin{aligned} D^{\dot{A}K}\psi^{\dot{B}}_{KL} + D^{\dot{B}K}\psi^{\dot{A}}_{KL} &= 2\kappa\phi_L^{\dot{A}\dot{B}} \\ D_{\dot{A}K}\phi_L^{\dot{A}\dot{B}} + D_{\dot{A}L}\phi_K^{\dot{A}\dot{B}} &= -2\kappa\psi^{\dot{B}}_{KL} \end{aligned} \tag{31.7}$$

but these do not give a second-order wave equation even in the limit of the force-free case. These can, in fact, be shown to describe particles of spin $\frac{3}{2}$, rest-mass κ *and* particles of spin $\frac{1}{2}$, rest-mass 2κ, and it is found that the total charge is no longer positive definite which prevents quantization consistent with the exclusion principle. (Cf. § 32.)

If, therefore, we are to retain equations (31.1) for the force-free case, the problem is seemingly to guarantee the self-consistency of the system under the substitution $\partial_{\dot{A}B} \to D_{\dot{A}B}$ by introducing suitable terms dependent on the field strengths. Pauli and Fierz approach the problem by noting the difficulty arising from the appearance of lower spin, higher rest-mass fields (cf. above), and so propose the artifice of introducing auxiliary spinors (or tensors) of all ranks lower than the original and deriving all equations from a generalized variation principle. By suitably choosing the parameters in the Lagrangian, it can then be arranged that the derived field equations are such that the auxiliary fields vanish in the absence of an external field, and the usual Lorentz-like (divergence) and symmetry conditions obtain automatically.

In any given case we must make some *Ansatz* for the Lagrangian and the procedure is so unwieldy in practice that it is doubtful if it can play a fundamental role, but it does serve to show that we cannot rule out higher-spin fields *just* on the grounds of the requirement of a Lagrangian origin. Moreover, it is certainly of interest to find that for spin 2 for which we make the *Ansatz*

$$\begin{aligned}\mathscr{L} = -\kappa^2 U_{\alpha\beta} U^{\alpha\beta} + U_{\alpha\beta,\gamma} U^{\alpha\beta,\gamma} - 2U_{\alpha\beta}{}^{\cdot\alpha} U^{\beta\gamma}{}_{,\gamma} \\ - \tfrac{3}{4}\kappa^2 C^2 - \tfrac{3}{8} C_{,\alpha} C^{\cdot\alpha} + U_{\alpha\beta}{}^{\cdot\beta} C^{\cdot\alpha}\end{aligned} \tag{31.8}$$

the equations obtaining in the limit $\kappa \to 0$ are exactly those of Einstein's *weak-field* gravitational approximation.[2]

§ 32. Spin and Statistics

(a) *Classification of Field Theories under the Proper Lorentz Group*

The field theories which we have considered thus far have had the common characteristic that the field variables or wave functions formed the components of a vector, of one of the *finite*-dimensional vector spaces $\mathfrak{R}(k, l)$, transforming according to the irreducible representation $\mathfrak{D}(k, l)$ of the proper Lorentz group, and we have noted the correlation between *indefinite* charge (energy) density and integral (half-integral) spin. It is therefore of interest that Pauli (40) has affirmed this correlation, assuming *only* that the field equations are linear and homogeneous in the field variables, and further, on general physical grounds, obtained a connection between spin and statistics.[1]

The field quantities $\psi(k, l)$ may be grouped into two natural classes, viz.

"I class"—$(k + l)$ integral $\to$ integral spins
"H class"—$(k + l)$ half-integral $\to$ half-integral spins

since, according to the rules for reduction of a product into irreducible constituents under the Lorentz group (cf. § 17), the product of two quantities of the I class or two quantities of the H class contains only quantities of the I class, whereas the product of a quantity of the I class with a quantity of the H class contains only quantities of the H class.

By the same argument, these are further separated into the respective subclasses

(1)	"$+1$ class"	k integral	l integral
(2)	"-1 class"	k half-integral	l half-integral
(3)	"$+\epsilon$ class"	k integral	l half-integral
(4)	"$-\epsilon$ class"	k half-integral	l integral

[2] Cf. Pauli-Fierz, *loc. cit.*; Chang (44).

[1] Although this beautiful result may be "the" connection, it is not possible to be so definite at the present stage of the theory of elementary particles. The introduction of either *infinite*-dimensional representations or an indefinite metric in the Hilbert vector space of quantized systems, however improbable on physical grounds, destroys the connection. On the other hand, if we are willing to make certain reasonable assumptions as to the permissible *form* of the field equations, it is also possible to prove that only fields corresponding to spin 0, 1, or $\frac{1}{2}$ are permissible (cf. § 35*c*).

which have the multiplication table:

	$+1$	-1	$+\epsilon$	$-\epsilon$
$+1$	$+1$	-1	$+\epsilon$	$-\epsilon$
-1	-1	$+1$	$-\epsilon$	$+\epsilon$
ϵ	ϵ	$-\epsilon$	$+1$	-1
$-\epsilon$	$-\epsilon$	$+\epsilon$	-1	$+1$

It is important to note that the complex conjugates $\bar{\psi}^{\pm}$, for which k and l are interchanged, belong to the same classes as $\psi^{\pm}$ under " I ", whereas under " H ", $\psi^{+\epsilon}$ and $\bar{\psi}^{-\epsilon}$ belong to the $+\epsilon$ class, while $\psi^{-\epsilon}$ and $\bar{\psi}^{+\epsilon}$ belong to the $-\epsilon$ class; hence, in the latter case the complex conjugate quantities must be considered explicitly.

(i) *Indefinite Character of the Charge-Current Density for Integral Spin Fields.*—The above classifications show that, solely on account of the requirement of invariance under the *proper* Lorentz group, the field equations must be of the prototype form

$$\left.\begin{aligned} k\psi^+ &= \psi^- \\ k\psi^- &= \psi^+ \end{aligned}\right\} \qquad (32.1)^2$$

which equations are to be understood as including *sums* of as many different terms of the same type as there are ψ^+'s and ψ^-'s (representing possibly many different particles—but all with integral spin).

By the same reasoning on the classes, it follows that the field tensors of even rank, such as the energy-momentum density tensor $\mathscr{T}$, which are quadratic or bilinear in the ψ's and thus constructed solely of quantities with even k and even l, will have the typical form (in the same sense as equations 32.1)

$$\mathscr{T} \sim \psi^+\psi^+ + \psi^-\psi^- + \psi^+k\psi^- \qquad (32.2)$$

Similarly, the odd-rank tensors, such as the charge-current density vector J, which are composed of quantities with half-integral k and half-integral l, will be of the typical form

$$J \sim \psi^+k\psi^+ + \psi^-k\psi^- + \psi^- \qquad (32.3)$$

Now we note that equations (32.1) are invariant under the substitution

$$\begin{aligned} &k_\mu \to -k_\mu && \\ &\psi^+ \to \psi^+ && \bar{\psi}^+ \to \bar{\psi}^+ \\ &\psi^- \to -\psi^- && \bar{\psi}^- \to -\bar{\psi}^- \end{aligned} \qquad (32.4)$$

but, correspondingly.

$$\begin{aligned} \mathscr{T} &\to +\mathscr{T} \\ J &\to -J \end{aligned} \qquad (32.5)$$

[2] We assume a plane wave so that $p_\mu \to k_\mu$, apart from a constant, and the presence of the propagation vector k_μ is indicated symbolically by k. It is clear that, since we do not assume that the equations are of the first order, a possible *even* number of factors k may occur on either side. These are not indicated, since under the above classifications their presence is clearly immaterial to the argument.

from which the indefinite character of the charge-current density and the total charge follows, since to every solution of the field equations there belongs another solution for which the components of J^{μ} reverse their sign. This indicates the impossibility of defining, in a physically satisfactory way, a definite *particle* density for integ al spin which transforms as the 4-component of a vector.

(ii) *Indefinite Character of the Energy-Density for Half-Integral Spin Fields.*—The case of half-integral spin fields is only slightly more involved, and corresponding to equations (32.1, 2, 3) we find:

$$k\psi^{+\epsilon} + k\bar{\psi}^{-\epsilon} = \psi^{-\epsilon} + \bar{\psi}^{+\epsilon} \qquad (32.6)$$
$$k\psi^{-\epsilon} + k\bar{\psi}^{+\epsilon} = \psi^{+\epsilon} + \bar{\psi}^{-\epsilon}$$

$$\begin{aligned}\mathcal{T} \sim \psi^{+\epsilon}\psi^{+\epsilon} + \bar{\psi}^{-\epsilon}\bar{\psi}^{-\epsilon} + \psi^{-\epsilon}\psi^{-\epsilon} + \bar{\psi}^{+\epsilon}\bar{\psi}^{+\epsilon} + \psi^{+\epsilon}\bar{\psi}^{-\epsilon} \\ + \psi^{-\epsilon}\bar{\psi}^{+\epsilon} + \psi^{+\epsilon}k\psi^{-\epsilon} + \bar{\psi}^{-\epsilon}k\psi^{-\epsilon} + \bar{\psi}^{+\epsilon}k\psi^{+\epsilon} + \bar{\psi}^{-\epsilon}k\bar{\psi}^{+\epsilon}\end{aligned} \qquad (32.7)$$

$$\begin{aligned}J \sim \psi^{+\epsilon}\psi^{-\epsilon} + \psi^{+\epsilon}\bar{\psi}^{+\epsilon} + \psi^{-\epsilon}\bar{\psi}^{-\epsilon} + \bar{\psi}^{-\epsilon}\bar{\psi}^{+\epsilon} + \psi^{+\epsilon}k\psi^{+\epsilon} \\ + \psi^{+\epsilon}k\bar{\psi}^{-\epsilon} + \psi^{-\epsilon}k\psi^{-\epsilon} + \psi^{-\epsilon}k\bar{\psi}^{+\epsilon} + \bar{\psi}^{-\epsilon}k\bar{\psi}^{-\epsilon} + \bar{\psi}^{+\epsilon}k\bar{\psi}^{+\epsilon}\end{aligned} \qquad (32.8)$$

Now, instead of (32.4), we consider the substitution

$$\begin{aligned} &k_{\mu} \to -k_{\mu} && \\ &\psi^{+\epsilon} \to i\psi^{+\epsilon} && \bar{\psi}^{-\epsilon} \to i\bar{\psi}^{-\epsilon} \\ &\psi^{-\epsilon} \to -i\psi^{-\epsilon} && \bar{\psi}^{+\epsilon} \to -i\bar{\psi}^{+\epsilon} \end{aligned} \qquad (32.9)$$

which is in accord with the algebraic requirement of passing to the complex conjugate, the requirement that quantities of the same class transform in the same way, and does not interfere with possible reality conditions of the type $\psi^{+\epsilon} = \bar{\psi}^{-\epsilon}$ or $\psi^{-\epsilon} = \bar{\psi}^{+\epsilon}$. The equations (32.6) are invariant to this substitution, but now we find the reverse effect to (32.5) on the field tensors:

$$\mathcal{T} \to -\mathcal{T}$$
$$J \to +J$$

from which the indefinite character of the energy-density and total energy follows, since, under the substitution (32.9), the energy density at every space-time point changes its sign.

These considerations do not prove that for integral spins there always exists a definite energy density and for half-integral spins a definite charge density, which is, indeed, not the case for spin > 1, as shown by Fierz (39); but there does exist a definite *total* energy for integral spins and a definite *total* charge for half-integral spins, in the c-number theory. It is rather interesting that Pauli's analysis does not require invariance with respect to space reflections, and thus includes the two-component spinor fields.

(b) *The Need for Second Quantization*

Apart from the previously mentioned limitations, this analysis indicates that a satisfactory interpretation within the limits of the one-particle problem is not possible, in accord with the fact that all relativistically invariant theories lead to annihilation and creation of particles, in pairs of opposite charge for charge-bearing particles, and singly for neutral particles.[3]

The process of quantization consists essentially of promoting the ψ's (and canonical conjugates) in the theory to the rank of operators, in the usual way, and the following argument devolves on only the most elementary bases of this method. The direct transcription of the canonical formalism, with its unnecessarily sharp distinction of the role of the time, is not applicable when there are supplementary conditions on the theory—as for spin > 1—so that we use the formalism of Jordan-Pauli (28).[4] The only new construct arising here is the generalized δ function, or invariant D function, which satisfies the second-order wave-equation and is given by

$$D(\mathbf{x}, x_4) = \frac{1}{V} \Sigma \exp [i(\mathbf{k} \cdot \mathbf{x})] \frac{\sin k_4 x_4}{k_4} \tag{32.10a}$$

in a periodicity volume V; or,

$$D(\mathbf{x}, x_4) = \frac{1}{(2\pi)^3} \int \frac{d^3k}{k_4} \exp [i(\mathbf{k} \cdot \mathbf{x})] \sin k_4 x_4 \qquad (32.10b)^5$$

$$k_4 = +\sqrt{(\mathbf{k}^2 + \kappa^2)}$$

in the limit $V \to \infty$. Further, the D function (which is even in the space coordinates and odd in the time coordinate) is *uniquely* determined by the conditions:

$$\left. \begin{aligned} (\Box + \kappa^2) D &= 0 \\ D(\mathbf{x}, 0) &= 0 \\ \left(\frac{\partial D}{\partial x_4}\right)_{x_4=0} &= \delta(\mathbf{x}) \end{aligned} \right\} \tag{32.11}$$

from which the general representation for D, obtained by Dirac (32), is

$$D(\mathbf{x}, x_4) = -\frac{1}{4\pi r} \frac{\partial}{\partial r} F(r, x_4)$$

$$F(r, x_4) = \left\{ \begin{array}{ll} J_0[\kappa(x_4{}^2 - r^2)^{1/2}] & (x_4 > r) \\ 0 & (r > x_4 > -r) \\ -J_0[\kappa(x_4{}^2 - r^2)^{1/2}] & (-r > x_4) \end{array} \right\} \tag{32.12}$$

that is, D vanishes in the region exterior, but not in the interior, to the light-cone, and the transition from $+$ to $-$ of the function F on the light-cone corresponds to the δ singularity of D,

$$[\delta(r - x_4) - \delta(r + x_4)]/4\pi r \tag{32.13}$$

[3] These general conclusions are in accord with elementary considerations of pair production and the uncertainty principle, $\Delta p \Delta x \geqslant h$, for a single particle. For $\Delta x \sim h/2mc$, $\Delta p \sim 2mc$, so that any attempt to locate a particle with an accuracy of the order h/mc generates an unknown number of pairs.

[4] The generalization of this method leads to the Dirac (32) *many-time* formalism which is the starting point of the more recent developments mentioned in the preface.

[5] The factor d^3k/k_4 is determined by its invariance on the hyperboloid of the four-dimensional momentum space $(\mathbf{k}, k_4)$. This may be seen by considering the invariant surface, defined by $k_4{}^2 = k^2 + \kappa^2$, in the $\mathbf{k}$-space, and noting that the surface element dS, which is given by $dS = \frac{\kappa}{k_4} d^3k$, has a Lorentz invariant meaning.

Apart from D, there is just one more such function, namely,

$$D_1(\mathbf{x}, x_4) = \frac{1}{(2\pi)^3}\int \frac{d^3k}{k_4} \exp\,[i(\mathbf{k}\cdot\mathbf{x})]\cos k_4 x_4 \tag{32.14}$$

which is given by

$$D_1(\mathbf{x}, x_4) = \frac{1}{4\pi r}\frac{\partial}{\partial r}F_1(r, x_4)$$

$$F_1(r, x_4) = \left\{\begin{array}{ll} N_0[\kappa(x_4^2 - r^2)^{1/2}] & (x_4 > r) \\ -iH_0^{(1)}[i\kappa(r^2 - x_4^2)^{1/2}] & (r > x_4 > -r) \\ N_0[\kappa(x_4^2 - r^2)^{1/2}] & (-r > x_4) \end{array}\right\} \tag{32.15}$$

Here the Bessel function is replaced by the Neumann and first Hankel cylinder function.[6]

One or other of these functions (and a finite number of derivatives thereof) will thus appear in the operator bracket expressions of all quantities which satisfy the second-order wave-equations, and the D_1 function is ruled out by the physical requirement that all measurements at space-time points separated by a space-like distance are compatible, i.e. the corresponding physical quantities commute. The conclusion on the use of the D function must be valid for both commutators and anticommutators, since otherwise it would follow that gauge-invariant quantities, such as the charge-density, constructed bilinearly from the field variables, would be non-commutable at points with space-like separation.

From the requirement of (proper) Lorentz invariance and the results of the preceding section on the classification of field quantities, we can infer directly the general form in which the *derivatives* of the D function may appear in the bracket expressions.[7] We assume the field variables ψ^r to be ordered in such a way that each field component is composed only of quantities of the same class, and consider first the bracket expression of a field component ψ^r with its own complex conjugate

$$[\psi^r(\mathbf{x}', x_4'),\ \bar{\psi}^r(\mathbf{x}'', x_4'')] \tag{32.16}$$

It is now necessary to distinguish between the two spin types, since for integral (half-integral) spin this expression transforms under Lorentz transformations as a tensor of even (odd) rank. Hence, for integral spin we have

$$[\psi^r(\mathbf{x}', x_4'),\ \bar{\psi}^r(\mathbf{x}'', x_4'')] = \text{even number of derivatives of } D(\mathbf{x}' - \mathbf{x}'', x_4' - x_4'') \tag{32.17a}$$

and for half-integral spin

$$[\psi^r(\mathbf{x}', x_4'),\ \bar{\psi}^r(\mathbf{x}'', x_4'')] = \text{odd number of derivatives of } D(\mathbf{x}' - \mathbf{x}'', x_4' - x_4'') \tag{32.17b}$$

both of which must again be understood to allow for a *sum* of expressions of the type indicated.

[6] The strongest singularity of D_1 on the surface of the light-cone is determined by $\frac{1}{2\pi^2}(r^2 - x_4^2)^{-1}$.

[7] N.B. The term " bracket expressions " *here* includes both types.

Now consider the expression

$$\mathfrak{X} \equiv [\psi^r(\boldsymbol{x}', x_4'), \bar{\psi}^r(\boldsymbol{x}'', x_4'')] + [\psi^r(\boldsymbol{x}'', x_4''), \bar{\psi}^r(\boldsymbol{x}', x_4')] \tag{32.18}$$

which is symmetrical in the two space-time points. Since the D function is even in the space coordinates and odd in the time coordinate, it follows from the symmetry of $\mathfrak{X}$ that $\mathfrak{X}$ equals an *even* number of space-like derivatives times an *odd* number of time-like derivatives of $D(\boldsymbol{x}' - \boldsymbol{x}'', x_4' - x_4'')$. This is consistent with (32.17*b*) for half-integral spins, but *contradicts* (32.17*a*) for integral spin—unless $\mathfrak{X}$ vanishes. Therefore, for integral spin,

$$\mathfrak{X} = 0 \tag{32.19}$$

The *bracket expression* thus far has not distinguished between the two cases of Bose-Einstein and Fermi-Dirac statistics, which correspond respectively to commutator and anticommutator. It is clear that with the latter (+ brackets) equation (32.19) leads to an algebraic contradiction, since the left-hand side is essentially positive for $x' = x''$ and cannot vanish unless both ψ^r and $\bar{\psi}^r$ vanish. This leads to the conclusions:

(*a*) *For integral spin, quantization according to the Fermi-Dirac statistics is not possible*;

(*b*) *For half-integral spin, quantization according to Bose-Einstein and Fermi-Dirac statistics is formally possible*;

but, according to our previous discussions, the total energy would not be positive in the former case, and since this is required on physical grounds it follows that we must quantize in accord with the Fermi-Dirac statistic (exclusion principle) and the interpretation of the Dirac " hole " theory.

Thus, a connection between spin and statistics is established by the two physical postulates:

(I) *Observables at points connected by a space-like vector are commutative*;

(II) *The total energy of the system is positive.*[8]

Finally, as implied in § 30*a* (i), it may be shown—in the quantized field theory—that the *different* requirement of invariance under time inversion *and* particle-antiparticle inversion (charge conjugation in the special case of quantum electrodynamics) also implies the above spin-statistic connection. Conversely, it may be said that all current orthodox quantized field theories appear to be invariant under these *two* requirements so that a theory which is not separately invariant under particle-antiparticle inversion will *necessarily* not be invariant under time inversion. The important system of the Maxwell-Dirac fields in interaction, which we speak of as quantum electrodynamics, is invariant under these requirements jointly *and* separately.

[8] Alternatively, that there is no infinite number of states of negative energy.

CHAPTER V

Relativistic Wave-Equations—Matrix-Algebraic Aspect

§ 33. The Wave Equations for Particles of Arbitrary Spin—Fundamentals

(*a*) *Introduction*

The problem of relativistic wave-equations for particles of arbitrary spin has been studied from two different standpoints, quite apart from the separation into field-particle aspects which we have seen, in several examples, to be essentially only a difference in formalism.

The investigations of Majorana (32), Dirac (36), Proca (36), Kemmer (38), Fierz (39), and Bhabha (45, 49), for example, represent, in various mathematical guise, the attempt to discover relativistically invariant linear equations whose solutions constitute a relativistically invariant manifold which describes the possible states of the system. This approach has doubtless been fruitful, *but* many of the proposed systems of equations have been subsequently found to be different mathematical formulations with the same *essential* content. Further, and more to the point, the few generally accepted physical conditions which arise from the necessity of a suitable interpretation do not suffice to limit the number of *apparently* allowed equations.[1]

The alternative approach, developed largely by Wigner (39, 47, 48),[2] is an invariant-theoretic method which attacks the broader problem of defining all relativistically invariant manifolds. The solution of this problem disposes of the question of *equivalent* systems of equations, so that a simplest prototype set can be *chosen* (*a posteriori*) to characterize each manifold, and at the same stroke uniquely delimit all possible relativistic equations. We shall consider this method further in § 40. Here we only outline briefly the underlying concepts—which are indeed simple—and the relation of this viewpoint to the preceding one.

Perhaps the most fundamental principle of quantum mechanics is that the wave functions, ψ, describing the possible states of a system form a linear manifold on which a positive definite, unitary inner product (ψ, ϕ) is defined for any two wave functions ψ, ϕ, i.e. a Hilbert space. The inner product usually involves an integration over the whole configuration- or momentum-space and, for particles of higher spin, a summation over spin indices.

The only indeterminate in the wave function is the unimodular phase factor, and the linear character of the manifold is equivalent to the statement of the super-

[1] We must, of course, accept the possibility that this situation is intrinsic to the interaction-free theory *per se*.

[2] Bhabha (45*b*, 47, 49), Lubanski (41*a*, *b*, 42*b*, *c*), and others have applied similar methods to the less general (and physical) problem of the homogeneous group. Cf. §§ 34, 35, 36, 38.

position principle. The squared modulus $|(\psi, \phi)|^2$ determines the transition probability from the state ψ on the state ϕ, or conversely, and is an invariant. By contrast, the description of states by wave functions is not, since the same state will be described by different wave-functions in different coordinate systems. If, for definiteness, we consider wave functions referring to a free particle, which satisfy a relativistic wave-equation, then there exists a correspondence between the wave functions ψ_l, $\psi_{l'}$ describing the same state in different Lorentz frames l, l':

$$\psi_{l'} = D(l', l)\psi_l \tag{33.1}$$

where $D(l', l)$ may be readily shown to be a linear unitary operator, determined up to a unimodular factor by the physical content of the theory. Apart from this constant, it is clear that the absence of real difference between frames of reference l, l', l'' requires that the operations $D(l', l)$ and $D(l'', l')$ be essentially identical, if l'' arises from l' by the same Lorentz transformation by which l' arises from l, so that in every Lorentz invariant linear theory the unitary operator $D(L) = D(l', l)$ is completely determined by the Lorentz transformation L carrying l over to $l' = Ll$. By successive transformations,

$$\begin{aligned} \psi_{l''} &= D(l'', l)\psi_l \\ &= D(l'', l')\, D(l', l)\psi_l \end{aligned} \tag{33.2}$$

it readily follows that

$$D(l'', l')\, D(l', l) = \omega D(l'', l) \tag{33.3a}$$

or

$$D(L_2)D(L_1) = \omega D(L_2 L_1) \tag{33.3b}$$

where ω is a unimodular factor, which may depend on L_2 and L_1. Thus the $D(L)$ form, up to a factor which can be shown to be normalizable to ± 1, a representation of the *inhomogeneous* Lorentz group by linear unitary operators (i.e. the operators $D(L)$ form a single- or double-valued representation). It is important to note that we consider here the group of *all* inhomogeneous Lorentz transformations, including the translations of the origin in space and time, in contradistinction to the homogeneous proper Lorentz group of all our previous work. It is well known that none of the finite representations of the latter, such as we have considered, is unitary (cf. § 17*b*).

From the linear manifold of states $\psi_1, \psi_2, \ldots$ we obtain the states $L\psi_1, L\psi_2, \ldots$ which arise from the original by a Lorentz transformation and, in the new coordinate system, have the same properties as the $\psi_1, \psi_2, \ldots$ in the original system. The existence of such states is precisely the content of relativistic invariance. Since all Lorentz frames are equivalent for the description of the system, it follows that together with ψ, $D(L)\psi$ is also a possible state—*viewed from the original Lorentz frame*, so that the (invariant) manifold contains with every ψ, all transforms $D(L)\psi$, where L is any Lorentz transformation. The $L\psi_i$ can be expressed in terms of the ψ_i, the expansion coefficients defining elements of a (unitary) representation of the group, in general of indefinite dimensionality, corresponding to the infinite number of possible wave functions.

The usual form of current theory is divided into the two distinct sections:

(i) the kinematic relations between various quantities at the same instant of time, such as the commutation rules, and

(ii) the specification of the causal relations between quantities at different times, such as the wave-equation. We see, therefore, that, since the inhomogeneous group includes " translations in time ", the representation can replace the latter *equations of motion* by giving the development of properties of the system in time—insofar as the initial properties are known.[3] It cannot replace the kinematic connections.

The intrinsic power of this invariant-theoretic approach is that *no* assumptions regarding the *nature* of the underlying equations are necessary, so that more general equations—such as quantized forms, insofar as they exist—are automatically included. The answer obtained in this methodology is as direct and unique as the problem posed, but by contrast it is stripped to the primitive essentials. No assertions regarding possible forms of interaction can be made without specific reference to the form and character of a particularly selected wave function or system (cf. § 40).

(*b*) *The Mathematical Conditions*

Apart from the assumptions on the continuity of space-time implicit in the use of Lorentz frames in the classical sense, it is all but trivial to state the requirement of relativistic invariance as the basic mathematical condition. *But* this is a condition without precise meaning unless the group involved is explicitly delimited. It is just at this point that we are forced to define the term *elementary particle*; or, more precisely, what entity we choose to consider an elementary particle. From the purely mathematical point of view it appears inevitable that this entity correspond to the simplest primitive form. The states ψ, which may be expressed as functions of the momenta p_μ and spin variables ξ, say $\psi(p, \xi)$, belong to an irreducible representation of the group, i.e. the manifold which is invariant and irreducible under all transformations of the group.

If we particularize the term " group " as the *inhomogeneous* Lorentz group, we have one definition; if as the *homogeneous* Lorentz group, quite another. Of course, all the equations for higher spin considered in the last chapter also define representations of the inhomogeneous Lorentz group, and their solutions the corresponding invariant manifolds, but it appears that, in general, these are not *irreducible* representations and manifolds of the inhomogeneous group. Though irreducible under the former group, they break up into inequivalent irreducible representations of the latter.[4]

On purely physical grounds, too, it appears more reasonable to define an elementary entity under this wider group, in which case it can be shown (cf. §§ 35*c*) that, for non-zero rest-mass, the only wave equations satisfying this and other simple physical requirements are those corresponding to spin 0, 1, or $\frac{1}{2}$. However, in the present state of our theoretical and, more particularly, experimental knowledge,

[3] If, for example, we knew the operator corresponding to the measurement of a physical quantity at the time $t = 0$, we could follow its development throughout time, since, in order to obtain the value at the time $t = t_1$, we need only transform the original wave-function ψ_l by $D(l', l)$ to a coordinate system whose time scale begins a time t_1 later.

[4] Cf. Bhabha (49); Bargmann (47, 48); Wigner (39, 47); also § 40.

there are insufficient grounds for excluding theories which couple several irreducible manifolds belonging to the wider group.

We shall, therefore, examine first the underlying principles of the theories which require only irreducibility under the homogeneous group *and* are cast in matrix-algebraic form. These considerations will be nearer the first of the above-mentioned approaches, and will serve *inter alia* to introduce some of the algebraic principles and methods which are of practical import.

We take as the basic equation the linear form

$$(\alpha^{\kappa}\partial_{\kappa} - i\kappa)\psi = 0 \tag{33.4}$$

where the matrices are required to be an irreducible representation of the corresponding symbols which generate an abstract algebra or ring (cf. §§ 37, 38) $\mathfrak{A}$.

(c) *The Physical Conditions*

The further physical assumptions concerning the particle devolve on whether or not the following conditions are satisfied, viz.

(i) the second-order equation $(\partial_{\kappa}\partial^{\kappa} + \kappa^2)\psi = 0$,

(ii) the total charge or total energy of the associated field is positive definite.

As previously discussed, the first condition ensures that the particle has a fixed rest-mass, while the second is required if the field is to be quantized. We shall classify theories for which (i) holds as type I and in the contrary case as type II, and investigate the general restrictions imposed on the α-matrices by the above conditions.

§ 34. Wave Equations of Type I

(a) *General Commutation Rules for the α-Matrices—The Minimal Equation*

The condition of Lorentz invariance requires

$$\begin{aligned} x^{\kappa} &\rightarrow L^{\kappa}{}_{\lambda}x^{\lambda} \\ \psi &\rightarrow T(L)\psi \\ T(L)\alpha^{\kappa}T^{-1}(L) &= L^{\kappa}{}_{\lambda}\alpha^{\lambda} \end{aligned} \tag{34.1}$$

so that the quantity $P = \alpha^{\lambda}k_{\lambda}$ (that is, with respect to plane-wave solutions) undergoes the transformation

$$P \rightarrow P' = \alpha^{\lambda}k_{\lambda}' = TPT^{-1} \tag{34.2}$$

As a matrix of finite degree, P satisfies a minimal equation[5] of finite degree whose coefficients must be polynomials in the k_{μ}, since P itself is a linear function of them. Moreover, (34.2) shows that the minimal equation of P' is the same as that of P, or that the minimal equation is Lorentz invariant and can therefore only contain the k_{μ} in an invariant form. The only such invariant combination is $k^2 = k_{\mu}k^{\mu}$, and since the minimal equation must be homogeneous in the k_{μ} it follows that it must contain only even or only odd powers of P. Every such equation can be written in the factorized form

$$(P^2 - a_1{}^2k^2)(P^2 - a_2{}^2k^2)\ldots = 0 \tag{34.3a}$$

[5] The algebraic equation of lowest degree satisfied by a matrix.

if its degree d is even, or P times this if its degree d is odd. Further, we can express this in the form

$$k_\kappa k_\lambda k_\mu k_\nu \ldots (\alpha^\kappa \alpha^\lambda - a_1{}^2 g^{\kappa\lambda})(\alpha^\mu \alpha^\nu - a_2{}^2 g^{\mu\nu}) \ldots = 0 \tag{34.3b}$$

and, since this is a polynomial of degree d which holds for arbitrary values of the numbers k_1, k_2, k_3, k_4, the coefficient of every term must vanish.

It follows that every set of α-matrices must satisfy an algebraic equation of degree d of the type (*the general commutation relations*)

$$\sum_{\mathscr{P}} \{(\alpha^\kappa \alpha^\lambda - a_1{}^2 g^{\kappa\lambda})(\alpha^\mu \alpha^\nu - a_2{}^2 g^{\mu\nu}) \ldots\} = 0 \tag{34.4}$$

if d is even, or α^η times (34.4) if d is odd, and $\sum_{\mathscr{P}}$ denotes a summation over similar terms with all possible permutations of the indices κ, λ, μ, ν. In particular, the minimal equation of any α^κ is obtained by putting all indices equal to κ, viz.

$$\{(\alpha^\kappa)^2 - a_1{}^2 g^{\kappa\kappa}\}\{(\alpha^\kappa)^2 - a_2{}^2 g^{\kappa\kappa}\} \ldots = 0 \qquad (34.5)^6$$

if d is even, or α^κ times this if d is odd.

For every solution of the equation $(\alpha^\kappa \partial_\kappa - i\kappa)\psi = 0$ we have

$$\begin{aligned} (P - \kappa)\psi &= 0 \\ P^n \psi &= (\kappa)^n \psi \end{aligned} \tag{34.6a}$$

and operating on ψ with (34.3a),

$$(\kappa^2 - a_1{}^2 k^2)(\kappa^2 - a_2{}^2 k^2) \ldots \tag{34.6b}$$

again with a factor κ depending on whether d is even or odd. Thus, the wave equation has plane-wave solutions of the type assumed only if k^2 has one of the values $\kappa^2/a_1{}^2$, $\kappa^2/a_2{}^2$, . . . where $\pm a_1$, $\pm a_2$, . . . are the non-zero eigenvalues of α^4. The rest-mass of a particle is defined as the positive root of k^2, and the values can only be real if all the eigenvalues of α_4 are real. Correspondingly, the equation (34.3a), as an identity in the k_μ, holds as an operator form in $-i\partial_\mu$ and then shows that every component of ψ satisfies a differential equation of order higher than the second—unless α_4 has *one and only one* pair of non-zero eigenvalues. By suitable normalization of κ and ψ, the non-zero value may be taken as $a_1 = 1$, so that the minimal equation—as the condition for a unique rest-mass—becomes

$$(\alpha_4)^r\{(\alpha_4)^2 - 1\}^s = 0 \qquad r > 0,\ s \geqslant 1 \qquad (34.7a)^7$$

As shown by Harish-Chandra (47a, c) it is actually sufficient that the minimal equation of α_4 be of the simpler form

$$(\alpha_4)^{n-2}\{(\alpha_4)^2 - 1\} = 0 \qquad n \geqslant 2 \tag{34.7b}$$

and we note, in passing, that the Dirac and Duffin-Kemmer matrices satisfy the simplest minimal equations of degree 2 and 3, respectively.

[6] For a more rigorous and detailed analysis, see, for example, Harish-Chandra (47a, b, c).

[7] For zero rest-mass it is readily shown that the corresponding wave equation is the same with the mass term promoted to a singular matrix, as in the example of § 30d.

(b) The Λ-Matrix

It appears to be essential for the construction of an energy-momentum tensor and current vector that the α-representation be equivalent to its Hermitian conjugate or adjoint; that is, there should exist a non-singular matrix Λ, which is often termed the *metric form*, such that

$$\alpha_\kappa{}^* = \Lambda \alpha_\kappa \Lambda^{-1} \tag{34.8a}$$ [8]

If such a matrix exists, it can always be chosen Hermitian

$$\Lambda = \Lambda^* \tag{34.8b}$$

since from (34.8*a*) it follows that $\Lambda^{*-1}\Lambda$ commutes with the α_κ which form an irreducible set, and is therefore a multiple of the unit matrix.

The representation $L \to T(L)$ of the homogeneous Lorentz group is uniquely determined by the condition

$$L^\kappa{}_\lambda \alpha^\lambda = T(L)\alpha^\kappa T^{-1}(L) \tag{34.9}$$

for if $L \to T'(L)$ is another representation satisfying (34.9), then $T^{-1}(L)T'(L)$ commutes with α^κ and therefore $T'(L) = c(L)T(L)$, where $c(L)$ is a number. Clearly $L \to c(L)$ is a one-dimensional representation of the Lorentz group; but the only possible one-dimensional representation is $L \to \pm 1$, where the $+$ sign is valid at least for all the proper transformations, so that in this case $T(L) = T'(L)$.

Now from (34.8, 9) and the reality of $L^\kappa{}_\lambda$, it follows that

$$L^\kappa{}_\lambda \alpha^\lambda = \Lambda^{-1}T^*(L)^{-1}\Lambda\alpha^\kappa\Lambda^{-1}T^*(L)\Lambda \tag{34.10}$$

and

$$L \to \Lambda^{-1}T^*(L)^{-1}\Lambda$$

is also a representation of the Lorentz group. Hence,

$$T(L) = \pm\Lambda^{-1}T^*(L)^{-1}\Lambda$$

or

$$T^*(L)\Lambda T(L) = \pm\Lambda \tag{34.11}$$

It is of basic importance that the choice of the $+$ sign is essential if the mathematical conditions resulting from the physical condition (ii) of § 33*c* are not to be violated (see next section).

(c) The Matrix Conditions Corresponding to the Positive Definiteness of the Total Charge or the Total Energy

The conditions (34.8) enable us to derive the wave equation by variation of the Lagrangian

$$\mathscr{L} = \frac{1}{2i}(\partial_\kappa\psi^+\alpha^\kappa\psi - \psi^+\alpha^\kappa\partial_\kappa\psi) + \kappa\psi^+\psi \tag{34.12}$$

[8] Without such a Λ it is not possible to construct the usual *c*-number positive definite unitary scalar product in momentum space; cf. § 40.

where $\psi^+ = \psi^* \Lambda$, and the energy-momentum tensor and current vector derived from (34.12) are

$$\mathscr{T}_{\kappa\lambda} = \tfrac{1}{2} i(\partial_\kappa \psi^+ \alpha_\lambda \psi - \psi^+ \alpha_\lambda \partial_\kappa \psi) \qquad (34.13a)$$

$$J_\kappa = e\psi^+ \alpha_\kappa \psi \qquad (34.13b)$$

We wish to show that the physical condition (ii) is equivalent to the condition that either $\Lambda\alpha_4^{m+1}$ (for positive definite total charge) or $\Lambda\alpha_4^m$ (for positive definite total energy), where m is the greatest even integer $\leq n - 1$ (cf. 34.7b), be a non-negative (eigenvalues $\geqslant 0$) matrix.

For this purpose decompose the field into plane waves, so that with (34.13) the total charge and total energy of the field are sums of terms each of which corresponds to a Fourier wave in the decomposition. Since it is assumed that $\kappa \neq 0$, every wave can be brought to rest by a suitable Lorentz transformation without affecting the value of the total charge or the sign of the total energy, and as the charge- and energy-densities for a particle at rest are independent of the spatial coordinates it becomes necessary and sufficient for our purpose to ensure that one or other of these densities is positive definite in the rest system.

The wave function of such a particle is given by

$$\psi = \phi_+(k)e^{i\kappa x_4} + \phi_-(k)e^{-i\kappa x_4} \qquad (34.14)$$

where, of course, ϕ_+ and ϕ_- are independent of x_μ. If we put

$$\phi = \phi_+ + \phi_- \qquad (34.15a)$$

$$\left.\begin{aligned} \psi_+ &= \phi_+ e^{i\kappa x_4} \\ \psi_- &= \phi_- e^{-i\kappa x_4} \end{aligned}\right\} \qquad (34.15b)$$

then from the wave equation

$$(1 \mp \alpha_4)\psi_\pm = 0 \qquad (34.16a)$$

$$(1 - \alpha_4^2)\psi = 0 \qquad (34.16b)$$

$$\psi_\pm = \tfrac{1}{2}(1 \pm \alpha_4)\psi \qquad (34.16c)$$

and the charge-density is given by

$$J_4 = e\psi^* \Lambda \alpha_4 \psi = e\phi^* \Lambda \alpha_4 \phi = e\phi^* \Lambda \alpha_4^{m+1} \phi \qquad (34.17)$$

where m is the greatest even integer $\leq n - 1$ (cf. the minimal equation 34.7b). Conversely, let ϕ be any arbitrary one-column matrix independent of x_μ. Then, since $m = n - 1$ or $n - 2$, it follows from (34.7b) that

$$(1 - \alpha_4^2)\alpha_4^m = 0 \qquad (34.18)$$

and therefore

$$\psi = \tfrac{1}{2}(1 + \alpha_4)\alpha_4^m \phi e^{i\kappa x_4} + \tfrac{1}{2}(1 - \alpha_4)\alpha_4^m \phi e^{-i\kappa x_4} \qquad (34.19)$$

is a solution of the wave equation (33.4) and represents a possible state of the particle at rest. Again we find that the charge density corresponding to (34.19) is (note that m is even and $\alpha_4^{m+2} = \alpha_4^m$)

$$J_4 = e\phi^* \Lambda \alpha_4^{m+1} \phi \qquad (34.20)$$

From (34.17) and (34.20) it is evident that the total charge is positive definite if

$$\phi^* \Lambda \alpha_4^{m+1} \phi > 0 \tag{34.21a}$$

for every ϕ such that

$$\alpha_4^m \phi \neq 0 \tag{34.21b}$$

This condition can be replaced by

$$\Lambda \alpha_4^{m+1} \phi \neq 0 \tag{34.21c}$$

since from $\Lambda \alpha_4^{m+1} \phi = 0$, it follows that

$$\alpha_4 \Lambda^{-1} \Lambda \alpha_4^{m+1} \phi = \alpha_4^{m+2} \phi = \alpha_4^m \phi = 0$$

and conversely. Now $\Lambda \alpha_4^{m+1}$ is Hermitian (cf. 34.18) and can therefore be transformed to diagonal form by a unitary transformation, so that from (34.21*a*, *c*) all the eigenvalues of $\Lambda \alpha_4^{m+1}$ must be $\geqslant 0$. Conversely, if $\Lambda \alpha_4^{m+1}$ is non-negative the conditions (34.21) obtain.

Similarly, we find that the energy-density corresponding to (34.14) takes the form

$$\mathscr{T}_{44} = \kappa \phi^* \Lambda \phi = \kappa \phi^* \Lambda \alpha_4^m \phi \tag{34.22}$$

and the total energy is positive definite if, and only if, $\Lambda \alpha_4^m$ is a non-negative matrix. If Λ does not satisfy the stated conditions, $T^*(L) \Lambda T(L) = +\Lambda$, it can be shown, by consideration of the reflection operator, that neither $\Lambda \alpha_4^m$ nor $\Lambda \alpha_4^{m+1}$ can be non-negative.[9]

(*d*) *Quantization*

We must now verify that the fields discussed can be quantized. Consider the commutation rules

$$[\psi_r^+(x'), \psi_s(x)]_\pm = \tfrac{1}{2} \left\{ \left(\frac{\alpha^\kappa \partial_\kappa}{i\kappa} \right)^m \left(1 + \frac{\alpha^\lambda \partial_\lambda}{i\kappa} \right) D(x - x') \right\}_{sr} \tag{34.23}$$

where $D(x - x')$ is the Jordan-Pauli relativistic delta-function of § 32*b*, and the $+$ or $-$ bracket is to be taken according as $\Lambda \alpha_4^{m+1}$ or $\Lambda \alpha_4^m$ is non-negative; that is, according as the total charge or total energy is positive definite.[10] These commutation rules are obviously relativistically invariant, and their consistency with the equation of motion (33.4) is assured by (34.4), so that

$$(i\alpha^\kappa \partial_\kappa + \kappa) \left(\frac{\alpha^{\kappa'} \partial_{\kappa'}}{i\kappa} \right)^m \left(1 + \frac{\alpha^\lambda \partial_\lambda}{i\kappa} \right) D = \frac{1}{\kappa} \left(\frac{\alpha^\kappa \partial_\kappa}{i\kappa} \right)^m (\kappa^2 + \partial_\lambda \partial^\lambda) D = 0$$

If now we transform (34.23) to the momentum space according to

$$\psi = \frac{1}{\sqrt{V}} \sum_{\boldsymbol{k}} a_{\boldsymbol{k}} e^{i(\boldsymbol{k} \cdot \boldsymbol{x})} \tag{34.24}$$

[9] Cf. Harish-Chandra (47*c*).

[10] Note that this is in accord with the formalism of § 30, where we considered the Dirac field, for which $n = 1$, $m = 0$.

and recall that (cf. § 32*b*)

$$D(x) = \frac{1}{V}\sum_{\boldsymbol{k}} \frac{\sin k_4 x_4}{k_4} e^{i(\boldsymbol{k}\cdot\boldsymbol{x})} \qquad k_4 = +\sqrt{(\boldsymbol{k}^2 + \kappa^2)}$$

it follows, on placing $\boldsymbol{k} = 0$ and $a_0 = a$, that

$$[(a_{\boldsymbol{k}}^* \Lambda \alpha_4)_r, a_s]_\pm = \begin{cases} (\alpha_4{}^m)_{sr} & \boldsymbol{k} = 0 \\ 0 & \boldsymbol{k} \neq 0 \end{cases} \tag{34.25}$$

Corresponding to (34.16*b*), we have

$$(1 - \alpha_4{}^2)a = 0 \tag{34.26}$$

and with (34.25)

$$[a_r^*, a_s]_\pm = (\alpha_4{}^{m-1}\Lambda^{-1})_{sr} \tag{34.27}$$

The minimal equations of $\alpha_4{}^m$ and $\alpha_4{}^{m+1}$ have no repeated factors, since

$$\left.\begin{aligned} \alpha_4{}^m(\alpha_4{}^m - 1) &= 0 \\ \alpha_4{}^{m+1}(\alpha_4{}^{2m+2} - 1) &= 0 \end{aligned}\right\} \tag{34.28}$$

and it is therefore possible to choose a representation in which $\alpha_4{}^m$ and $\alpha_4{}^{m+1}$ are diagonal and Hermitian. Therefore, from (34.8), Λ commutes with $\alpha_4{}^m$ and $\alpha_4{}^{m+1}$ in this representation and can be diagonalized without affecting $\alpha_4{}^m$ and $\alpha_4{}^{m+1}$. Now, for the case of positive definite total charge, $\Lambda\alpha_4{}^{m+1}$ is non-negative as is also $\alpha_4{}^{m+1}\Lambda^{-1} = \Lambda\alpha_4{}^{m+1}\Lambda^{-2}$. The diagonal elements $(\alpha_4{}^{m+1}\Lambda^{-1})_{ss} = 0$ correspond to $(\alpha_4{}^m)_{ss} = 0$, since $\alpha_4\alpha_4{}^{m+1} = \alpha_4{}^m$. Hence, for $s = r$ the right-hand side of (34.27) is >0 unless $(\alpha_4{}^m)_{ss} = 0$, in which case $a_s = 0$ by (34.26). The commutation rules (34.27) are therefore consistent in the case of Fermi-Dirac statistics, and the validity of (34.23) follows by Lorentz transformation. As we have previously found (§ 32*b*), the case corresponding to Bose-Einstein statistics does not require detailed investigation.

The conditions imposed on the α's are quite restrictive, so that if we could show that only the Dirac (spin $\frac{1}{2}$) and Duffin-Kemmer (spin 0, 1) cases satisfy them, this would mean that the existence of higher-spin elementary particles is inconsistent with the basic assumptions of this theory. As Bhabha (49) and Harish-Chandra (47*c*) have shown by construction of counter examples, this conclusion, which would appear so desirable on grounds of simplicity, is not so easily obtained. However, both of these examples implicitly contain *subsidiary conditions* (cf. § 35*c*) deriving from (33.4), in consequence of the properties assigned the α-matrices; and if such subsidiary conditions are ruled out, it can be shown that (for non-zero rest-mass) only particles of spin 0, 1, or $\frac{1}{2}$ are allowed. To give a general proof we shall consider the analysis of the structure of equation (33.4) on the lines given by Bhabha (45*a*, *b*, 46) and Wild (47). The argument is essentially a different, and more restrictive, mathematical form of the preceding, *but* with the important difference that subsidiary conditions are excluded.

§ 35. Analysis of the Structure of the First-order Wave-Equations

(a) *The Relation of the Nucleus of the Lorentz Group to the Algebra of the α-Matrices*

From the conditions of relativistic invariance (34.1), considered for the infinitesimal transformations, we obtain

$$[\alpha^\kappa, \mathfrak{T}^{\lambda\mu}]_- = g^{\kappa\lambda}\alpha^\mu - g^{\kappa\mu}\alpha^\lambda \tag{35.1}$$

Together with the fundamental commutation rules of the group nucleus

$$[\mathfrak{T}^{\kappa\lambda}, \mathfrak{T}^{\mu\nu}]_- \equiv g^{\kappa\nu}\mathfrak{T}^{\lambda\mu} - g^{\kappa\mu}\mathfrak{T}^{\lambda\nu} + g^{\lambda\mu}\mathfrak{T}^{\kappa\nu} - g^{\lambda\nu}\mathfrak{T}^{\kappa\mu} \tag{35.2}$$

this resolves the problem of finding all equations invariant under the proper Lorentz group into the problem of finding all $\mathfrak{T}^{\kappa\lambda}$, that is, representations, and α^κ satisfying equations (35.1, 2). This again is the invariant-theoretic aspect previously discussed.

It is readily verified that the solution of (35.1) reduces to the solution of the system:

$$[\alpha^4, \mathfrak{T}^{kl}]_- = 0 \tag{35.3a}$$
$$\big[[\alpha^4, \mathfrak{T}^{41}], \mathfrak{T}^{41}\big]_- = \alpha^4$$

the remaining operators being defined by

$$\alpha^k = [\alpha^4, \mathfrak{T}^{4k}]_- \tag{35.3b}$$

Although thus far we have required that the α's form an irreducible set, the matrices $\mathfrak{T}^{\mu\nu}$ will in general be a reducible representation of the nucleus of the proper Lorentz group, and may always be written in the reduced form:

$$\mathfrak{T}^{\mu\nu}(k_1 l_1) \oplus \mathfrak{T}^{\mu\nu}(k_2, l_2) \oplus \dots \tag{35.4a}$$

or

$$\mathfrak{T}^{\mu\nu} = \left\| \begin{array}{c|c|c} \mathfrak{T}^{\mu\nu}(k_1, l_1) & & \\ \hline & \mathfrak{T}^{\mu\nu}(k_2, l_2) & \\ \hline & & \ddots \end{array} \right\| \tag{35.4b}$$

where our notation indicates that we are restricting our considerations to the finite representations (§ 17). Further, since the wave equation and the $\mathfrak{T}^{\mu\nu}(k, l)$ are required to be invariant under reflections, there must appear in the decomposition the associated $\mathfrak{T}^{\mu\nu}(l, k)$ if $k \neq l$ (cf. § 17). Corresponding to this decomposition, the wave function also reduces to a direct sum of irreducible component functions:

$$\psi = \psi(k_1, l_1) \oplus \psi(k_2, l_2) \oplus \dots \tag{35.4c}$$

The wave equation $(\alpha^\kappa \partial_\kappa - i\kappa)\psi = 0$ then states that the α's connect the various irreducible parts of ψ, but it is immediately evident that, since p_μ transforms according to $\mathfrak{D}(\frac{1}{2}, \frac{1}{2})$, an irreducible component of ψ belonging to $\mathfrak{D}(k, l)$ can, at most, be coupled to components belonging to $\mathfrak{D}(k + \frac{1}{2}, l + \frac{1}{2})$, $\mathfrak{D}(k + \frac{1}{2}, l - \frac{1}{2})$, $\mathfrak{D}(k - \frac{1}{2}, l + \frac{1}{2})$ and $\mathfrak{D}(k - \frac{1}{2}, l - \frac{1}{2})$—corresponding to the reduction of $\mathfrak{D}(\frac{1}{2}, \frac{1}{2}) \times \mathfrak{D}(k, l)$.

All the irreducible representations are now known, so that it is a comparatively straightforward calculation to determine the explicit form of possible α-matrices. For this purpose we shall again find the spinor formalism necessary, and we adopt the relations given by (17.14), viz.

$$\left.\begin{aligned} \mathfrak{K}_1 &= \tfrac{1}{2}(i\mathfrak{T}^{23} + \mathfrak{T}^{41}) \\ \mathfrak{L}_1 &= \tfrac{1}{2}(i\mathfrak{T}^{23} - \mathfrak{T}^{41}) \end{aligned}\right\} \tag{35.5}$$

(cyclically), and relate $\mathfrak{T}^{\mu\nu}$ to two symmetric spinors $\mathfrak{K}_{AB}$, $\mathfrak{L}_{\dot{A}\dot{B}}$, according to (cf. 13.20, 21)

$$2\mathfrak{T}_{\mu\nu} = S_{\mu\nu AB}\mathfrak{K}^{AB} + \bar{S}_{\mu\nu\dot{A}\dot{B}}\mathfrak{L}^{\dot{A}\dot{B}} \tag{35.6a}$$

$$\mathfrak{K}_{AB} = \tfrac{1}{4}\mathfrak{T}^{\mu\nu}S_{\mu\nu AB}$$

$$\mathfrak{L}_{\dot{A}\dot{B}} = \tfrac{1}{4}\mathfrak{T}^{\mu}\bar{S}_{\mu\nu\dot{A}\dot{B}}$$

or

$$\mathfrak{K}_1{}^1 = -\mathfrak{K}_2{}^2 = \mathfrak{K}_3 \qquad \mathfrak{K}_1{}^2 = \mathfrak{K}^- \qquad \mathfrak{K}_2{}^1 = \mathfrak{K}^+ \tag{35.6b}$$

$$\mathfrak{L}_{\dot{1}}{}^{\dot{1}} = -\mathfrak{L}_{\dot{2}}{}^{\dot{2}} = -\mathfrak{L}_3 \qquad \mathfrak{L}_{\dot{1}}{}^{\dot{2}} = -\mathfrak{L}^+ \qquad \mathfrak{L}_{\dot{2}}{}^{\dot{1}} = -\mathfrak{L}^-$$

Further, we require the spinor-matrices u^A, v^A for which the equations (17.36) again obtain here:

$$u^B(k + \tfrac{1}{2})v_B(k + \tfrac{1}{2}) = v^B(k)u_B(k) = (2k + 1)(-1)^{2k+1} \tag{35.7a}$$ [1]

$$u^B(k + \tfrac{1}{2})u_B(k) = v^B(k)v_B(k + \tfrac{1}{2}) = 0 \tag{35.7b}$$

$$v_B(k + \tfrac{1}{2})u^C(k + \tfrac{1}{2}) = \{\mathfrak{K}_B{}^C(k) + (k + 1)\,\delta_B{}^C\}(-)^{2k+1} \tag{35.7c}$$

$$u_B(k)v^C(k) = \{k\,\delta_B{}^C - \mathfrak{K}_B{}^C(k)\}(-)^{2k+1} \tag{35.7d}$$

and from which we may readily verify the useful relations

$$u_A(k)\mathfrak{K}_{BC}(k - \tfrac{1}{2}) - \mathfrak{K}_{BC}(k)u_A(k) = \tfrac{1}{2}\epsilon_{BA}u_C(k) + \tfrac{1}{2}\epsilon_{CA}u_B(k) \tag{35.8a}$$

$$v_A(k)\mathfrak{K}_{BC}(k) - \mathfrak{K}_{BC}(k - \tfrac{1}{2})v_A(k) = \tfrac{1}{2}\epsilon_{BA}v_C(k) + \tfrac{1}{2}\epsilon_{CA}v_B(k) \tag{35.8b}$$

When the representations of $\mathfrak{K}^{BC}(k)$ and $\mathfrak{K}^{BC}(k - \frac{1}{2})$ are given, these equations also serve in place of equations (35.7) to determine the u's and v's, up to a multiplicative factor.

Now, $\alpha^\mu = \frac{1}{2}g^{\mu AB}\alpha_{AB}$ and conversely $\alpha^{AB} = g^{\mu AB}\alpha_\mu$, and we require the commutator of α_{AB} with $\mathfrak{K}_{CD}$ and $\mathfrak{L}_{\dot{C}\dot{D}}$. These are readily obtained, since with (35.6a)

$$\begin{aligned} [\alpha_{\dot{A}B}, \mathfrak{K}_{CD}]_- &= \tfrac{1}{4}[g_{\kappa\dot{A}B}\alpha^\kappa, \mathfrak{T}^{\lambda\mu}S_{\lambda\mu CD}]_- \\ &= \tfrac{1}{4}[\alpha^\kappa, \mathfrak{T}^{\lambda\mu}]_- g_{\kappa\dot{A}B}S_{\lambda\mu CD} \end{aligned}$$

Then, noting (35.1) and the relation

$$g^\mu{}_{\dot{A}B}S_{\mu\nu CD} = -\epsilon_{BC}g_{\nu\dot{A}D} - \epsilon_{BD}g_{\nu\dot{A}C}$$

[1] The corresponding relations in the dotted spinors with k, $\mathfrak{K}_B{}^C$ replaced by l, $\mathfrak{L}_{\dot{B}}{}^{\dot{C}}$, etc., have precisely the same form but for a change of sign in each equation.

which follows from (13.4), we find

$$[\alpha_{\dot{A}B}, \mathfrak{K}_{CD}]_- = \tfrac{1}{2}(\epsilon_{CB}\alpha_{\dot{A}D} + \epsilon_{DB}\alpha_{\dot{A}C}) \tag{35.9a}$$

and similarly

$$[\alpha_{\dot{A}B}, \mathfrak{L}_{\dot{C}\dot{D}}]_- = \tfrac{1}{2}(\epsilon_{\dot{C}\dot{A}}\alpha_{\dot{D}B} + \epsilon_{\dot{D}\dot{A}}\alpha_{\dot{C}B}) \tag{35.9b}$$

In the reduction indicated by equations (35.4) the matrices $\mathfrak{K}^{AB}(k)$, $\mathfrak{L}^{\dot{C}\dot{D}}(l)$ also split up into submatrices $\mathfrak{K}^{AB}(k_s)$, $\mathfrak{L}^{\dot{C}\dot{D}}(l_s)$ corresponding to the component $\mathfrak{T}^{\kappa\lambda}(k_s, l_s)$, and similarly the submatrices of the $\alpha^{\dot{A}B}$ take the form $(k_r, l_r \mid \alpha^{\dot{A}B} \mid k_s, l_s)$. Corresponding to this reduction, equations (35.9) read as

$$\begin{aligned}(k_r, l_r \mid \alpha_{\dot{A}B} \mid k_s, l_s)\,\mathfrak{K}_{CD}(k_s) &- \mathfrak{K}_{CD}(k_r)(k_r, l_r \mid \alpha_{\dot{A}B} \mid k_s, l_s)\\ &= \tfrac{1}{2}\epsilon_{CB}(k_r, l_r \mid \alpha_{\dot{A}D} \mid k_s, l_s) + \tfrac{1}{2}\epsilon_{DB}(k_r, l_r \mid \alpha_{\dot{A}C} \mid k_s, l_s)\end{aligned} \tag{35.10a}$$

$$\begin{aligned}(k_r, l_r \mid \alpha_{\dot{A}B} \mid k_s, l_s)\mathfrak{L}_{\dot{C}\dot{D}}(l_s) &- \mathfrak{L}_{\dot{C}\dot{D}}(l_r)(k_r, l_r \mid \alpha_{\dot{A}B} \mid k_s, l_s)\\ &= \tfrac{1}{2}\epsilon_{\dot{C}\dot{A}}(k_r, l_r \mid \alpha_{\dot{D}B} \mid k_s, l_s) + \tfrac{1}{2}\epsilon_{\dot{D}\dot{A}}(k_r, l_r \mid \alpha_{\dot{C}B} \mid k_s, l_s)\end{aligned} \tag{35.10b}$$

Since the indices $\dot{A}$, B of $\alpha_{\dot{A}B}$ are unaffected in (35.10*a*) and (35.10*b*) respectively, we may infer that the general solution of these equations is of the form

$$(k_r, l_r \mid \alpha_{\dot{A}B} \mid k_s, l_s) = (k_r \mid \alpha_B \mid k_s) \times (l_r \mid \alpha_{\dot{A}} \mid l_s) \tag{35.11}$$

which is the *direct* product of the two rectangular matrices with $2k_r + 1$ rows and $2k_s + 1$ columns, and $2l_r + 1$ rows and $2l_s + 1$ columns. Then (35.10*a*), for example, reduces to

$$\begin{aligned}(k_r \mid \alpha_B \mid k_s)\mathfrak{K}_{CD}(k_s) &- \mathfrak{K}_{CD}(k_r)(k_r \mid \alpha_B \mid k_s)\\ &= \tfrac{1}{2}\epsilon_{CB}(k_r \mid \alpha_D \mid k_s) + \tfrac{1}{2}\epsilon_{DB}(k_r \mid \alpha_C \mid k_s)\end{aligned} \tag{35.12}$$

which is essentially the form (35.8) if $k_s = k_r \pm \tfrac{1}{2}$, and it remains to prove that $(k_r \mid \alpha_B \mid k_s) = 0$ otherwise (by a previous remark this is, of course, the case).

Recalling the matrix forms of $\mathfrak{K}_3$, $\mathfrak{K}^+$, $\mathfrak{K}^-$ (cf. 17.15) and the relations (35.6*b*), equation (35.12) for $B = C = 1$, $D = 2$, reads

$$(k_r \mid \alpha_1 \mid k_s)\mathfrak{K}_3(k_s) - \mathfrak{K}_3(k_r)(k_r \mid \alpha_1 \mid k_s) = \tfrac{1}{2}(k_r \mid \alpha_1 \mid k_s) \tag{35.13a}$$

or

$$(k_r, m_r \mid \alpha_1 \mid k_s, m_s)\{m_s - m_r - \tfrac{1}{2}\} = 0 \tag{35.13b}$$

Because m_r, m_s have the usual range of values with respect to k_r, k_s, $\{m_s - m_r - \tfrac{1}{2}\}$ can only vanish if k_r and k_s differ by a half-integer, and the only non-zero elements are then $(k_r, m_r \mid \alpha_1 \mid k_s, m_r + \tfrac{1}{2})$. If $k_s \neq k_r + \tfrac{1}{2}n$, the corresponding submatrix of α_1 is zero.

Now let $k_s = k_r + \tfrac{1}{2}n$, n being an odd integer, and set $B = C = D = 1$ in (35.12). We obtain

$$\begin{aligned}(k_r \mid \alpha_1 \mid k_s)\mathfrak{K}^-(k_s) &- \mathfrak{K}^-(k_r)(k_r \mid \alpha_1 \mid k_s) = 0\\ &= (k_r, m_r \mid \alpha_1 \mid k_s, m_r + \tfrac{1}{2})[(k_s + m_r + \tfrac{3}{2})(k_s - m_r - \tfrac{1}{2})]^{1/2}\\ &\quad - [(k_r + m_r + 1)(k_r - m_r)]^{1/2}(k_r, m_r + 1 \mid \alpha_1 \mid k_s, m_r + \tfrac{3}{2})\end{aligned} \tag{35.14a}$$

or

$$\begin{aligned}(k, m \mid \alpha_1 \mid k + \tfrac{1}{2}n, m + \tfrac{1}{2})&[(k + \tfrac{1}{2}n + m + \tfrac{3}{2})(k + \tfrac{1}{2}n - m - \tfrac{1}{2})]^{1/2}\\ &= [(k + m + 1)(k - m)^{1/2}(k, m + 1 \mid \alpha_1 \mid k + \tfrac{1}{2}n, m + \tfrac{3}{2})\end{aligned} \tag{35.14b}$$

where we have set $k_s = k_r + \frac{1}{2}n$ and omitted the indices. Consideration of the consistency requirement on this equation for $m = k$ gives the condition

$$(k + n + k + \tfrac{3}{2})(k + \tfrac{1}{2}n - k - \tfrac{1}{2}) = 0$$

which requires $n = 1$. For this case (35.12) reduces to (35.8*b*), so that $(k \mid \alpha_1 \mid k + \frac{1}{2})$ is, up to a factor, the matrix $v_1(k + \frac{1}{2})$, and from (35.14*b*) the explicit expression for the matrix elements of $(k \mid \alpha_1 \mid k + \frac{1}{2})$, or v_1, is

$$(k, m \mid \alpha_1 \mid k + \tfrac{1}{2}, m + \tfrac{1}{2}) = c[k + m + 1]^{1/2} \tag{35.15a}$$

where c is an arbitrary constant (however, cf. § 35*c*). Similarly, with $B = 1$, $C = D = 2$ we find

$$(k, m \mid \alpha_2 \mid k + \tfrac{1}{2}, m - \tfrac{1}{2}) = c[k - m + 1]^{1/2} \tag{35.15b}$$

A similar argument for n negative shows that it must be -1, equation (35.12) takes the form of (35.8*a*), and

$$(k \mid \alpha_A \mid k - \tfrac{1}{2}) = du_A(k) \tag{35.15c}$$

Thus the spinor components of α_μ must be of the form (omitting the indices r, s which indicate the representations coupled)

$$\left.\begin{aligned} (k, l \mid \alpha_{AB} \mid k + \tfrac{1}{2}, l - \tfrac{1}{2}) &= cv_B(k + \tfrac{1}{2})u_A(l) \\ (k + \tfrac{1}{2}, l - \tfrac{1}{2} \mid \alpha_{AB} \mid k, l) &= du_B(k + \tfrac{1}{2})v_A(l) \end{aligned}\right\} \tag{35.16a}$$

$$\left.\begin{aligned} (k, l \mid \alpha_{AB} \mid k + \tfrac{1}{2}, l + \tfrac{1}{2}) &= cv_B(k + \tfrac{1}{2})v_A(l) \\ (k + \tfrac{1}{2}, l + \tfrac{1}{2} \mid \alpha_{AB} \mid k, l) &= du_B(k + \tfrac{1}{2})u_A(l + \tfrac{1}{2}) \end{aligned}\right\} \tag{35.16b}$$

which must obviously be interpreted in the sense of *direct* product, and the two sets of matrix elements correspond to coupling of irreducible components of the types $(k, l) \rightleftharpoons (k + \frac{1}{2}, l - \frac{1}{2})$ and $(k, l) \rightleftharpoons (k + \frac{1}{2}, l + \frac{1}{2})$, respectively.

. . . .

It is easy to show that, by suitable transformation, every irreducible representation of the α's can be brought to a form such that the space-like α^k are skew-Hermitian and the time-like α^4 is Hermitian: $\alpha^{k*} = -\alpha^k$, $\alpha^{4*} = \alpha^4$. Indeed, in a suitably chosen representation, the Λ-matrix (§ 34*a*) is simply the total space-reflection operator R and, in fact, usually appears in this form in physical applications ($\Lambda_\gamma = \gamma^4$, $\Lambda_\beta = 1 - 2(\beta^4)^2$, for electron, meson equations, respectively).[2]

While considering here the problem of the *general* form of the α-matrices, we wish to mention, in passing, one *special* case which will be of interest for our later work, though we leave the detailed discussion to § 36 *et seq.* We note that *if* it happens that there exists a connection between the α^κ and the infinitesimal Lorentz gene-

[2] From this and the connection with the 5-dimensional Lorentz group (cf. § 36*a*), it follows that Λ is a numerical multiple of the transformation which reverses the directions of the 4,5-axes, that is, the rotation of angle π in the 4,5-plane. Hence $\Lambda \sim \exp i\pi\alpha_4$, the various polynomial forms of Λ following by virtue of the characteristic equation satisfied by α_4 for the various systems of linear wave-equations.

rators $\mathfrak{T}^{\kappa\lambda}$ such that, apart from an inessential numerical factor, $\mathfrak{T}^{\kappa\lambda} = [\alpha^\kappa, \alpha^\lambda]$, then with equations (35.1–3) we obtain the *special* cyclic relations:

$$[\alpha^4, \alpha^k] = \mathfrak{T}^{4k}$$
$$[\alpha^k, \mathfrak{T}^{4k}] = \alpha^4$$
$$[\mathfrak{T}^{4k}, \alpha^4] = -\alpha^k$$

that is, essentially the commutation rules for any three components of angular-momentum. If, then, we consider α^4 (or any one of the α^κ) in diagonal form, it has eigenvalues, say, $n, n-1, \ldots, -n$ of respective multiplicities $\mu_n, \mu_{n-1}, \ldots \mu_{-n}$ $(\mu_n = \mu_{-n})$,[3] and we may assume a representation so chosen that all equal eigenvalues occur together in sequence; namely, α_4 has the form

$$\| \alpha_4 \| = \mathrm{diag}\, \| n, n-1, \ldots, -n \|$$

where the multiplying unit matrices of corresponding dimension are, of course, tacitly understood. It follows from the above *special* cyclic commutators that in an irreducible representation the other two members in the cycle, α^k and $\mathfrak{T}^{4k}$, have *at most* non-vanishing matrix elements labelled (for convenience at the moment), though incompletely, by eigenvalues of α_4 differing by ± 1. Of course, the same holds true for every reducible representation comprising a sum of irreducible representations, in each of which the maximum α_4 eigenvalue is one of the set $n, n-1, \ldots, -n$. Thus, the matrices of the α^k *under this specialization* are of the form

$$\| \alpha^k \| = \left\| \begin{matrix} 0 & (n|\alpha^k|n-1) & 0 & \cdots & \\ (n-1|\alpha^k|n) & 0 & (n-1|\alpha^k|n-2) & \cdots & \\ 0 & (n-2|\alpha^k|n-1) & 0 & \cdots & \\ \cdots & \cdots & \cdots & \cdots & (-n+1|\alpha^k|-n) \\ \cdots & \cdots & \cdots & (-n|\alpha^k|-n+1) & 0 \end{matrix} \right\| \tag{35.16c}$$

and for any n, $(n \mid \alpha^k \mid n-1) = -(n-1 \mid \alpha^k \mid n)^*$. The corresponding splitting of ψ is into the sets $\psi_n \oplus \psi_{n-1} \oplus \ldots \oplus \psi_{-n}$ in $\mu_n, \mu_{n-1}, \ldots, \mu_{-n}$ components.

(b) *Classification of Coupling Schemes*

There are now two possibilities in the general coupling scheme: either it is possible to pass from any component $\mathfrak{T}^{\kappa\lambda}(k, l)$ to every other through a succession of couplings of the type described, or there are at most two unconnected sets, in each of which it is possible to pass from one $\mathfrak{T}^{\kappa\lambda}(k, l)$ to any other. In the latter case the two sets which are unconnected by matrix elements of the α's must pass into each other under a reflection, since any other possibility would lead to reducible representations of the α's and $\mathfrak{T}$'s. Clearly both sets will have the same structure and only one need be considered explicitly.

If we choose $d = -c$ in (35.16), which we are free to do, then corresponding to

[3] The n here is one of the two numbers (both integral or half-integral) labelling the irreducible representation $\mathfrak{D}^{(5)}(n, m)$ of the 5-dimensional Lorentz group, which is considered in §§ 36*a*, *b*, *et seq.* For the present consideration, however, this connection is not essential.

the simplest coupling scheme of the type $(k, l) \rightleftharpoons (k + \frac{1}{2}, l - \frac{1}{2})$ with but one link, the wave equation takes the form

$$\partial_{AB} \begin{Vmatrix} 0 & cv^B(k+\frac{1}{2})u^A(l) \\ -cu^B(k+\frac{1}{2})v^A(l) & 0 \end{Vmatrix} \cdot \begin{Vmatrix} \psi(k, l) \\ \psi(k+\frac{1}{2}, l-\frac{1}{2}) \end{Vmatrix} = \kappa \begin{Vmatrix} \psi(k, l) \\ \psi(k+\frac{1}{2}, l-\frac{1}{2}) \end{Vmatrix} \tag{35.17}$$

Clearly the matrix factors on the left are to be understood in the sense of *direct* products, and as such have the appropriate dimensions.

The general Dirac equations (28.11, 12) are of this type, being equivalent to

$$\left.\begin{aligned} \partial^{AB} v_B(k+\tfrac{1}{2})\psi(k+\tfrac{1}{2}, l-\tfrac{1}{2}) - \kappa\left(\frac{2k+1}{2l}\right)^{1/2} v^A(l)\psi(k, l) = 0 \\ \partial_{AB} v^A(l)\psi(k, l) + \kappa\left(\frac{2l}{2k+1}\right)^{1/2} v_B(k+\tfrac{1}{2})\psi(k+\tfrac{1}{2}, l-\tfrac{1}{2}) = 0 \end{aligned}\right\} \tag{35.18a}$$

which may be transcribed to the form (35.17) on using the properties of the u's and v's. However, these equations imply the subsidiary or supplementary conditions:

$$\left.\begin{aligned} \partial^{AB} v_A(l-\tfrac{1}{2})v_B(k+\tfrac{1}{2})\psi(k+\tfrac{1}{2}, l-\tfrac{1}{2}) = 0 \\ \partial_{AB} v^B(k)v^A(l)\psi(k, l) = 0 \end{aligned}\right\} \tag{35.18b}$$

which involve only one or other of the component parts of ψ, and are required to ensure the validity of the second-order equation. As usual, such conditions cannot be obtained as equations of motion in the quantized system, but must be imposed as conditions on the possible states. The only exception is the coupling scheme $(0, \frac{1}{2}) \rightleftharpoons (\frac{1}{2}, 0)$, i.e. the Dirac electron equation, for which both subsidiary conditions vanish.

By a straightforward calculation, with the use of (35.6a), it is found that this exception is also the only equation of the simple coupling type $(k, l) \rightleftharpoons (k + \frac{1}{2}, l - \frac{1}{2})$, for which

$$\mathfrak{T}^{\mu\nu} \sim [\alpha^\mu, \alpha^\nu]_- \tag{35.19}$$

that is, the simplest possible relation between the $\mathfrak{T}$'s and α's.[4]

Similarly it is easily shown that, of the simple coupling type $(k, l) \rightleftharpoons (k + \frac{1}{2}, l + \frac{1}{2})$, only the scalar meson equation, corresponding to $(0, 0) \rightleftharpoons (\frac{1}{2}, \frac{1}{2})$, leads to this simple connection (35.19).

These are, of course, more complicated systems involving either open or closed chains of connections of either or both types. For example, the open chain

$$(k, l) \rightleftharpoons (k+\tfrac{1}{2}, l-\tfrac{1}{2}) \rightleftharpoons (k+1, l-1) \rightleftharpoons \ldots \rightleftharpoons (k+n+\tfrac{1}{2}, l-n-\tfrac{1}{2})$$

for which repeated application of the preceding results show that the validity of (35.19) requires $k = 0$, $l = n - \frac{1}{2}$ so that the chain is of the form

$$(0, \lambda) \rightleftharpoons (\tfrac{1}{2}, \lambda - \tfrac{1}{2}) \rightleftharpoons \ldots (\lambda, 0)$$

with constant $\lambda = k + l$. The vector meson equation is of this type and corresponds to the scheme $(0, 1) \rightleftharpoons (\frac{1}{2}, \frac{1}{2}) \rightleftharpoons (1, 0)$, so that there are two equations of the type

[4] Cf. Harish-Chandra (47*a*, *c*); Bhabha (45, 49).

(35.17) and one equation involving all three component functions $\psi(0, 1)$, $\psi(\frac{1}{2}, \frac{1}{2})$, and $\psi(1, 0)$.

We shall leave for later consideration all the higher-spin equations of this open coupling type as well as the more complicated equations based on closed coupling with the typical link

$$\begin{array}{ccc} (k, l) & \rightleftharpoons & (k + \frac{1}{2}, l - \frac{1}{2}) \\ \upharpoonleft\downharpoonright & & \upharpoonleft\downharpoonright \\ (k - \frac{1}{2}, l - \frac{1}{2}) & \rightleftharpoons & (k, l - 1) \end{array}$$

in which the horizontal connections are of the first type $(k, l) \rightleftharpoons (k + \frac{1}{2}, l - \frac{1}{2})$ and the vertical connections of the type $(k, l) \rightleftharpoons (k + \frac{1}{2}, l + \frac{1}{2})$. These do not lead to (35.19), and in general are not in the category of what we have called *equations of type I* (cf. § 33*c*), since the second-order equation is not satisfied. By the same token, they represent composite systems, at least in the sense that states of higher spin and mass occur, although in the absence of interaction with other fields they are not usually coupled.

(c) *Further Considerations on Spin and Statistics*

We are now in a position to consider the previously mentioned delimitation to the cases of spin 0, 1, and $\frac{1}{2}$ which results from the added requirement that no subsidiary[5] conditions occur in the theory. The problem may be formulated as the determination of the allowed forms of α_4 which, for half-integral spin, give eigenvalues of $\Lambda\alpha_4$ of one sign only, and for integral spin satisfy the condition $\Lambda\alpha_4 = \alpha_4$, where Λ is the metric operator.[6]

For a representation of the α's in which $\Lambda\alpha_4$ (which is Hermitian by virtue of (34.8*a*) and the choice of Λ) is diagonal, those equations of the set

$$(\partial_\kappa \Lambda\alpha^\kappa - i\kappa\Lambda)\psi = 0 \tag{35.20}$$

corresponding to zero values of the diagonal elements, say λ_r, constitute a set of inhomogeneous equations which, if solvable, define the redundant components; and if Λ is non-degenerate, the wave equation follows from (35.20). The canonical commutation rules can be expressed as

$$[\bar{\psi}_r(\mathbf{x}, x_4), \psi_s(\mathbf{x}', x_4)]_\pm = \pm \frac{1}{\lambda_r} \delta_{rs}\, \delta(\mathbf{x} - \mathbf{x}') \tag{35.21a}$$

$$[\psi_r(\mathbf{x}, x_4), \psi_s(\mathbf{x}', x_4)]_\pm = 0$$

[5] Those components of ψ whose time-derivatives are given explicitly in terms of *all* the components and their spatial derivatives are called the *essential* components; those which are defined in terms of the *spatial* derivatives of these essential components are termed redundant—and can be eliminated. In the electron theory there are no redundant components in contrast to the meson fields which for spin 1 and 0 have four (of ten) and three (of five) redundant components, respectively. For the elimination of these see § 39*d*.

By contrast, *subsidiary* conditions impose restrictions on the permissible solutions for the essential components.

[6] For each non-zero eigenvalue of α_4, there is a solution of the wave equation in the rest system for which, if Λ and α_4 are both in diagonal form, the total energy has the same sign as the corresponding eigenvalue of Λ. Hence, for the total energy to be positive definite, it is necessary that the eigenvalues of α_4 corresponding to negative eigenvalues of Λ be zero. Since all the eigenvalues of Λ are ± 1, the above condition for integral spin follows.

or in the matrix form

$$\Lambda\alpha_4[\psi, \psi^*]\Lambda\alpha_4 = \delta(\mathbf{x} - \mathbf{x}')\Lambda\alpha_4 \tag{35.21b}$$

where

$$[\psi, \psi^*]_{rs} = [\psi_r(\mathbf{x}, t), \bar{\psi}_s(\mathbf{x}', t)]$$

In order to study the properties of α_4 and Λ, it is convenient to define a standard representation. We choose the representation of the α's such that the representation of the Lorentz group afforded by the components of ψ is the direct sum of its irreducible components, all equivalent components are equal, and the contained representation of the rotation subgroup is unitary. Each irreducible representation of the Lorentz group, which we denote by $\{k, l\}$ for convenience, is to be the direct sum of the two irreducible representations (k, l) and (l, k) of the proper Lorentz group. The spatial reflection $R(x^k \to -x^k)$ is to be represented by

$$\begin{Bmatrix}\psi(k, l)\\ \psi(l, k)\end{Bmatrix} \to a \begin{Bmatrix}\psi(l, k)\\ \psi(k, l)\end{Bmatrix} \qquad (k \neq l) \tag{35.22}$$

$$\psi_{rs}(k, k) \to b\psi_{sr}(k, k)$$

where a takes the value 1 or i for $k + l$ half-integral, 1 for $k + l$ integral, b the value $+1$ or -1, and the indices r, s *here* refer to $(k, 0)$, $(0, k)$, respectively; (k, l) being given by the direct product of $(k, 0)$ and $(0, l)$.

The condition $T^*(L)\Lambda T(L) = \Lambda$ (cf. 34.11) considered for the infinitesimal transformations reads

$$\Lambda\mathfrak{T}_{\kappa\lambda} + \mathfrak{T}_{\kappa\lambda}{}^*\Lambda = 0 \tag{35.23a}$$

or, for the present choice ($\mathfrak{T}_{kl}{}^* = -\mathfrak{T}_{kl}$, $\mathfrak{T}_{4k}{}^* = \mathfrak{T}_{4k}$),

$$\begin{aligned}\Lambda\mathfrak{T}_{kl} &= \mathfrak{T}_{kl}\Lambda\\ \Lambda\mathfrak{T}_{4k} &= -\mathfrak{T}_{4k}\Lambda\end{aligned} \tag{35.23b}$$

and, in general,

$$\begin{aligned}R\mathfrak{T}_{kl} &= \mathfrak{T}_{kl}R\\ R\mathfrak{T}_{4k} &= -\mathfrak{T}_{4k}R\\ R\Lambda &= \Lambda R\end{aligned} \tag{35.24}$$

Note also that $R\alpha_4 R^{-1} = \alpha_4$, $R\alpha_k R^{-1} = -\alpha_k$.

If we label by r, s, etc., the irreducible representations of the Lorentz group which occur in the representation of the α_κ, then equations (35.23, 24) give

$$R_r^{-1}\Lambda_{rs}(\mathfrak{T}_{\kappa\lambda})_s = (\mathfrak{T}_{\kappa\lambda})_r R_r^{-1}\Lambda_{rs} \qquad \text{(no summation)} \tag{35.25}$$

for the component of Λ which couples two such representations, so that $R_r^{-1}\Lambda_{rs}$ is a multiple of the unit matrix or zero (and hence $\Lambda_{rs} = 0$) according as r and s are, or are not, equivalent. $R^{-1}\Lambda$ can therefore be diagonalized by a unitary transformation which does not affect the representation of the Lorentz group, whence

$$\Lambda_{rs} = \lambda_r R_r\,\delta_{rs} \tag{35.26a}$$

By the further substitution

$$\psi_r \to \sqrt{|\lambda_r|}\,\psi_r \qquad (\lambda_r \neq 0)$$

all constants in (35.26) can be made equal to 1, 0, -1 for single-valued representations, or $\pm i$ for two-valued representations. Further, $\Lambda^2 = 1$ and $\Lambda\alpha_\kappa{}^* = \alpha_\kappa\Lambda$, and since R commutes with α_4 it follows that $R^{-1}\Lambda$ commutes with $\alpha_4{}^* + \alpha_4$ and anticommutes with $\alpha_4{}^* - \alpha_4$. Therefore, since $R^{-1}\Lambda$ is diagonal (in the standard representation),

$$(\alpha_{4st})^* = (\alpha_4{}^*)_{ts} \qquad \text{or} \qquad (\alpha_{4st})^* = -(\alpha_4{}^*)_{ts} \tag{35.26b}$$

according as $R_s{}^{-1}\Lambda_s$ and $R_r{}^{-1}\Lambda_r$ have the same or opposite signs.

To complete the analysis we require, besides the spinor-matrices u^A, v^A, P^{At} and R^{At} as defined in §17b(ii), the spinor-matrices

$$\begin{aligned} Q^{A_t} &= Q^{A_1 \cdots A_{2l}}(j, l, k) \qquad\qquad (35.27a)\\ &= v^{A_1}(j + \tfrac{1}{2}) \ldots v^{A_s}(j + \tfrac{1}{2}s)u^{A_{s+1}}(j + \tfrac{1}{2}s) \ldots u^{A_{2l}}(k + \tfrac{1}{2})\\ &\quad s = k + l - j \qquad k + l \geqslant j \qquad j + l \geqslant k \end{aligned}$$

which are evidently symmetric in the sets of indices $A_1 \ldots A_s$ and $A_{s+1} \ldots A_{2l}$, and have the properties:

$$\begin{aligned} &\Sigma(\text{Perm } A_t) Q^{A_r}(j, l, k) Q_{A_t}(k, l, j')\\ &\qquad = (-)^{l(4k+2l-1)} \frac{(k + l - j)!\,(j + l - k)!\,(k + l + j + 1)!}{(2j + 1)(k + j - l)!} \delta_{jj'} \end{aligned} \tag{35.27b}$$

$$Q^{A_t}(j, k, 0) Q_{A_t}(0, k, j') = (-)^{[j]} \frac{(k + j + 1)!\,(k - j)!}{(2j + 1)} \delta_{jj'} \tag{35.27c}$$ [7]

where $[j]$ denotes the integral part of j. Here the convention is used that when the umbral index A appears in two places with different subscripts such as A_r and A_t, then the indices represented by A_t are an arbitrary permutation of those represented by A_r, and contraction over repeated indices[8] is implied. The sum $\Sigma(\text{Perm } A_t)$ then extends over all such permutations. When the umbral index has the same subscript in both places, the indices represented by it are to be identically ordered, and contraction is again implied.

For the allowed form of the spinor components of the α_κ, we write

$$\begin{array}{lccc} (k_r, l_r \mid \alpha_{AB} \mid k_s', l_s') & k' & l' & \\ c_{rs}\left\{\begin{array}{l} v_B(k + \frac{1}{2}) \times v_A(l + \frac{1}{2}) \\ v_B(k + \frac{1}{2}) \times u_A(l) \\ u_B(k) \times u_A(l) \\ u_B(k) \times v_A(l + \frac{1}{2}) \end{array}\right\} & \begin{array}{c} k + \frac{1}{2} \\ k + \frac{1}{2} \\ k - \frac{1}{2} \\ k - \frac{1}{2} \end{array} & \begin{array}{c} l + \frac{1}{2} \\ l - \frac{1}{2} \\ l - \frac{1}{2} \\ l + \frac{1}{2} \end{array} & \end{array} \tag{35.28}$$

where c_{rs} again denotes an arbitrary complex number, and for consistency with the standard representation the representation of the $u(k)$ and $v(k)$ must be chosen such that

$$u^A(k) = u_A(k) \qquad v^A(k) = v_A(k) \tag{35.29a}$$

$$v^A(k) = (-)^{2k+1} u_A^*(k) \tag{35.29b}$$

[7] For the detailed evaluation refer to the original article by Wild (47).

[8] In the following the summation convention is also extended to apply to a repeated index which appears once as a dotted and once as an undotted index.

The first condition (35.29a) ensures the consistency of the representation of R, and the second (35.29b) the unitary property of the transformation $U(k)$ (cf. 17.34) and therefore that of the representation of the rotation group. Further, if r, t and s, u denote two pairs of inequivalent irreducible representations (of the proper Lorentz group), such that r goes into t and s into u by reflection, it follows from (35.29) and that fact that R commutes with α_4, that $c_{rs} = -c_{tu}$, and from (35.26b, 29) $c_{rs} = \pm\bar{c}_{sr}(-)^{2k+1}$, according as $R_r^{-1}A_r$ and $R_s^{-1}A_s$ have the same or opposite signs.

To examine the sign of the eigenvalues of $A\alpha_4$, we transform to a representation in which the representation of the rotation subgroup contained in each irreducible representation of the proper Lorentz group is the direct sum of its irreducible constituents, equivalent representations again being equal. The point of this transformation is that in the new representation the matrices A and α_4, since they commute with the $\mathfrak{T}_{kl}$, will have no components which couple inequivalent representations of the rotation subgroup and will be represented by multiples of the unit matrix for components which couple equivalent (equal) representations. Conversely, this criterion suggests the unitary transformation, say S, which will effect the reduction. S is defined (for $k \geqslant l$) by

$$(j_r \mid S \mid k_s, l_s) = N(j, l, k)Q^{A_t}(j, l, k) \times R^{A_t}(l)\, \delta_{rs} \tag{35.30}$$

$$(k_r, l_r \mid S^{-1} \mid j_s) = (-)^{4l(k+l)} N(j, l, k) Q_{A_t}(k, l, j) \times P_{A_t}(l)\, \delta_{rs}$$

$$N(j, l, k) = \left[\frac{(2j+1)(k+j-l)\,!}{(k+l-j)\,!\,(j+l-k)\,!\,(k+l+j+1)\,!}\right]^{1/2}$$

and for $l > k$, the indices l and k, dotted and undotted indices are interchanged, and the defining forms multiplied by $(-)^{2k}$. With the help of the properties of the matrices u, v, P, R, Q it is readily verified that these forms are indeed reciprocal and that S is unitary. That the required reduction is effected thereby follows if we can show that this transformation brings A and α_4 to the required diagonal form. This has been proved by Wild (47). For the (j, j') component of the former in the representation $\{k, l\}$ of the complete Lorentz group, we obtain:[9]

$k \neq l$:

$$\begin{aligned}
&(-)^{4l(k+l)+2l} Q^{A_r}(j, l, k) Q_{\dot{B}_t}(k, l, j') \times R^{A_t}(l) P_{B_t}(l) \\
&= (-)^{4l(k+l)+2l} Q^{A_r}(j, l, k) Q_{B_t}(k, l, j') \times R_{A_r}(l) P^{B_t}(l) \\
&= (-)^{l(4k+2l-1)} \Sigma(\text{Perm } A_t) Q^{A_r}(j, l, k) Q_{A_t}(k, l, j') \\
&= \delta_{jj'}
\end{aligned} \tag{35.31a}$$

$k = l$:

$$\begin{aligned}
&(-)^{8k} Q^{A_r}(j, k, k) P_{\dot{B}_t}(k) \times R^{A_r}(k) Q_{B_t}(k, k, j') \\
&= (-)^{2k} Q^{A_r}(j, k, k) P^{B_t}(k) \times R_{A_r}(k) Q_{B_t}(k, k, j') \\
&= (-)^{2k} Q^{A_s}(j, 2k, 0) Q_{A_s}(0, 2k, j') \\
&= (-)^{2k+1}\, \delta_{jj'}
\end{aligned} \tag{35.31b}$$

[9] In the following equations (35.31) and (35.32) the normalizing factors N are omitted except in the final result.

For the (j, j') component of the latter, which couples the representations $r\,(=(k, l))$ and $s\,(=(k', l'))$, we obtain:

$$(j \mid \alpha_4 \mid j') = (j \mid S \mid k, l)(k, l \mid g_{4\dot{A}B}\,\alpha^{\dot{A}B} \mid k', l')(k', l' \mid S^{-1} \mid j')$$

$$\left.\begin{array}{l}(k, l) \rightleftharpoons (k - \tfrac{1}{2}, l - \tfrac{1}{2}),\ k \geqslant l: \\ (-)^{k+l+j} c_{rs}[(k + l + j + 1)(k + l - j)]^{1/2}\,\delta_{jj'}\end{array}\right\} \tag{35.32a}$$

$$\left.\begin{array}{l}(k, l) \rightleftharpoons (k - \tfrac{1}{2}, l + \tfrac{1}{2}),\ k > l + \tfrac{1}{2} \\ c_{rs}[(k + j - l)(j + l - k + 1)]^{1/2}\,\delta_{jj'}\end{array}\right\} \tag{35.32b}$$

$$\left.\begin{array}{l}(k + \tfrac{1}{2}, k) \rightleftharpoons (k, k + \tfrac{1}{2}) \\ (-)^{[j]+1} c_{rs}(j + \tfrac{1}{2})\,\delta_{jj'}\end{array}\right\} \tag{35.32c}$$

$$\left.\begin{array}{l}(k, k) \rightleftharpoons (k - \tfrac{1}{2}, k + \tfrac{1}{2}) \\ (-)^{2k-1+j} c_{rs}[j(j + 1)]^{1/2}\,\delta_{jj'}\end{array}\right\} \tag{35.32d}$$

from which, with (35.29), the coupling components for all other values of k and l can be obtained.[10]

These relations, together with those previously noted for associated values of c_{rs}, give the correct commutation relations of α_4 with R and Λ. Further, in all representations $r\ (=\{k, k\})$ of the Lorentz group, R must have the same sign for any one representation (e.g. $\mathfrak{D}(0)$) of the rotation subgroup; and if $s\,(=(k + \tfrac{1}{2}, k - \tfrac{1}{2}))$ and $t\,(=(k - \tfrac{1}{2}, k + \tfrac{1}{2}))$ are representations which go into each other by reflection and are coupled to r, then $c_{sr} = \mp (-)^{2k} c_{tr}$, according as the sign (of R) is positive or negative.

The results which are essential to the present issue are the dependence of sign on j in (35.31b) and (35.32c), and the non-occurrence of zero (j, j) components in the range $k + l \geqslant j \geqslant |k - l|$, $k' + l' \geqslant j \geqslant |k' - l'|$.

Half-integral Spin

The α-algebra may contain any representation $\{k, l\}$ of the complete Lorentz group such that $k + l$ is half-integral, and we require the forms of α_4 which give eigenvalues of $\Lambda\alpha_4$ of one sign only.

Consider first the case that includes a representation $\{k, l\}$ with $k \neq l + \tfrac{1}{2}$, or a representation $\{k + \tfrac{1}{2}, k\}$ in which the constituent $(k + \tfrac{1}{2}, k)$ and $(k, k + \tfrac{1}{2})$ are not coupled in α_4. Then, in the *standard* representation, premultiplication by Λ interchanges the rows corresponding to the constituents (k, l) and (l, k), and the corresponding diagonal elements of $\Lambda\alpha_4$ are zero. Hence, $\Lambda\alpha_4$ is a Hermitian matrix with zero diagonal elements at the intersection of some non-zero rows and columns, and by considering the unitary transformation which carries it from diagonal form to the form considered, it is readily shown that $\Lambda\alpha_4$ cannot be non-negative.

Therefore we need consider only the case in which there appear representations $\{k + \tfrac{1}{2}, k\}$ whose two constituents are coupled in α_4. Then from (35.31, 32), after

[10] The component $(k', l') \rightleftharpoons (k, l)$ is $(-)^{2k+2l}$ times that of $(k, l) \rightleftharpoons (k', l')$. The component of $(l, k) \rightleftharpoons l', k')$ is the negative of that of $(k, l) \rightleftharpoons (k', l')$ except when $k = l$ or $k' = l'$. This exceptional case is given by (35.32d), which corresponds, on reflection, to the inverse of (35.32b) for $k' = l'$.

the transformation (35.30), the diagonal elements of $\Lambda\alpha_4$ arising from this coupling which correspond to representations $\mathfrak{D}(j + \frac{1}{2})$ with j even and j odd are opposite in sign (provided $k \neq 0$). $\Lambda\alpha_4$ is therefore a Hermitian matrix with diagonal elements of both signs, and it follows as before that it cannot be non-negative.

Thus, the only permissible representation[11] is $\{\frac{1}{2}, 0\}$, and the corresponding wave equation is that for the electron (spin $\frac{1}{2}$), since the α_4 (and hence α_κ) are determined up to a numerical factor and correspond to the Dirac equation.

Integral Spin

For integral spin the α-algebra may contain any representation $\{k, l\}$ of the complete Lorentz group such that $k + l$ is integral, and the condition to be satisfied is $\Lambda\alpha_4 = \alpha_4$.

Consider again the case which includes a representation $\{k', l'\}$ with $k' \neq l'$. Again in the standard representation, premultiplication by Λ interchanges the rows which correspond to the two constituent representation of the proper Lorentz group, and the condition $\Lambda\alpha_4 = \alpha_4$ can only be satisfied for components of α_4 which couple both constituents to the *same* representation of the proper Lorentz group. From (35.28), this jointly held representation must be of the type (k, k).

The only permissible couplings, then, are of the type $\{k, k\} \rightleftharpoons \cdots \{k', l'\}$, but from (35.31*b*) the condition on Λ, α_4 is not satisfied unless the couplings are such that after the transformation (35.30) they have a non-zero component in only one representation of the rotation subgroup. The possibilities are therefore reduced to $\{\frac{1}{2}, \frac{1}{2}\} \rightleftharpoons \{1, 0\}$ and $\{\frac{1}{2}, \frac{1}{2}\} \rightleftharpoons \{0, 0\}$, which may not occur together. In each case, only one irreducible wave equation results, namely, the meson field of spin 1 and spin 0, respectively.[12]

§ 36. Wave Equations of Type II

(*a*) *Introduction*

Before considering the more formal aspects of the properties of the α-matrices and the algebras generated by them, we wish to examine briefly the characteristics of the wave equations classed as type II in § 33*c*. For reasons already noted, and others to be indicated later, this class appears less likely to be of physical importance. The present considerations are none the less of interest because of the connection between equations of this type and the orthogonal group in 5 dimensions—developed principally by Lubanski (41*a*, *b*; 42*a*, *b*), Bhabha (45*a*, *b*; 46) and Madhava Rao (47); a connection which, moreover, also obtains in the physically important cases of the electron and meson fields and thereby presents an interesting aspect of their structure and the couplings discussed in §§ 35*a*, *b*.

[11] If this representation appears more than once, the system is reducible.

[12] There remain the possibilities that α_4 is: (i) not Hermitian in the standard representation; (ii) diagonalizable but possesses complex eigenvalues; (iii) not diagonalizable. As shown by Wild, the first case leads to the same result as above, the second does not lead to definite energy density, and the third leads to subsidiary conditions, since the number of spatially constant solutions of the wave equation is then less than the number of essential components.

This connection follows, in the first instance, from the assumption that the matrices $\mathfrak{T}_{\kappa\lambda}$ of the nucleus of the proper Lorentz group are given in terms of the α_κ by the particular relation

$$\mathfrak{T}_{\kappa\lambda} = [\alpha_\kappa, \alpha_\lambda]_- \tag{36.1}$$

apart from a possible numerical factor which is not physically important. It is seen that this *Ansatz* is consistent with the commutation relations (35.2), and satisfies the equation

$$\big[[\alpha_\kappa, \alpha_\lambda]_-, \mathfrak{T}_{\mu\nu}\big]_- = -g_{\kappa\mu}[\alpha_\lambda, \alpha_\nu]_- + g_{\kappa\nu}[\alpha_\lambda, \alpha_\mu]_- + g_{\lambda\mu}[\alpha_\kappa, \alpha_\nu]_- - g_{\lambda\nu}[\alpha_\kappa, \alpha_\mu]_- \tag{36.2}$$

which follows from (35.1). We must then study the algebra of the four α's and six $\mathfrak{T}$'s, satisfying equations (35.1, 2) and (36.2), for which it is readily shown (cf. Bhabha, *loc. cit.*) that

$$\alpha^4,\ i\alpha^1,\ i\alpha^2,\ i\alpha^3,\ \mathfrak{T}^{4k},\ i\mathfrak{T}^{12},\ i\mathfrak{T}^{23},\ i\mathfrak{T}^{31}$$

all satisfy the same characteristic equation

$$\{X^2 - n^2\}\{X^2 - (n-1)^2\} \ldots = 0$$

the last factor being X^2 or $\{X^2 - (\tfrac{1}{2})^2\}$ according as n is an integer or half-integer.

Then, if we introduce an index 5, and define

$$\mathfrak{T}^{\kappa 5} = -\mathfrak{T}^{5\kappa} = \alpha^\kappa \tag{36.3a$\dagger$}$$
$$g^{55} = -1 \qquad g^{\kappa 5} = 0 \qquad (\kappa \neq 5)$$

equations (35.1, 2) and (36.1) are all combined into the one equation

$$[\mathfrak{T}_{\kappa\lambda}, \mathfrak{T}_{\mu\nu}]_- = -g_{\kappa\mu}\mathfrak{T}_{\lambda\nu} + g_{\kappa\nu}\mathfrak{T}_{\lambda\mu} - g_{\lambda\nu}\mathfrak{T}_{\kappa\mu} + g_{\lambda\mu}\mathfrak{T}_{\kappa\nu} \tag{36.3b$\dagger$}$$

where the indices range from 1 to 5. The ten matrices $\mathfrak{T}^{\kappa\lambda}$, with this index range, therefore satisfy the commutation rules of the infinitesimal transformations of the Lorentz group in 5 dimensions;[1] that is, the fundamental quadratic form is again of *inertial index* 1, corresponding to one time-like and four space-like variables with the metric signature $(- - - + -)$. All finite irreducible representations of this group are known, so that in *this* theory every such representation determines an irreducible set of α-matrices and a (presumed) possible wave-equation for the elementary particles so defined.

(*b*) *The Lorentz and Real Orthogonal Groups in 5 Dimensions*

Every finite representation of the Lorentz group in five dimensions, which we denote for the moment by $\mathfrak{D}^{(5)}(n, m)$, can be labelled by two numbers n, m such that $n \geqslant m \geqslant 0$, and both are either integers (including zero) or half-integers, correspond-

† Where this range is used we retain the usual tensor index labelling, but mark by † the equations for which this connection is to be understood.

[1] Note that, since every finite irreducible representation can be transformed to a form in which the matrices $\mathfrak{T}_{\kappa\lambda}$ are anti-Hermitian for $\kappa \neq \lambda \neq 4$ and Hermitian for κ or $\lambda = 4$, any of the matrices can be brought to diagonal form and, being on equal footing, all ten (with multiplication by i if $\kappa, \lambda \neq 4$) have the same eigenvalues and characteristic equation.

ing, respectively, to the one-valued (tensor) and two-valued (spinor) representations.[2] Further, it is well known (cf. Murnaghan (38), Weyl (35, 39), that from every finite irreducible representation of the orthogonal group in n dimensions, we can obtain an irreducible representation of the, so to speak, attached Lorentz group—by suitable change in the reality conditions on the parameters of the group, the so-called *unitary trick*.[3] The advantage gained thereby stems from the compactness (and unitarity) of the real orthogonal group, which allows us to draw on important results obtaining from the methods of group integration.

An alternative, though essentially equivalent, formulation of this theory has been given by Kramers, Belinfante, and Lubanski (41), taking the wave-function corresponding to spin s to be a 4-spinor of rank $N = 2s$, the α^μ being represented by

$$\alpha^\mu = \sum_{r=1}^{N} \gamma^\mu_{(r)} \tag{36.4a}$$

$$\gamma^\mu_{(r)} = I \times \ldots \times I \times \underset{(r)}{\gamma^\mu} \times I \times \ldots \times I \tag{36.4b}$$

$$\gamma^\mu_{(r)} \gamma^\nu_{(t)} = \gamma^\nu_{(t)} \gamma^\mu_{(r)} \qquad (r \neq t) \tag{36.4c}$$

$$\gamma^\mu_{(r)} \gamma^\nu_{(r)} + \gamma^\nu_{(r)} \gamma^\mu_{(r)} = 2g^{\mu\nu}$$

This representation of the α^μ's is identical with the general spinor representation of the orthogonal group discovered by Cartan (13, 38) and investigated by Brauer and Weyl (35) in the n-dimensional case.

Thus, from either point of view, the fundamental basis of this theory is the concept of the spinor representations of the orthogonal group (and hence the attached Lorentz group) in n dimensions, to which we now turn our attention. In these considerations we find more definite basis for the preceding schematic picture, and at the same time a logical point of entry to the formal domain of algebraic concepts. Apart from the natural development of the fundamental principles *per se*, we find, in a very simple way, the general basis for many of the results obtained in previous chapters through more special and elementary arguments.

§ 37. Elementary Algebraic Concepts

(a) *The Spinor Representations of the Orthogonal Group in n Dimensions*

We begin with the familiar matrices

$$P_1 = \begin{Vmatrix} 0 & i \\ i & 0 \end{Vmatrix} \quad P_2 = \begin{Vmatrix} 0 & -1 \\ 1 & 0 \end{Vmatrix} \quad P_3 = \begin{Vmatrix} i & 0 \\ 0 & -i \end{Vmatrix} \quad P_4 = \begin{Vmatrix} i & 0 \\ 0 & i \end{Vmatrix} \tag{37.1a}$$

[2] For the moment we only note that, since $\mathfrak{T}^{45}$ and $\mathfrak{T}^{12}$ commute, we can choose the basic vectors of the representation space such that they are simultaneous eigenvectors of these two operators. If n, m denote their respective maximum eigenvalues in this representation, the two numbers uniquely determine the irreducible representation denoted by $\mathfrak{D}^{(5)}(n, m)$. Cf. § 38.

[3] That this device does not yield *all possible* representations of the Lorentz group (particularly with reference to the infinite case) is technically described as due to the different topological structure of the two groups; in particular, the *aperiodic* nature of the *temporal rotations* of the Lorentz group which, therefore, unlike the orthogonal group, is not compact. Cf. Murnaghan, *loc. cit.*, Chapter XII.

which satisfy the commutation rules

$$\left.\begin{aligned} P_r P_s + P_s P_r &= 2g_{rs} I_{(2)} \\ P_1 P_2 P_3 &= i P_4 \end{aligned}\right\} \qquad (g_{11} = g_{22} = g_{33} = -1)$$

For our later work it will be convenient to denote them alternatively by the symbols

$$\mathscr{P},\ \mathscr{Q},\ \mathscr{I}',\ \mathscr{I} \tag{37.1b}$$

respectively.

Any matrix of two rows and columns can be expressed linearly in terms of P_1, P_2, P_3, P_4 or, alternatively stated, these four matrices constitute a basis for the full matrix algebra of dimension 2 and order 4. In this connection a matrix algebra $\mathfrak{A}(A)$ appears as a set of elements A for which three operations are defined: addition $A + B$ and multiplication AB of two elements A, B and multiplication λA of an element by a number[1] λ.

If we start with an arbitrary set of matrices $\mathfrak{A}$, we can first form its so-called multiplicative closure consisting of all finite products ΠA (A in $\mathfrak{A}$), which is a semi-group (that is, closed with respect to multiplication but, in general, lacking the group properties that: (i) the unit matrix is contained, (ii) every element is non-singular, (iii) A^{-1} is a member of the set together with A), and pass to the linear closure $[\mathfrak{A}]$ consisting of all finite linear combinations $c^r A_r$ with coefficients from the field of complex numbers. These two steps provide the so-called *enveloping* algebra $\mathfrak{A}(A)$, which is a fancy term meaning simply the collection of all matrices obtained by forming *all possible* linear combinations of matrices in the set: $c^r A_r + c^{rs} A_r A_s + c^{rst} A_r A_s A_t + \ldots$.

An abstract algebra $\mathfrak{a}$ is simply a set of abstract elements obeying the above-defined properties and, if we have any representation of an abstract algebra, the enveloping algebra of the representation is termed a representation of the enveloping algebra.

After this brief preliminary digression on algebra, let us return to the matrices (37.1) and the problem of the spinor representations of the orthogonal group. The procedure we follow is essentially the Brauer-Weyl (35) generalization of Dirac's basic paper. We consider the linearization of a quadratic form

$$-(x_1^2 + x_2^2 + \ldots + x_n^2) = (\gamma^r x_r)^2 \qquad (r = 1, 2, \ldots, n) \tag{37.2}$$

where the summation convention applies, and the signature on the left-hand side is conventional, being chosen here because, with our world-metric signature $(- - - +)$, the transition to the attached Lorentz group (in 4 dimensions) corresponds to changing the last sign $+$ to $-$ and, therefore, multiplying by i all tensors having the index 4, and substituting hyperbolic rotations for orthogonal rotations of the k, 4-axes. Similarly, it is convenient to recognize formally covariant and contravariant vectors, even though the difference disappears under the orthogonal group.

The γ^r satisfy

$$\gamma^r \gamma^s + \gamma^s \gamma^r = 2g^{rs} \qquad (\text{all } g^{rr} = -1) \tag{37.3}$$

[1] Properly, we must define the field of numbers in which we operate.

that is, a simple generalization to n dimensions of the relation (14.3) for 4-spinors. The quantities γ^r generate an associative algebra Π of order 2^n consisting of all linear combinations of the 2^n units

$$e^{A_1 A_2 \ldots A_n} = \gamma_1{}^{A_1} \ldots \gamma_n{}^{A_n} \qquad (A_1, \ldots, A_n, \text{ integers mod. } 2)$$

whose multiplication table follows from (37.3). The algebra Π is well known in the form of the second quantization for Fermi-Dirac systems, which connection also yields a definite representation $\gamma_r \to \Gamma_r$ by $n = 2\nu$ (considering even dimensionality first) matrices of dimensions 2^ν, each of which is the Kronecker product of ν matrices selected from the set $\mathscr{P}$, $\mathscr{Q}$, $\mathscr{I}'$, $\mathscr{I}$. We have (a common multiplicative factor i is implied for ν even)

$$\gamma_r \to \Gamma_r = \mathscr{I}' \times \ldots \times \mathscr{I}' \times \underset{(r)}{\mathscr{P}} \times \mathscr{I} \times \ldots \times \mathscr{I} \qquad (r = 1, 2, \ldots, \nu) \tag{37.4}$$

$$\gamma_{\nu+r} \to \Gamma_{\nu+r} = \mathscr{I}' \times \ldots \times \mathscr{I}' \times \underset{(r)}{\mathscr{Q}} \times \mathscr{I} \times \ldots \times \mathscr{I}$$

and noting that $(A_1 \times A_2 \times \ldots \times A_n)(B_1 \times B_2 \times \ldots \times B_n) = (A_1 B_1 \times A_2 B_2 \times \ldots)$

$$\Gamma_r \Gamma_s + \Gamma_s \Gamma_r = 2 g_{rs} I_{(2^\nu)} \tag{37.5}$$

The algebra Π is of the same order $2^{2\nu} = (2^\nu)^2$ as the algebra consisting of all matrices in the 2^ν-dimensional space, the relation $x \rightleftharpoons X$ between elements x of the algebra and their matrix image X is a 1–1 isomorphic mapping of Π upon the full matrix algebra of the 2^ν-dimensional *spin-space*. The proof of this assertion requires only that it be proved that the collection of $2^{2\nu}$ matrices $\Gamma_1{}^{A_1} \Gamma_2{}^{A_2} \ldots \Gamma_n{}^{A_n}$, where each of the A's takes on the two values 0, 1, constitutes a basis for the full matrix algebra of dimension 2^ν; that is, the algebra generated by the matrices $\Gamma_1, \ldots, \Gamma_{2\nu}$ has the full 2^ν-dimensional matrix algebra as its enveloping algebra. It is readily shown that from these elements we can construct $(2^\nu)^2$ distinct matrices such that each has but a single non-zero element which is unity (cf. Weyl, Murnaghan, *loc. cit.*), which therefore constitute an orthonormal set in the $2^{2\nu}$-dimensional vector space.

In particular, for $\nu = 2$, we find:

$$\Gamma_1 = \begin{Vmatrix} 0 & 0 & i & 0 \\ 0 & 0 & 0 & i \\ i & 0 & 0 & 0 \\ 0 & i & 0 & 0 \end{Vmatrix} \qquad \Gamma_2 = \begin{Vmatrix} 0 & i & 0 & 0 \\ i & 0 & 0 & 0 \\ 0 & 0 & 0 & -i \\ 0 & 0 & -i & 0 \end{Vmatrix} \tag{37.6}$$

$$\Gamma_3 = \begin{Vmatrix} 0 & 0 & -1 & 0 \\ 0 & 0 & 0 & -1 \\ 1 & 0 & 0 & 0 \\ 0 & 1 & 0 & 0 \end{Vmatrix} \qquad \Gamma_4 = \begin{Vmatrix} 0 & -1 & 0 & 0 \\ 1 & 0 & 0 & 0 \\ 0 & 0 & 0 & 1 \\ 0 & 0 & -1 & 0 \end{Vmatrix}$$

and, according to the prescription for the transition to the Lorentz group and metric signature $(- - - +)$, the corresponding set will be simply $(\Gamma_1, \Gamma_2, \Gamma_3, i\Gamma_4)$. The latter have the same properties as the $\gamma^{\mu a}{}_b$ (14.4b), and, in fact,

$$\Gamma_\mu = U\gamma_\mu U^{-1} \tag{37.7}$$

$$U = \frac{1}{\sqrt{2}}\begin{Vmatrix} 0 & 0 & 1 & -i \\ -1 & i & 0 & 0 \\ i & -1 & 0 & 0 \\ 0 & 0 & -i & 1 \end{Vmatrix}$$

The *necessity* of the equivalence of this representation to the previous one in the $\gamma^{\mu a}{}_b$ may be viewed as a consequence of the fundamental lemma: *the full matrix algebra allows only inner automorphisms*; that is, every automorphism $A \to D(A) = UAU^{-1}$ is *on* transforming matrices U which themselves belong to the set $\{A\}$, hence the term *inner*. To return to the general case, we say, therefore, that if we can construct a new set Γ_r' satisfying the basic relations (37.5), the correspondence $\Gamma_r \to \Gamma_r'$ must be an inner automorphism, there being a matrix S of dimension 2^ν which transforms [2] Γ_r into Γ_r', every r. Again considering the orthogonal group O_n, the matrices defined by the correspondence

$$\Gamma_r \to \Gamma_r' = o_{rs}\Gamma_s \tag{37.8a}$$

(through rotations $o = \| o_{rs} \|$ in the n-dimensional space) evidently also satisfying the relations (37.5), and by the fundamental proposition on automorphisms of the algebra Π, this representation is equivalent to the original. Hence there exists a non-singular matrix $S(o)$ determined, but for a numerical (or gauge) factor, by

$$\Gamma_r' = S(o)\Gamma_r S(o)^{-1} \tag{37.8b}$$

$S(o)$ is to be interpreted not as an affine transformation of the 2^ν-dimensional vector space, but in the homogeneous sense of a *collineation* [3] representation of the projective space consisting of its rays. Once the gauge factor for two rotations o, o' and their product $o'o$ is fixed (in an arbitrary manner), there necessarily exists a relation of the type

$$S(o'o) = \kappa S(o')S(o) \tag{37.9}$$

so that we have a collineation representation of degree 2^ν of the rotation group; the so-called *spin representation* $\Delta : o \to S(o)$.

The quantities ψ^r and ϕ_r, transforming according to the representation Δ and the contragredient representation Δ^0, are termed contravariant and covariant spinors, respectively. By comparison of (37.8, 9) for the case $\nu = 1$ and (10.6a, b) of § 10b, this is seen to be in accord with our previous notation and nomenclature.

(b) *General Connection between Covariant and Contravariant Spinors*

Before considering the slight modification involved for the case of odd dimensionality, we examine the general connection between covariant and contravariant spinors. This devolves on the *equivalence* of the representation Δ^0 to the repre-

[2] The proof is elementary and need not be developed here. See, for example, the preceding references or Brauer-Weyl (35).

[3] The significance of the terms homogeneous and collineatory is simply that only the *ratios* of the spinor components have meaning. If the collineation representation is suitably normalized, we obtain an ordinary, though double-valued, representation. Cf. § 37b.

sentation Δ, as known to us already for special cases. Observe that, since the transposed matrices $\tilde{\Gamma}_s$ satisfy (37.5), there must exist a definite matrix C, such that for all s

$$\tilde{\Gamma}_s = C\Gamma_s C^{-1} \tag{37.10a}$$

Now since $\tilde{\mathscr{P}} = \mathscr{P}$ and $\tilde{\mathscr{Q}} = -\mathscr{Q}$, it follows that

$$\tilde{\Gamma}_r = \Gamma_r \qquad \tilde{\Gamma}_{\nu+r} = -\Gamma_{\nu+r} \qquad (r = 1, 2, \ldots, \nu) \tag{37.10b}$$

which property directly determines C; for, the product $\gamma_1 \ldots \gamma_\nu$ commutes (anticommutes) with the γ_r and anticommutes (commutes) with the $\gamma_{\nu+r}$, if ν is odd (even), so that we can take

$$c \sim \gamma_1 \ldots \gamma_\nu \quad \text{or} \quad c \sim \gamma_{\nu+1} \ldots \gamma_{2\nu}$$

according as ν is odd or even. In this way we obtain in both cases

$$C = \mathscr{P} \times \mathscr{Q} \times \mathscr{P} \times \ldots \qquad (\nu \text{ factors}) \tag{37.11}$$

Corresponding to (37.8),

$$\tilde{\Gamma}_r \to \tilde{\Gamma}_r' = o_{rs}\tilde{\Gamma}_s \tag{37.12a}$$

is expressed by

$$\tilde{\Gamma}_r \to \tilde{S}(o)^{-1}\tilde{\Gamma}_r\tilde{S}(o) = S^0(o)\tilde{\Gamma}_r S^0(o)^{-1} \tag{37.12b}$$

and, since the transformation $\tilde{\Gamma}_r = C\Gamma_r C^{-1}$ is performed by $CS(o)C^{-1}$, there must obtain an equation of the form

$$CS(o)C^{-1} = \rho(o)S^0(o) \qquad (\rho, \text{ a numerical factor}) \tag{37.12c}$$

On multiplication of $S(o)$ by λ, $S^0(o)$ is multiplied by $1/\lambda$ and $\rho \to \rho\lambda^2$, hence the arbitrary gauge factor in S may be chosen such that $\rho \to 1$:

$$S^0(o) = CS(o)C^{-1} \tag{37.13a}$$

Then

$$(\det S)^2 = 1 \tag{37.13b}$$

and $S(o)$ is uniquely determined, apart from sign. After normalizing this sign for two rotations o, o' and their product $o'o$, in an arbitrary manner, the factor κ in (37.9) becomes ± 1—the matrices $X = S(o')S(o)$ and $X = S(o'o)$ both satisfy the normalizing condition $X^0 = CXC^{-1}$. Δ is then an ordinary, though double-valued, representation.

Equation (37.13a) states the general relation between covariant and contravariant spinors:

$$\phi_r = C_{rs}\psi^s \tag{37.14}$$

For the important case $\nu = 2$ we find

$$C = \mathscr{P} \times \mathscr{Q} = \begin{Vmatrix} 0 & 0 & 0 & -i \\ 0 & 0 & i & 0 \\ 0 & -i & 0 & 0 \\ i & 0 & 0 & 0 \end{Vmatrix} \tag{37.15}$$

which, by (37.7, 10), goes into the metric 4-spinor $\varepsilon = \tilde{U}CU$ in the representation of § 14f.

Finally, if the dimensionality is odd, $n = 2\nu + 1$, we simply adjoin a further quantity $\gamma_{2\nu+1}$ whose square is again -1 and which anticommutes with the previous γ_s. The representation $\gamma_s \to \Gamma_s$ $(s = 1, 2, \ldots, 2\nu)$ can be extended by establishing the correspondence

$$\gamma_n \to \Gamma_n = (\mathscr{I}' \times \mathscr{I}' \times \ldots \times \mathscr{I}') \qquad (n = 2\nu + 1) \tag{37.16}$$

The quantity $u = \backepsilon\gamma_1 \ldots \gamma_n$, with $\backepsilon = 1$ or i according as ν is even or odd, satisfies $u^2 = -1$ and commutes with all elements of the algebra. It can therefore only be represented by $\pm$ a multiple of i, and corresponding to these two possibilities there are two inequivalent representations of degree 2^ν, one obtaining from the other by mere reversal of sign of all γ's.

The order $2(2^\nu)^2$ of the algebra Π is now twice as large as the order of the algebra of all matrices in the 2^ν-dimensional spin-space. Since we do not require further properties of this case, we refer to the article by Weyl (35) for detailed discussion. Let us note, however, that since

$$(-)^\nu \tilde{\Gamma}_{2\nu+1} = C\Gamma_{2\nu+1}C^{-1} \tag{37.17}$$

the matrix C can be used for the purpose of connecting covariant and contravariant spinors only if ν is even. In the contrary case we must replace C by $C\Gamma_{2\nu+1}$, for which $C\Gamma_s C^{-1} = -\tilde{\Gamma}_s$, all s. In both cases the equation (37.13*a*) obtains after appropriate normalization, and $\Delta^0 \sim \Delta$ which allows the explicit connection between the two types of spinors.

(c) *The Spinor Representations of the Lorentz Group in n Dimensions*

In the general case we may have a fundamental indefinite ground form

$$-(x^1)^2 - (x^2)^2 - \ldots - (x^t)^2 + (x^{t+1})^2 + \ldots + (x^n)^2 \tag{37.18}$$

of so-called inertial index $(n - t)$, corresponding to t space-like and $(n - t)$ time-like variables. The $\Gamma_1, \ldots, \Gamma_t$ keep their previous significance, while the $\Gamma_{t+1}, \ldots, \Gamma_n$ assume the factor i. Since, for the previous Γ_s,

$$\bar{\Gamma}_s = -\tilde{\Gamma}_s \qquad (s = 1, \ldots, 2\nu) \tag{37.19a}$$

we now have

$$\bar{\Gamma}_s = \begin{cases} -\tilde{\Gamma}_s & s = 1, \ldots, t \\ \tilde{\Gamma}_s & s = t+1, \ldots, n \end{cases} \tag{37.19b}$$

and there must again exist a matrix B which, for all s, transforms the $\bar{\Gamma}_s$ into the $\tilde{\Gamma}_s$:

$$\begin{aligned} \tilde{\Gamma}_s &= B\bar{\Gamma}_s B^{-1} \\ -\tilde{\Gamma}_s &= B\bar{\Gamma}_s B^{-1} \end{aligned} \tag{37.20}$$

according as $(n - t)$ is even or odd. For the important case $\nu = 2$, $(n - t) = 1$, we have (choosing B skew-Hermitian)

$$B = i\,\Gamma_1\Gamma_2\Gamma_3 = \begin{Vmatrix} 0 & i & 0 & 0 \\ i & 0 & 0 & 0 \\ 0 & 0 & 0 & i \\ 0 & 0 & i & 0 \end{Vmatrix} \tag{37.21}$$

We shall not pursue the development, which clearly parallels that for the orthogonal group, except to note that from (37.20) there follows

$$B\bar{S}(l)\,B^{-1} = \rho(l)\,S^0(l) \tag{37.22}$$

or

$$\tilde{S}(l)\,B\bar{S}(l) = \rho B$$

As in the definite case, we infer that $\rho(l) = \pm 1$ and $\rho(l'l) = \rho(l')\rho(l)$, and it is readily shown that $\rho(l)$ coincides with the *character* $\sigma_-(l)$ which is the *temporal signature* of the temporal minor of the Lorentz transformation $\| l^r{}_s \|$.

Any representation $l \to G(l)$ of the Lorentz group gives rise to another one, $l \to \sigma_-(l)G(l)$, and equation (37.22), or

$$\bar{S}(l) = \sigma_-(l)\,B^{-1}S^0(l)\,B \tag{37.23}$$

then proves the equivalence

$$\bar{\Lambda} \sim \sigma_-\Lambda^0$$

It follows that, insofar as we restrict ourselves to Lorentz transformations which *do not* reverse the sense of time ($\sigma_- = 1$), the transformation B changes the conjugate of a contravariant spinor ψ into a covariant spinor $\phi : \phi = B\bar{\psi}$. The form of B (37.21) in the representation of §§ 14, 30 is then given by (cf. 37.7)

$$\tilde{U}B\bar{U} = \begin{Vmatrix} 0 & 0 & -i & 0 \\ 0 & 0 & 0 & -i \\ -i & 0 & 0 & 0 \\ 0 & -i & 0 & 0 \end{Vmatrix} \tag{37.24}$$

that is, the charge-conjugator $\pounds_{ab}$.

(*d*) *The Infinitesimal Description*

Since infinitesimal methods play such a significant role, it is of interest to examine the corresponding correlate to the preceding finite description derived by algebraic means. For this purpose it is more convenient to express the invariant form (37.2) as

$$-(z_1y_1 + z_2y_2 + \ldots + z_\nu y_\nu) \tag{37.25}$$

the z_r, y_r being the $n = 2\nu$ variables. Hence, the linearization now involves the quantities

$$\alpha_r = \tfrac{1}{2}(\gamma_r - i\gamma_{\nu+r}) \qquad \beta_r = \tfrac{1}{2}(\gamma_r + i\gamma_{\nu+r}) \tag{37.26a}$$

to which correspond the relations

$$\left.\begin{aligned} \alpha_r\beta_s + \beta_s\alpha_r &= g_{rs} \\ \alpha_r\alpha_s + \alpha_s\alpha_r &= 0 \\ \beta_r\beta_s + \beta_s\beta_r &= 0 \end{aligned}\right\} \qquad (r,\, s = 1, 2, \ldots, \nu) \tag{37.26b}$$

$$\alpha_r \to \mathfrak{A}_r = \mathscr{I}' \times \ldots \times \mathscr{I}' \times \begin{Vmatrix} 0 & i \\ 0 & 0 \end{Vmatrix} \times \mathscr{I} \ldots \times \mathscr{I} \qquad \text{(cf. 37.4)} \tag{37.26c}$$

$$\beta_r \to \mathfrak{B}_r = \mathscr{I}' \times \ldots \times \mathscr{I}' \times \begin{Vmatrix} 0 & 0 \\ i & 0 \end{Vmatrix} \times \mathscr{I} \times \ldots \times \mathscr{I}$$

All infinitesimal rotations are linear combinations of rotations of the types

$$\delta z_r = z_r \qquad \delta y_r = -y_r \tag{37.27a}$$

$$\delta z_r = z_s \qquad \delta y_s = -y_r \qquad (r < s) \tag{37.27b}$$ [4]

all other increments being zero. These are represented in Δ by the infinitesimal transformations

$$\tfrac{1}{2}\mathfrak{U} = \tfrac{1}{2}(-i)^{\nu}\,(\mathscr{I} \times \ldots \times \mathscr{I} \times \mathscr{I}' \times \mathscr{I} \times \ldots \times \mathscr{I}) \tag{37.28a}$$

$$-\mathfrak{A}_r\mathfrak{B}_s = (-)^{\nu+1}\left(I \times \ldots \times I \times \underset{(r)}{\begin{Vmatrix} 0 & -1 \\ 0 & 0 \end{Vmatrix}} \times \begin{Vmatrix} 1 & 0 \\ 0 & -1 \end{Vmatrix} \times \ldots \times \begin{Vmatrix} 1 & 0 \\ 0 & -1 \end{Vmatrix} \times \underset{(s)}{\begin{Vmatrix} 0 & 0 \\ 1 & 0 \end{Vmatrix}} \times I \times \ldots \times I\right) \tag{37.28b}$$

the proof of which requires only verification of the equations

$$\delta\mathfrak{X} = \tfrac{1}{2}[\mathfrak{U}, \mathfrak{X}] = \begin{Bmatrix} \mathfrak{A}_r \\ -\mathfrak{B}_r \end{Bmatrix} \qquad \begin{matrix} \mathfrak{X} = \mathfrak{A}_r \\ \mathfrak{X} = \mathfrak{B}_r \end{matrix} \tag{37.29a}$$

$$\delta\mathfrak{X} = [-\mathfrak{A}_r\mathfrak{B}_s, \mathfrak{X}] = \begin{Bmatrix} \mathfrak{A}_r \\ -\mathfrak{B}_s \end{Bmatrix} \qquad \begin{matrix} \mathfrak{X} = \mathfrak{A}_s \\ \mathfrak{X} = \mathfrak{B}_r \end{matrix} \tag{37.29b}$$ [5]

It follows from (37.28a) that the finite rotation o:

$$z_1 \to e^{i\theta} z_1 \qquad y_1 \to e^{-i\theta} y_1$$

is associated with the operation $S(o)$ multiplying the variable in the spin-space (which may be labelled by the indices $\sigma_1 \ldots \sigma_\nu$, $[\sigma_r = \pm 1]$) by $e^{\frac{1}{2}\sigma_1\theta}$, showing again that the representation Δ is double-valued.

§ 38. The α-Algebras

(a) *Group-theoretical Method of Deriving and Decomposing Representations*

With the results of the two preceding sections before us, the problem of deriving the irreducible representations of the matrix-algebra or -ring $\mathfrak{A}_N(\alpha)$ generated by the α^μ (represented as in equations 36.4) in the wave equation

$$(\alpha^\mu \partial_\mu - iN\kappa)\Psi = 0 \qquad (N = 2s) \tag{38.1}$$ [1]

is referred back to the study of the irreducible representations $D^{(5)}(n, m)$ of the 5-dimensional orthogonal group. The definition (36.4) of the ring $\mathfrak{A}_N(\alpha)$ assumes that the representations of the algebras generated by the α^μ's are contained in the direct product

$$\Delta \times \Delta \times \ldots \times \Delta \times \Delta \tag{38.2a}$$

[4] Note that in (37.27b) we may exchange, independently of each other, $z_r \rightleftharpoons y_r$ and $z_s \rightleftharpoons y_s$.

[5] This follows directly from (37.26b) on writing

$$[\mathfrak{A}_r\mathfrak{B}_s, \mathfrak{X}]_- = \mathfrak{A}_r[\mathfrak{B}_s, \mathfrak{X}]_+ - [\mathfrak{X}, \mathfrak{A}_r]_+ \mathfrak{B}_s.$$

[1] Cf. the example of the meson equations, § 30b.

where Δ is the basic spinor representation of the rotation group in 5 dimensions,[2] and the number of factors is $2s$ or $2s+1$ according as the spin is s or $s+\frac{1}{2}$. For $N>1$, the ring $\mathfrak{A}_N(\alpha)$ is reducible (cf. *infra*), so that the basic problem of obtaining irreducible representations is simply the decomposition of the direct product of the group representation Δ (or $D^{(5)}(\frac{1}{2},\frac{1}{2})$) taken with itself a certain number of times.[3]

The solution of this problem is well known in group theory, and is given by

$$D^{(5)}(\tfrac{1}{2},\tfrac{1}{2}) \times \ldots (N \text{ factors}) \ldots \times D^{(5)}(\tfrac{1}{2},\tfrac{1}{2}) \sim \sum_{n=\epsilon}^{N/2} \sum_{m=\epsilon}^{n} C^N_{n,m} D^{(5)}_{(n,m)} \qquad (38.2b)$$

where ϵ is 0 or $\frac{1}{2}$ according as N is even or odd, and $C^N_{n,m}$ denotes the multiplicity of *equivalent* irreducible components $D^{(5)}(n,m)$. In turn, this decomposition obtains by mere repetition of the basic formula

$$D^{(5)}(\tfrac{1}{2},\tfrac{1}{2}) \times D^{(5)}(n,m) \sim D^{(5)}(n+\tfrac{1}{2}, m+\tfrac{1}{2}) \oplus D^{(5)}(n+\tfrac{1}{2}, m-\tfrac{1}{2})$$
$$\oplus D^{(5)}(n-\tfrac{1}{2}, m+\tfrac{1}{2}) \oplus D^{(5)}(n-\tfrac{1}{2}, m-\tfrac{1}{2}) \qquad (38.3)$$

where it is understood that a term for which $n-\frac{1}{2}<m+\frac{1}{2}$ is to be omitted. While it would not be profitable to consider here the problem in complete detail, it is worth noting that the essential results are quite easily obtained. Indeed, we need merely note that the operator

$$B(n,m) = -\tfrac{1}{2}\mathfrak{T}^{\kappa\lambda}(n,m)\mathfrak{T}_{\kappa\lambda}(n,m) \qquad (38.4)^\dagger$$

is an invariant of O_5 and has the value $n(n+3)+m(m+1)$ in $D^{(5)}(n,m)$.[4] Then, as for the 3-dimensional rotation group and 4-dimensional Lorentz group, the invariant operator

$$A(n,m) = -\tfrac{1}{2}\mathfrak{T}^{\kappa\lambda}(\tfrac{1}{2},\tfrac{1}{2}) \times \mathfrak{T}_{\kappa\lambda}(n,m) \qquad (38.5)^\dagger$$

commutes with the infinitesimal transformations

$$\mathfrak{T}^{\kappa\lambda}(\tfrac{1}{2},\tfrac{1}{2}) \times I_d + I_{d'}' \times \mathfrak{T}^{\kappa\lambda}(n,m) \qquad (38.6)^\dagger$$

of the nucleus of the product representation, I_d and $I_{d'}'$ denoting the unit matrices of dimensions $d(n,m)$ and $d(\frac{1}{2},\frac{1}{2})$ of $\mathfrak{R}^{(5)}(n,m)$ and $\mathfrak{R}^{(5)}(\frac{1}{2},\frac{1}{2})$, respectively. The decomposition may therefore again be referred to finding matrices, which will now be 4-spinor matrices and four in number, such that A is brought to diagonal form. It follows from (38.4–38.6) that $A(n,m)$ satisfies a fourth-degree equation corresponding to eigenvalues

$$\tfrac{1}{2}(n+m),\ \tfrac{1}{2}(n-m-1),\ -\tfrac{1}{2}(n-m+3),\ -\tfrac{1}{2}(n+m+4) \qquad (38.7a)$$

with respective multiplicities

$$d(n+\tfrac{1}{2}, m+\tfrac{1}{2}),\ d(n+\tfrac{1}{2}, m-\tfrac{1}{2}),\ d(n-\tfrac{1}{2}, m+\tfrac{1}{2}),\ d(n-\tfrac{1}{2}, m-\tfrac{1}{2}) \qquad (38.7b)$$

[2] The case of n (arbitrary) dimensions is also amenable to this approach, as will be indicated in § 39, but is of little physical interest except for certain special considerations on 6-dimensional spaces. (Cf. § 39*d*.)

[3] See for example, Murnaghan, *loc. cit.*

[4] In the general case of $n(2\nu$ or $2\nu+1)$ dimensions, the irreducible representations $D^{(n)}(n_1, n_2 \ldots n_\nu)$ are labelled by ν parameters, and we may show that

$$-\tfrac{1}{2}\mathfrak{T}^{\kappa\lambda}\mathfrak{T}_{\kappa\lambda} = \sum_{r=1}^{\nu} n_r(n_r+n-2r)$$

where $d(n, m) = \frac{1}{6}(2n + 3)(2m + 1)(n + m + 2)(n - m + 1)$. The check by degrees gives $4d(n, m) = d(n + \frac{1}{2}, m + \frac{1}{2}) + \ldots + d(n - \frac{1}{2}, m + \frac{1}{2})$. It is to be noted that the vanishing of $d(n - \frac{1}{2}, m + \frac{1}{2})$ for $n = m$ automatically eliminates the spurious component $D^{(5)}(n - \frac{1}{2}, m + \frac{1}{2})$ for $n - \frac{1}{2} < m + \frac{1}{2}$, and, as required by the vanishing of the spur of A—which corresponds to a symmetric second-rank 4-spinor—the sum of the roots multiplied by their multiplicities is zero.

The decomposition (38.2*b*) into irreducible constituents leads to a direct sum of all possible wave-equations, of this type, for particles of spin s *and* those of lower spin. The wave-functions so obtained may be further decomposed into parts according to the three characteristics: (i) the irreducible representation of the $\mathfrak{A}_N(\alpha)$, (ii) the different spin values, (iii) the different rest-mass values. Except for the cases $N = 1, 2$, the resulting equations appear necessarily to be interpreted as describing composite systems of particles—corresponding to the various values of spin and mass, i.e. the component states nominally called *excited* states of spin and mass. The interested reader may find full details in the references listed.

Because it is unlikely that the *higher-spin composite* cases correspond to anything in Nature, we shall not consider the general Ψ's of equation (38.1) which are 4-spinors $\Psi^{a_1 \cdots a_N}$ without special symmetry properties and form the basis for *reducible* representations of the 5-dimensional group and, hence, the ring of matrices. The higher-spin equations of Dirac, Fierz, and Pauli are probably no more likely of physical realization, but it is of interest to note, in passing, that these obtain as a special case of the present formalism when $\Psi^{a_1 \cdots a_N}$ is a completely symmetric 4-spinor (cf. § 40). It is not difficult to verify that in this instance each component satisfies the second-order wave-equation, and likewise that there appears *but one* spin value (cf. § 40). Indeed, the corresponding representation $D^{(5)}\left(\frac{N}{2}, \frac{N}{2}\right)$ is irreducible, the simplest case being the spin 1 equation ($N = 2$) of the meson field discussed in §§ 15*b*, 30*b*, *c*.

The world-tensor equivalents of a second-rank symmetric 4-spinor, as given in § 15*b*, illustrate nicely the detailed transition: 4-dimensional Lorentz → 5-dimensional Lorentz → 5-dimensional Orthogonal group, the (pseudo) 4-vector and 6-vector combining to form a 5-dimensional skew-tensor or 10-vector with the previously noted reinterpretation of tensor components with index 4. Conversely, we have the reduction $G^{\dagger}{}_{\kappa\lambda} \to G_{\kappa\lambda} \oplus G_{\kappa 5}$ and similarly $F^{\dagger}{}_{\kappa} \to F_{\kappa} \oplus F_5$, which examples illustrate the fact that every irreducible representation of the Orthogonal group (or Lorentz group, depending on the reality conditions on the parameters) in five dimensions provides an, in general, reducible representation of the Lorentz group in four dimensions—if we consider only rotations, keeping the 5-axis fixed.

The reduction above corresponds to the decomposition

$$
\begin{aligned}
D^{(5)}(1, 1) &\to \{1, 1\} \oplus \{1, 0\} \qquad (38.8a)\\
D^{(5)}(1, 0) &\to \{1, 0\} \oplus \{0, 0\}
\end{aligned}
$$

while the higher-spin coupling case mentioned in § 35 corresponds to

$$D^{(5)}(\tfrac{3}{2}, \tfrac{1}{2}) \to \{\tfrac{3}{2}, \tfrac{1}{2}\} \oplus \{\tfrac{1}{2}, \tfrac{1}{2}\} \qquad (38.8b)$$

The decomposition of the wave functions discussed in §§ 35*a*, *b* then follows directly from the relation (cf. Murnaghan, *loc. cit.*)

$$D^{(5)}(n, m) \to \sum_{p=m}^{n} \sum_{q=\epsilon}^{m} \{p, q\} \tag{38.9a}$$

or, in terms of the constituent irreducible representations of the *proper* Lorentz group,

$$D^{(5)}(n, m) \to \sum_{p=m}^{n} \sum_{q=-m}^{m} \mathfrak{D}(p, q) \tag{38.9b}$$

where $\{n; m\}$ is connected to the usual $\mathfrak{D}(k, l)$ by the relation

$$\{n, m\} = \left\{ \begin{array}{ll} \mathfrak{D}\left(\frac{n+m}{2}, \frac{n-m}{2}\right) \oplus \mathfrak{D}\left(\frac{n-m}{2}, \frac{n+m}{2}\right) & m \neq 0 \\ \mathfrak{D}\left(\frac{n}{2}, \frac{n}{2}\right) & m = 0 \end{array} \right\} \tag{38.9c}$$

that is, $2k = p + q$, $2l = p - q$. A detailed proof of this result is scarcely required, since it is readily inferred by considering the first few values of (n, m), particularly for symmetric 4-spinors,[5] and the check by degrees.

$$\sum_{p=m}^{n} \sum_{q=-m}^{m} (p + q + 1)(p - q + 1) = d(n, m)$$

Returning to the more formal algebraic aspects, we can say that the $\alpha^\mu(n, m)$ are matrices of degree $d(n, m)$ of the *irreducible* representations of the ring $\mathfrak{A}_N(\alpha)$ afforded by $D^{(5)}(n, m)$. It is most important to observe that from (38.2*b*) we obtain the number of *independent* elements of the ring

$$Z(N) = \sum_{n=\epsilon}^{N/2} \sum_{m=\epsilon}^{n} d^2(n, m) \tag{38.10a}$$

as well as the number of inequivalent irreducible representations

$$\eta(N) = \binom{s+2}{s} \tag{38.10b}$$ [6]

For example, for $N = 1, 2, 3, 4$ we find, respectively,

$Z(N)$	$\eta(N)$
16	1
126	3
672	3
2772	6

(38.11)

[5] From the 2-spinor composition, it follows that a totally symmetric 4-spinor of rank N determines a series of symmetric Nth rank 2-spinors, of which the prototype is symmetric in its $(\frac{1}{2}N + q)$ covariant indices and $(\frac{1}{2}N - q)$ contravariant conjugated indices, the index q taking on the $N + 1$ values $(\frac{1}{2}N \ldots -\frac{1}{2}N)$. For given N and q the number of independent components is evidently $(\frac{1}{2}N + q + 1)(\frac{1}{2}N - q + 1)$ and their sum for all q is just $d(\frac{1}{2}N, \frac{1}{2}N.)$

[6] The total number of representations, which is evidently the same for spins s and $s + \frac{1}{2}$, is just the number of ways of choosing the representation indices n, m such that $n = 0, \ldots, \nu$ and $m = 0, \ldots, n$. In the general case of n (arbitrary) dimensions, we find $\binom{\nu + s}{s}$.

For the third, or spin $\frac{3}{2}$, case, for example, it follows that the triple direct product of Δ contains the 20th order representation once, the 16th order representation twice, and the 4th order (Dirac) representation three times, since the corresponding third-rank 4-spinor has 64 components, which gives the check by degrees: $64 = 20 + 2.16 + 3.4$. This illustrates the general result that the representations of the algebra for spin s or $s + \frac{1}{2}$ contain those which belong to the lower spins $(s - 1)$ and $(s - \frac{1}{2})$ respectively. Otherwise stated, this means that the α's corresponding to the *lower* spins *also* satisfy the commutation rules of the respective α's for *higher* spins.

(*b*) *Algebraic Methods of Deriving and Decomposing Representations*

While the general theory of abstract algebras, *per se*, can be made as complicated as desired, the concepts and theorems deriving therefrom—which have found use in the theory of elementary particles—are essentially elementary. Indeed, many of the concepts are carried over from the generically related theory of groups, so that the proof of many *corresponding* theorems is virtually self-evident. With the advantage gained from the preceding group-theoretical analysis, it will best suit our needs to consider the essential[7] algebraic definitions and theorems first with respect to the Dirac and Duffin-Kemmer algebras $\mathfrak{A}(\gamma)$, $\mathfrak{A}(\beta)$ which occur most often in physical applications, and *together* illustrate and comprise *all* the necessary fundamentals of the algebraic aspect. In general we shall now adopt the convention that corresponding types, such as A and a, $\mathfrak{A}$ and $\mathfrak{a}$ of the upper and lower case, mark the transition from matrices to abstract elements, the matrix algebra $\mathfrak{A}$ being a *faithful representation* of the abstract algebra $\mathfrak{a}$.

Apart from the basic properties of a ring or algebra already defined in § 37*a*, an algebra is said to contain *the* unit e (which is unique) if for all elements a in $\mathfrak{a}$ ($a \varepsilon \mathfrak{a}$ or $a \subset \mathfrak{a}$) : $ae = ea = a$. Naturally enough, it is further termed a *division* algebra if every element $a \neq 0$ possesses an inverse $a^{-1} : aa^{-1} = a^{-1}a = e$. All the α-algebras which we consider possess the unit. None of them is a division algebra although, for half-integral spins, the generating elements $\alpha_1, \ldots, \alpha_4$ do possess inverses. As a consequence, the transition from the linear wave-equation to the Hamiltonian form only follows by simple multiplication by $\alpha_4{}^{-1}$ for half-integrals pins. This difference corresponds, *inter alia*, to the appearance of the eigenvalue 0 in the generating α's for integral spin (from which it follows that the α's are singular and their reciprocals do not exist), and to the previously mentioned occurrence of redundant components of Ψ for integral spins (in particular the meson equations). It has the further awkward consequence that certain physical quantities which are identically zero may be *represented* by an operator which is not identically zero—though all its expectation values are zero, as is to be expected. (Cf. Kemmer (39), Heitler (43)). None the less, a *reduced* Hamiltonian form can be obtained, simultaneously with the circumvention of these awkward features, by the so-called Peirce decomposition, which we shall apply to the meson case in the next section.

[7] We do not seek to introduce the most general methods possible, nor do we use concepts which are not material for application in field theory. The few basic principles may be found, for example, in Weyl (31, 39). More general aspects of specifically algebraic context may be found in the works of van der Waerden (31), Artin (46), and Albert (37).

When dealing with a representation $\mathfrak{a}(a) \to \mathfrak{A}(A)$ of an algebra $\mathfrak{a}$ containing the unit element e, there will correspond to e an idempotent matrix $J(e)$, that is $JJ = J$. In the vector space P of the representation we can construct subspaces P_1, P_2 of those vectors Ψ_1, Ψ_2 for which

$$\begin{aligned}(1 - J)\Psi_1 &= 0 \\ J\Psi_2 &= 0\end{aligned} \tag{38.12a}$$

The splitting $P \to P_1 \oplus P_2$ gives linearly independent subspaces, since $\Psi = \Psi_1 \oplus \Psi_2$ implies $J\Psi = \Psi_1$ and leads to the unique decomposition

$$\begin{aligned}\Psi_1 &= J\Psi \\ \Psi_2 &= (1 - J)\Psi\end{aligned} \tag{38.12b}$$

In a coordinate system adapted to this decomposition, J is the unit matrix rimmed by zeros:

$$\| J \| = \begin{Vmatrix} I & 0 \\ 0 & 0 \end{Vmatrix}$$

and every A $(= A(a))$ carries a rim of zeros of the same width in consequence of $AJ = JA = A$ which images the relation $ae = ea = a$. Making use of this idempotent, we can decompose the ring $\mathfrak{A}$ according to

$$\mathfrak{A} = J\mathfrak{A}J + J\mathfrak{A}(1 - J) + (1 - J)\mathfrak{A}J + (1 - J)\mathfrak{A}(1 - J) \tag{38.13}$$

or

$$\mathfrak{A} = J\mathfrak{A}J + J\mathfrak{B} + \mathfrak{C}J + \mathfrak{D} \tag{38.14}$$

where $\mathfrak{B}$ and $\mathfrak{C}$, respectively, are termed the *left* and *right ideal* and $\mathfrak{D}$ is defined by $\mathfrak{D} = [\mathfrak{B}, \mathfrak{C}]_-$. With respect to a ring $\mathfrak{A}$, which need not be a matrix ring, a subset $\mathfrak{B}$ is said to be a *left ideal* if (i) $\mathfrak{B}$ is a subgroup (relative to addition in $\mathfrak{A}$) and (ii) $\mathfrak{A}\mathfrak{B} \subseteq \mathfrak{B}$. For a *right ideal* $\mathfrak{C}$, (ii) is replaced by (ii′) $\mathfrak{C}\mathfrak{A} \subseteq \mathfrak{C}$, and a *two-sided ideal* such as the commutator $\mathfrak{D}$ satisfies (i), (ii), and (ii′). Its elements X evidently satisfy $JX = XJ = 0$.

The components in (38.13) or (38.14) constitute subrings, and if we confine ourselves to the subring $J\mathfrak{A}J$, then J is the unit element by virtue of $JA = AJ = A$ for every element A in $J\mathfrak{A}J$.

A ring $\mathfrak{a}$ can be considered in the dual role of vector space as well as operator domain, i.e. as an $\mathfrak{a}$-vector space, with ring multiplication from the left being the operation of $\mathfrak{a}$ (as ring) on $\mathfrak{a}$ (as space).[8] In this way we associate with each element a of $\mathfrak{a}$ the linear transformation

$$(a) : x \to x' = ax$$

which is to say, the left translation by a as x ranges over $\mathfrak{a}$. The correspondence $(\mathfrak{a}) : a \to (a)$ defines, as in group theory, the fundamental *regular representation* of degree equal to the order, say h, of the algebra.

[8] The usual conditions on a vector space as an Abelian group with the ring as operator domain reduce in this instance to certain of the ring postulates. Taking a matrix ring, purely for convenience of notation, the conditions that a subset $\mathfrak{B}$ of $\mathfrak{A}$ shall be a subspace are (i) $\mathfrak{B}$ is an Abelian subgroup, and (ii) $\mathfrak{A}\mathfrak{B} \subset \mathfrak{B}$, that is, exactly the conditions that $\mathfrak{B}$ be a left ideal.

Of the various types of algebras, we are concerned only with the so-called simple and semisimple algebras which, in the least sophisticated sense, are such that the first are capable of only one irreducible representation (apart from equivalence), while the second admit several irreducible representations, the corresponding regular representations being fully reducible.[9] We must necessarily sharpen these definitions, but it is a brief and elementary step from the above to the statement: *if* $\ni_r$ *denotes the matrix representing the rth independent element of a semisimple algebra* $\mathfrak{a}$ *in the regular representation, and* $\lambda_{rs} = \text{spur}\,(\ni_r \ni_s)$, *then* $\det \| \lambda_{rs} \| \neq 0$. This condition may be verified for the Dirac and Duffin-Kemmer algebras, but such a laborious procedure is not required since for both of these there holds the connection $\mathfrak{I}_{\kappa\lambda} \sim [\alpha_\kappa, \alpha_\lambda]_-$ and from this connection with the orthogonal group O_{n+1}, all of whose representations are fully reducible, it follows that every representation of these algebras is also fully reducible. In particular, this is true for the corresponding regular representations, so that both are semisimple.[10] It is important to note that though a semisimple algebra, *in abstracto*, ordinarily has several inequivalent irreducible representations ($\mathfrak{a}(a) \to \mathfrak{A}(A)$), any one of these is itself a simple matrix algebra. For the rigorous definition of simple and semisimple rings we require the concepts of *nilpotent ideal, radical,* and *ring with minimum condition.* An element of a ring $\mathfrak{R}$ is called nilpotent if a power $A^m = 0$, and similarly an ideal $\mathfrak{L}$ is nilpotent if a power $\mathfrak{L}^m$ is the null ideal. The *radical* of a ring is defined as the sum (hence the largest) of all nilpotent left ideals of $\mathfrak{R}$, and from this definition it may be shown that the radical $\mathfrak{N}$ is a nilpotent two-sided ideal, containing *all* nilpotent right and left ideals of $\mathfrak{R}$. Finally, $\mathfrak{R}$ is said to be a ring with minimum condition *on left ideals* if $\mathfrak{R}$, considered as a left $\mathfrak{R}$-space, has minimum condition on subspaces. The left $\mathfrak{R}$-subspaces are just the left ideals, and the existence of minimum condition for ring or space (the latter is more general) is simply the statement that there exists a *minimal member* of the descending chain of subspaces: $P \supset P_1 \supset P_2 \ldots$ which must therefore terminate.

By definition, both *simple* and *semisimple* rings satisfy the minimum condition on left ideals, the former having no two-sided ideals other than $\mathfrak{R}$ and the null ideal, the latter having radical zero. A minimal two-sided ideal of a semisimple ring can be shown to be a simple ring, thus justifying the name *simple ideal.* The term semisimple ring is justified by the structure of such a ring relative to its simple ideals as stated in the *principal* theorem: *A semisimple ring* $\mathfrak{R}$ *contains only a finite number of simple ideals and as a direct sum.*[11]

We obtain the equivalent statement: *Every semisimple algebra is expressible as a direct sum of simple algebras*:

$$\mathfrak{R}(\alpha) = \mathfrak{R}_1(\alpha) \oplus \mathfrak{R}_2(\alpha) \oplus \ldots \oplus \mathfrak{R}_r(\alpha) \tag{38.15}$$

where $\mathfrak{R}_s(\alpha)$ ($s = 1, 2, \ldots, r$) are all simple and the decomposition is unique, apart

[9] The former are termed normal simple algebras by some writers, indicating that only the unit element commutes with all elements of the algebra. Cf. *infra*, the concept of *centrum*.

[10] It may be shown that for n (arbitrary) generating elements α_μ, corresponding to n-dimensional space, the Dirac algebraic is also *normal* for even dimensionality. Cf. Kemmer (43); Harish-Chandra (47*a*, *c*).

[11] For the proof see Artin (46), Chapter IV.

from the order of the $\mathfrak{R}_s(\alpha)$ in the decomposition. Since, therefore, every irreducible representation of an abstract semisimple algebra is a faithful representation of a simple subalgebra (the other subalgebras being represented by zero matrices in this representation), it follows that we can specify an abstract algebra in such a way that it only has any one representation that we wish to select. This again is related to the Peirce decomposition, and in the next section we shall see how, in the example of spin 1, we can so specify the algebra—in addition to the Duffin-Kemmer commutation rules—that it possesses only one of the two (non-trivial) representations known to us.

The number of inequivalent irreducible representations of $\mathfrak{R}$ is r, the number of simple subalgebras in the decomposition (38.15), corresponding to which the unity quantity of $\mathfrak{R}$ takes the form

$$e = e_1 + \ldots + e_r \tag{38.16}$$

where $e_s \,\varepsilon\, \mathfrak{R}_s$ is the unity quantity of $\mathfrak{R}_s$. The $(e_1, \ldots, e_r)$ obviously form a linearly independent normal set $e_r e_s = \delta_{rs} e_s$, and are known as the *primitive*[12] *normal* idempotent generators of the *centrum* of $\mathfrak{R}$, that is, the set of elements commuting with *all* elements of $\mathfrak{R}$. Clearly the centrum is a *commutative* subring. It is variously referred to as the commutator or commutator algebra of $\mathfrak{R}$. From this we obtain the well-known theorem: *The number of inequivalent irreducible representations is also the order of the centrum.*[13] By examining the requirement of invariance of every irreducible representation $\mathfrak{A}(\alpha)$ under the orthogonal group (including reflections) it can be further shown that it is only necessary to obtain all scalar or pseudoscalar quantities in $\mathfrak{A}$ which commute with the α's, and to choose from them a linearly independent set $c_1(\alpha), \ldots, c_r(\alpha)$. Moreover, it follows that an irreducible representation is reflection-invariant only if the pseudoscalar elements are zero.[14]

As in group theory, the regular representation contains each irreducible representation a number of times equal to its dimensionality, and the connection (38.10*a*) between the order of the algebra and the sum of the squares of the irreducible representation follows in precisely the same way.[15]

We shall also require three additional theorems to the effect that (i) *every automorphism of a simple algebra is inner*, (ii) *every automorphism of a semisimple algebra (that is a direct sum of simple algebras) which leaves the elements of the centrum invariant is inner, and* (iii) *the direct product* $\mathfrak{A}(\alpha) = \mathfrak{A}(\gamma) \times \mathfrak{A}(\beta)$ *of two simple algebras is again a simple algebra.*

[12] A primitive or indecomposable idempotent admits only the trivial decomposition $e = e + 0$. In terms of these primitive idempotent elements of the centrum, the principal theorem on semisimple rings takes the form $\mathfrak{A} = \mathfrak{A}e_1 + \mathfrak{A}e_2 + \ldots$, the direct sum of left ideals.

[13] As previously remarked, when the centrum contains only multiples of the unit element, the algebra $\mathfrak{R}$ is normal (simple) and has but one representation.

[14] Cf. Kemmer (39), Harish-Chandra (47).

[15] Cf., for example, van der Waerden, *loc. cit.*; Weyl (31), esp. pp. 302–319.

§ 39. Derivation of Irreducible Representations by the Methods of Generating Idempotents, Algebra Products, and Peirce Decomposition

(a) *Introduction*

In this section we wish to illustrate, by definite examples, the principles and methods discussed. The examples chosen are at the same time the algebras which appear most frequently in physical applications; these include the rings associated with the names of Pauli, Dirac, Duffin-Kemmer, Chandra, and Bhabha, or, equivalently, spins ($\frac{1}{2}$), ($\frac{1}{2}$), (0, 1), ($\frac{3}{2}$, $\frac{1}{2}$) and ($\frac{3}{2}$, $\frac{1}{2}$), respectively.[1] The methods of derivation are here selected so as to illustrate the fundamental principles in the natural setting of developing these algebras and their individual properties.

The determination of matrix representations (irreducible or not) of the α-algebras is unavoidable as soon as we wish to make explicit calculations of physical quantities, but, with the exception of the case of spin ($\frac{1}{2}$), the obvious direct method becomes prohibitively difficult because of the rapid increase of dimensionality. Further, the spurs of the α's and their multiple products are important in many physical applications,[2] and even when the desired representation is explicitly known, the computation of spurs is tedious.

Both problems can be materially simplified through combination of various general techniques,[3] of which we consider first the *method of generating idempotents.*

(b) *The Pauli Ring*

The simplest ring of four independent elements (σ_1, σ_2, σ_3, 1) becomes the ring (σ, 1) of Pauli operators when subject to the commutation rule

$$\left.\begin{aligned} \sigma_i\sigma_j + \sigma_j\sigma_i &= 2\delta_{ij} \\ i\sigma_k &= \tfrac{1}{2}\epsilon_{klm}\sigma^l\sigma^m \end{aligned}\right\} (i, j, \ldots = 1, 2, 3) \tag{39.1}$$

Now, if for any α-ring we can find a complete set of *generating* idempotents (ϵ_1, ϵ_2, . . . , ϵ_t), which are functions of the α's satisfying

$$\left.\begin{aligned} 1 &= \epsilon_1 + \epsilon_2 + \ldots + \epsilon_t \\ \epsilon_r\epsilon_s &= \epsilon_s\delta_{rs} \end{aligned}\right\} \tag{39.2}$$

[1] The two algebras proposed for the composite system ($\frac{3}{2}$, $\frac{1}{2}$), that is, which includes states corresponding to both spins, differ in that the first is associated with but one rest-mass value and the second with two rest-mass values. We have already considered, from the point of view of the coupling scheme (§§ 35*a*, *b*, *c*), the algebra proposed by Bhabha. However, as already noted, neither system can be considered as defining an elementary particle—at least under the requirement of only one rest-mass *and* absence of subsidiary conditions. Cf. *infra*.

[2] Cf., for example, Wilson (40*a*, *b*).

[3] In this section we take, for convenience, the pseudo-Euclidean metric of purely positive signature corresponding to the choice $x_4 = ict$. This is useful in the present study because so many recent papers follow this convention. The transition to our more customary metric, when desirable, is trivial.

then ϵ_r may be represented by a matrix, whose sole non-zero element is the one in the rth row and column; and if quantities ϵ_{rs} are determined such that

$$\epsilon_r \alpha^\mu \epsilon_s = (\alpha^\mu)_{rs} \epsilon_{rs} \qquad (39.3a)$$

$$\epsilon_{rs} \epsilon_{st} = \epsilon_{rt}$$

the $(\alpha^\mu)_{rs}$ are the matrix elements of α^μ and

$$\epsilon_r A \epsilon_s = A_{rs} \epsilon_{rs} \qquad (39.3b)$$

for any element A in the α-ring. Clearly these generating idempotents are related to, but not in general equal (*per se* or in numbers) to the *primitive* normal idempotents (which commute with all elements and cannot therefore generate anything), and, correspondingly, the representation obtained is usually reducible. The simplest application of this method is to the above σ-ring, and we denote the engendered representation—which in this case is irreducible—as the Σ-algebra: $\mathfrak{A}(\sigma) \to \mathfrak{A}(\Sigma)$. Let us note first that the σ-ring is *semisimple*, since the regular representation generated by σ-ring operating on σ-space, in the above order of elements, represents the σ's by (cf. § 38*b*):

$$\sigma_1 \to \backepsilon_1 = \begin{Vmatrix} 0 & 0 & 0 & 1 \\ 0 & 0 & i & 0 \\ 0 & -i & 0 & 0 \\ 1 & 0 & 0 & 0 \end{Vmatrix} \qquad \sigma_2 \to \backepsilon_2 = \begin{Vmatrix} 0 & 0 & -i & 0 \\ 0 & 0 & 0 & 1 \\ i & 0 & 0 & 0 \\ 0 & 1 & 0 & 0 \end{Vmatrix} \qquad (39.4)$$

$$\sigma_3 \to \backepsilon_3 = \begin{Vmatrix} 0 & i & 0 & 0 \\ -i & 0 & 0 & 0 \\ 0 & 0 & 0 & 1 \\ 0 & 0 & 1 & 0 \end{Vmatrix} \qquad \sigma_4 = 1 \to \backepsilon_4 = \begin{Vmatrix} 1 & 0 & 0 & 0 \\ 0 & 1 & 0 & 0 \\ 0 & 0 & 1 & 0 \\ 0 & 0 & 0 & 1 \end{Vmatrix}$$

and $\det \| \lambda_{rs} \| = \det \| \text{spur } \backepsilon_r \backepsilon_s \| = 16$.

From equations (39.3), it is clear that in the present application *the* σ which appears in the definition of the generating idempotents (*the*, because there are now but two) will be represented by a diagonal matrix. To obtain the familiar matrix representation of the σ's, which is of course not essential, we therefore take the generators

$$\epsilon_1 = \tfrac{1}{2}(1 + \sigma_3) \qquad (39.5a)$$

$$\epsilon_2 = \tfrac{1}{2}(1 - \sigma_3)$$

As stated, $\epsilon_1 \sigma_3 \epsilon_2 = \epsilon_2 \sigma_3 \epsilon_1 = 0$, and, for example,

$$\epsilon_1 \sigma_1 \epsilon_2 = \sigma_1 \epsilon_2 \qquad (39.5b)$$

$$\epsilon_2 \sigma_1 \epsilon_1 = \sigma_1 \epsilon_1$$

so that we *may* choose $\epsilon_{12} = \sigma_1 \epsilon_2$ and $\epsilon_{21} = \sigma_1 \epsilon_1$, whence $\epsilon_{11} = \epsilon_{12} \epsilon_{21} = \epsilon_1$, $\epsilon_{22} = \epsilon_{21} \epsilon_{12} = \epsilon_2$. It readily follows that the representation so defined is

$$\sigma_1 \to \Sigma_1 = \begin{Vmatrix} 0 & 1 \\ 1 & 0 \end{Vmatrix} \qquad \sigma_2 \to \Sigma_2 = \begin{Vmatrix} 0 & -i \\ i & 0 \end{Vmatrix} \qquad \sigma_3 \to \Sigma_3 = \begin{Vmatrix} 1 & 0 \\ 0 & -1 \end{Vmatrix} \qquad (39.6)$$

and this is the single irreducible representation of the σ-ring.[4]

[4] Note that the centrum consists only of multiples of the one *primitive* idempotent 1, and the dimensionality—ring order relation is satisfied.

(c) *The Dirac Ring*

$\mathfrak{A}(\Sigma)$ is simple, so that by the main theorem on *direct* or *algebra* products of simple algebras, $\mathfrak{A}(\Gamma) = \mathfrak{A}(\Sigma) \times \mathfrak{A}(\Sigma)$ is also simple and the representation of the Dirac algebra $\mathfrak{A}(\gamma)$ so obtained is irreducible. In particular, if we set (note $\Sigma_4 = I_{(2)}$)

$$\Gamma_1 = \Sigma_1 \times \Sigma_4 = \begin{Vmatrix} 0 & 0 & 1 & 0 \\ 0 & 0 & 0 & 1 \\ 1 & 0 & 0 & 0 \\ 0 & 1 & 0 & 0 \end{Vmatrix} \qquad \Gamma_2 = \Sigma_3 \times \Sigma_1 = \begin{Vmatrix} 0 & -i & 0 & 0 \\ i & 0 & 0 & 0 \\ 0 & 0 & 0 & i \\ 0 & 0 & -i & 0 \end{Vmatrix} \tag{39.7a}$$

$$\Gamma_3 = \Sigma_2 \times \Sigma_4 = \begin{Vmatrix} 0 & 0 & -i & 0 \\ 0 & 0 & 0 & -i \\ i & 0 & 0 & 0 \\ 0 & i & 0 & 0 \end{Vmatrix} \qquad \Gamma_4 = \Sigma_3 \times \Sigma_2 = \begin{Vmatrix} 0 & -i & 0 & 0 \\ i & 0 & 0 & 0 \\ 0 & 0 & 0 & i \\ 0 & 0 & -i & 0 \end{Vmatrix}$$

we have

$$\Gamma_\mu \Gamma_\nu + \Gamma_\nu \Gamma_\mu = 2\delta_{\mu\nu} \tag{39.7b}$$

and this is precisely the same (apart from metric signature) as the algebra Π (§ 37*a*) which appeared in the development of the basic spinor representation of the orthogonal group in four dimensions. In terms of the abstract $\mathfrak{A}(\gamma)$, the result (39.7) is equivalent to the statement that the γ^μ's can be expressed in terms of two commuting triads of Pauli operators, σ_k, σ_k', and the above representation corresponds to the *particular* correlation

$$\begin{aligned} \gamma_1 &= \sigma_1 \sigma_4' \\ \gamma_2 &= \sigma_3 \sigma_1' \\ \gamma_3 &= \sigma_2 \sigma_4' \\ \gamma_4 &= \sigma_3 \sigma_2' \end{aligned} \tag{39.8}$$

Precisely the irreducible representation (39.7*a*) is obtained by taking as generating idempotents:

$$\begin{aligned} \epsilon_1 &= \tfrac{1}{4}(1 + \sigma_3)(1 + \sigma_3') = \tfrac{1}{4}(1 - i\gamma_1\gamma_3 - i\gamma_2\gamma_4 + \gamma_1\gamma_2\gamma_3\gamma_4) \\ \epsilon_2 &= \tfrac{1}{4}(1 + \sigma_3)(1 - \sigma_3') = \tfrac{1}{4}(1 - i\gamma_1\gamma_3 + i\gamma_2\gamma_4 - \gamma_1\gamma_2\gamma_3\gamma_4) \\ \epsilon_3 &= \tfrac{1}{4}(1 - \sigma_3)(1 + \sigma_3') = \tfrac{1}{4}(1 + i\gamma_1\gamma_3 - i\gamma_2\gamma_4 - \gamma_1\gamma_2\gamma_3\gamma_4) \\ \epsilon_4 &= \tfrac{1}{4}(1 - \sigma_3)(1 - \sigma_3') = \tfrac{1}{4}(1 + i\gamma_1\gamma_3 + i\gamma_2\gamma_4 + \gamma_1\gamma_2\gamma_3\gamma_4) \end{aligned} \tag{39.9}$$

In the work of § 37 we have, in effect, already determined that the *matrix systems* γ, $\tilde{\gamma}$, $\bar{\gamma}$ and γ^* are equivalent. This conclusion, which we have also seen to be of physical importance, is again an instance of the critical dependence of the results of the Dirac electron-positron field on the dimensionality of the problem,[5] since it follows from the main lemma on automorphisms of the full matrix algebra with which $\mathfrak{A}(\Gamma)$ coincides in four dimensions. It is clear that for general dimensionality n (and hence generating elements γ_r of the algebra) every scalar element is again

[5] Even more demanding on the dimensionality and signature is the requirement from the Maxwell-Dirac theory that the positive definite charge density $\Psi^*\Psi$ be the temporal component of a *vector*. For inertial index $(n - t) \neq 1$ (cf. § 37*c*), this property is possessed not by the vector contained in $\Delta \times \bar{\Delta}$ but by the *temporal* component $J_{t+1 \ldots n}$ of the *tensor* of rank $(n - t)$ with components $\Psi^* B \Gamma_r \Gamma_s \ldots \Psi$ $(r, s, \ldots,$ all distinct).

simply a multiple of the unit quantity, since in any such function of the γ's all indices are contracted and, by the commutation rules, may be brought adjacent to one another—each pair giving a factor $\gamma_r \gamma^r = n$. By the same argument it is readily seen that every pseudoscalar quantity must be a multiple of, say, $\omega_n = \frac{i^{n/2}}{n!} \epsilon^{rs\cdots} \gamma_r \gamma_s \ldots$, which is necessarily linearly independent of 1 and commutes or anticommutes with all γ's according as n is odd or even. Thus, the basis of the centrum of the Dirac algebra is $(1, \omega)$ or (1) in these two cases, and correspondingly there will be two or one inequivalent irreducible representations. Although the Dirac algebra corresponding to arbitrary n is only of limited physical interest, it seems desirable to make these considerations more precise, since a certain amount of misunderstanding has arisen concerning the correlation between the cases of odd and even dimensionality—and the *number* of *generating* elements—particularly in the analogous case of the Duffin-Kemmer algebra (cf. *infra*).

It is clear that the general algebra is of order $N = 2^n$, having irreducible representations of dimensions $d_1 = 2^\nu$ and $d_1^2 + d_2^2 = 2^{2\nu+1}$ according as $n = 2\nu$ or $n = 2\nu + 1$. The two representations of $\mathfrak{A}_{2\nu+1}(\gamma)$ can be constructed from the single representation of $\mathfrak{A}_{2\nu}(\gamma)$ by forming the element $\omega_{2\nu} = i^\nu \gamma_1 \ldots \gamma_{2\nu}$ (which can replace any element γ_r in the basic commutation rules),[6] and defining $\gamma_{2\nu+1}$ by $\gamma_{2\nu+1} = \pm\omega_{2\nu}$. In this way we obtain the two *inequivalent* representations of $\mathfrak{A}_{2\nu+1}(\gamma)$ of dimensions $d_1 = d_2 = 2^\nu$, the result already known from the analysis of the basic spinor representations of the orthogonal group.

(*d*) *The Duffin-Kemmer Ring*

The Duffin-Kemmer algebra $\mathfrak{A}(\beta)$, with *four* generating elements β_μ subject to the commutation rules

$$\beta_\kappa \beta_\lambda \beta_\mu + \beta_\mu \beta_\lambda \beta_\kappa = \beta_\kappa \delta_{\lambda\mu} + \beta_\mu \delta_{\kappa\lambda} \qquad (39.10)$$

is the most elementary example, of physical importance, in which we find illustrated all the defining characteristics of semisimple algebras. The number of independent elements, 126, is finite, and the basis of the centrum will be seen to consist of three *primitive* idempotents e_1, e_2, e_3 whence, because the algebra has the unit element, it follows that $\mathfrak{A}(\beta)$ is the *direct* sum of the left ideals $\mathfrak{A}(\beta)e_1$, $\mathfrak{A}(\beta)e_2$, $\mathfrak{A}(\beta)e_3$, and therefore semisimple. As we already know, this fixes the number of inequivalent irreducible representations as three, of dimensions 1, 5, and 10.

Our main interest here will be the derivation of these results and their generalization to an arbitrary number of generating elements subject to (39.10), and the Peirce decomposition of $\mathfrak{A}(\beta)$—in preparation for the application in § 39*d* to the reduction of the meson equations.

Granting, for the moment, the above-mentioned properties of $\mathfrak{A}_4(\beta)$, let us consider briefly the derivation of the irreducible representations—for which purpose the method of generating idempotents illustrates no new principles and itself now becomes tediously long. The simplest realization of these representations is, of course, precisely in the context in which they were first obtained, i.e. the connection

[6] This point appears to cause the difficulty. The correct interpretation has been given in detail by Kemmer (39, 43), Schrödinger (43), and others.

with meson theory. As we have, in effect, already done for the electron case, we need only transcribe the well-known tensor *field* equations[7] (neglecting the dual cases)

$$\partial_\mu U_\nu - \partial_\nu U_\mu = \kappa G_{\mu\nu} \qquad \text{(cf. 26.21)} \qquad (39.11)$$
$$\partial^\nu G_{\mu\nu} = \kappa U_\mu$$

for spin 1, and

$$\partial_\mu \phi = \kappa \chi_\mu \qquad \text{(cf. 26.12)} \qquad (39.12)$$
$$\partial^\mu \chi_\mu = \kappa \phi$$

for spin 0, to the linear wave-equation—or particle-like—form

$$(\partial^\mu \beta_\mu + \kappa)\psi = 0 \qquad (39.13)$$

In particular, with the correspondence

$$\begin{aligned} (G_{41},\ G_{42},\ G_{43}) &\to -i(\psi_1,\ \psi_2,\ \psi_3) \\ (G_{23},\ G_{31},\ G_{12}) &\to -(\psi_4,\ \psi_5,\ \psi_6) \\ (U_1,\ U_2,\ U_3,\ -iU_4) &\to (\psi_7,\ \psi_8,\ \psi_9,\ \psi_{10}) \end{aligned} \qquad (39.14)$$

we read off directly the familiar and very useful form of the ten-dimensional matrices (given by Kemmer (39)) (39.15a)

$$\beta_1 = \left\|\begin{array}{ccc|ccc|ccc|c} 0&0&0&0&0&0&0&0&0&-1\\ 0&0&0&0&0&0&0&0&0&0\\ 0&0&0&0&0&0&0&0&0&0\\ \hline 0&0&0&0&0&0&0&0&0&0\\ 0&0&0&0&0&0&0&0&-1&0\\ 0&0&0&0&0&0&0&1&0&0\\ \hline 0&0&0&0&0&0&0&0&0&0\\ 0&0&0&0&0&1&0&0&0&0\\ 0&0&0&0&-1&0&0&0&0&0\\ \hline -1&0&0&0&0&0&0&0&0&0 \end{array}\right\| \qquad \beta_2 = \left\|\begin{array}{ccc|ccc|ccc|c} &&&&&&&&&0\\ &&&&&&&&&-1\\ &&&&&&&&&0\\ \hline &&&&&&0&0&1&\\ &&&&&&0&0&0&\\ &&&&&&-1&0&0&\\ \hline &&&0&0&-1&&&&\\ &&&0&0&0&&&&\\ &&&1&0&0&&&&\\ \hline 0&-1&0&&&&&&& \end{array}\right\|$$

$$\beta_3 = \left\|\begin{array}{ccc|ccc|ccc|c} &&&&&&&&&0\\ &&&&&&&&&0\\ &&&&&&&&&-1\\ \hline &&&&&&0&-1&0&\\ &&&&&&1&0&0&\\ &&&&&&0&0&0&\\ \hline &&&0&1&0&&&&\\ &&&-1&0&0&&&&\\ &&&0&0&0&&&&\\ \hline 0&0&-1&&&&&&& \end{array}\right\| \qquad \beta_4 = \left\|\begin{array}{ccc|ccc|ccc|c} &&&&&&-i&0&0&\\ &&&&&&0&-i&0&\\ &&&&&&0&0&-i&\\ \hline &&&&&&&&&\\ &&&&&&&&&\\ &&&&&&&&&\\ \hline i&0&0&&&&&&&\\ 0&i&0&&&&&&&\\ 0&0&i&&&&&&&\\ \hline &&&&&&&&& \end{array}\right\|$$

[7] We note again that in this entire section (39) we use the pseudo-Euclidean metric, though formally retaining the convention of contracting only on repeated indices which appear once in covariant form and once in contravariant form. Though the transcription of all the results of this section to our more usual formalism is trivial, it is important to bear in mind the minor differences relative to our former equations and, particularly, in the definition and calculation of various field or particle density tensors.

and similarly from (39.12), the five-dimensional matrices

$$\beta_1 = \left\| \begin{array}{c|ccc|c} & & & & 0 \\ \hline & & & & 1 \\ & & & & 0 \\ & & & & 0 \\ \hline 0 & 1 & 0 & 0 & 0 \end{array} \right\| \quad \beta_2 = \left\| \begin{array}{c|ccc|c} & & & & 0 \\ \hline & & & & 0 \\ & & & & 1 \\ & & & & 0 \\ \hline 0 & 0 & 1 & 0 & 0 \end{array} \right\| \quad \beta_3 = \left\| \begin{array}{c|ccc|c} & & & & 0 \\ \hline & & & & 0 \\ & & & & 0 \\ & & & & 1 \\ \hline 0 & 0 & 0 & 1 & 0 \end{array} \right\| \quad \beta_4 = \left\| \begin{array}{c|ccc|c} & & & & -i \\ \hline & & & & 0 \\ & & & & 0 \\ & & & & 0 \\ \hline i & 0 & 0 & 0 & 0 \end{array} \right\| \qquad (39.15b)\dagger$$

The irreducibility of these matrix sets is also verified in that any matrix of similar dimension can be expressed as a linear combination of the β_μ's (more properly the B_μ's of $\mathfrak{A}(\beta) \to \mathfrak{A}(B)$) and their multiple products; equivalently, we can construct from them the corresponding matrices, each with but a single distinct non-zero unit matrix element, which are the orthonormal basis vectors of linear vector spaces of dimensions 100 and 25.

Clearly, we may also derive equivalent irreducible representations directly from the basic commutation rules—and the knowledge of the eigenvalues of the β_μ's and their multiplicities. We shall apply this method in § 39*f* to the system proposed by Bhabha and designated spin $(\frac{3}{2}, \frac{1}{2})$.

The remaining (and perhaps simplest) method amounts to reading off the β_μ's from the results of the general analysis of § 35*a* of the possible (2-spinor) forms of these matrices. From equations (35.16*a*, *b*, *c*) *et seq.*, we have (spin 1)

$$\beta_{AB} = \left\| \begin{array}{ccc} 0 & v_B(\frac{1}{2})u_A(1) & 0 \\ -u_B(\frac{1}{2})v_A(1) & 0 & v_B(1)u_A(\frac{1}{2}) \\ 0 & -u_B(1)v_A(\frac{1}{2}) & 0 \end{array} \right\| \qquad (39.16)^8$$

and (spin 0)

$$\beta_{AB} = \sqrt{2} \left\| \begin{array}{cc} 0 & -u_B(\frac{1}{2})u_A(\frac{1}{2}) \\ v_B(\frac{1}{2})v_A(\frac{1}{2}) & 0 \end{array} \right\| \qquad (39.17)$$

the β^μ's following from $\beta^\mu = \frac{1}{2}g^{\mu\dot{A}B}\beta_{AB}$, with the $g^{\mu\dot{A}B}$ given by (8.10).

Returning to the underlying algebraic problem, we always require the order of the ring or the number of linearly independent quantities among the generating elements and their multiple products. In the present case it is convenient for this purpose to introduce the auxiliary quantities

$$\eta_\mu = 2\beta_\mu^{\ 2} - 1 \qquad (39.18)$$

of which, as is readily verified, η_4 is the operator representing reflection in the origin.[9]

† The divisions in these sets of matrices merely denote the splitting of the two wave functions with respect to the subgroup of spatial rotations; that is, three 3-vectors and a scalar and one 3-vector and two scalars, respectively.

[8] N.B. We use here the u, v spinor-matrices satisfying the conditions (cf. § 17*b*(ii)) $v^A(k) = (-)^{2k+1}u^*{}_A(k)$ to which the explicit forms (17.39′, 39″) correspond. If the factor $(-)^{2k+1}$ is not included, as is often the case in the literature, the explicit forms given must be changed in sign and, correspondingly, the second *row* of matrices in (39.16*a*) takes on a minus sign. The net result is, of course, the same in either case.

[9] That is, up to a unimodular factor depending on the metric signature and whether we are dealing with the vector and scalar fields or their respective duals.

Then, writing out (39.10) in full, we have

$$\left.\begin{array}{ll} \beta_\kappa^3 = \beta_\kappa & \\ \beta_\kappa\beta_\lambda^2 + \beta_\lambda^2\beta_\kappa = \beta_\kappa & (\kappa \neq \lambda) \\ \beta_\kappa\beta_\lambda\beta_\mu + \beta_\mu\beta_\lambda\beta_\kappa = 0 & (\kappa \neq \lambda,\ \mu \neq \lambda) \\ \beta_\kappa\beta_\lambda\beta_\kappa = 0 & (\kappa \neq \lambda) \end{array}\right\} \qquad (39.19a)$$

$$\left.\begin{array}{ll} \eta_\mu^2 = 1 & \\ \eta_\mu\eta_\nu - \eta_\nu\eta_\mu = 0 & \\ \eta_\mu\beta_\nu + \beta_\nu\eta_\mu = 0 & (\mu \neq \nu) \\ \beta_\mu = \eta_\mu\beta_\mu = \beta_\mu\eta_\mu & \end{array}\right\} \qquad (39.19b)$$

and we can now readily enumerate the types and respective numbers of the 126 elements (note, multiple products such as $\beta_\kappa\beta_\lambda\beta_\mu$ are *here* condensed to $\beta_{\kappa\lambda\mu}$):

Element	Number	Element	Number
1	1	$\eta_\kappa\beta_{\lambda\mu\nu}$	12
β_κ	4	$\eta_{\kappa\lambda}$	6
$\beta_{\kappa\lambda}$	12	$\eta_{\kappa\lambda}\beta_\mu$	12
$\beta_{\kappa\lambda\mu}$	12	$\eta_{\kappa\lambda}\beta_{\mu\nu}$	12
$\beta_{\kappa\lambda\mu\nu}$	6	$\eta_{\kappa\lambda\mu}$	4
η_κ	4	$\eta_{\kappa\lambda\mu}\beta_\nu$	4
$\eta_\kappa\beta_\lambda$	12	$\eta_{\kappa\lambda\mu\nu}$	1
$\eta_\kappa\beta_{\lambda\mu}$	24		

(39.20)

The elements forming a basis for the centrum are the linearly independent complete scalar set (cf. *infra*)

$$\begin{aligned} &I \\ &M = \sum_\mu \eta_\mu - \sum_{\mu<\nu} \eta_{\mu\nu} \\ &N = \prod_\mu \eta_\mu(1 - \sum_\nu \eta_\nu) = \prod_\mu \eta_\mu - \sum_{\lambda<\mu<\nu} \eta_{\lambda\mu\nu} \end{aligned} \qquad (39.21a)$$

which are thus the only such elements, apart from linear combinations, as seen from the relations

$$\begin{aligned} M^2 &= 10 - 6M + 6N \\ N^2 &= 5 - 2M \\ MN &= 3M - 4N \end{aligned} \qquad (39.21b)$$

With the help of these we find the three *primitive* idempotents

$$\begin{aligned} 16e_1 &= 1 - M + N \\ 16e_2 &= 10 + 2M + 2N \\ 16e_3 &= 5 - M - 3N \end{aligned} \qquad (39.21c)$$

The *commuting elements* (39.21*a*) may be obtained by the direct procedure of successively postulating commutability with each of the quantities in the list (39.20); or, as we shall show presently, by appeal to general principles.

In the case of arbitrary n, the corresponding basis for the centrum is the sequence of elements

$$I, \quad \sum_r \eta_r - \sum_{r<s} \eta_{rs}, \quad \sum_{r<s<t} \eta_{rst} - \sum_{r<s<t<u} \eta_{rstu}, \quad \ldots \tag{39.22}$$

in which the last of the $\frac{1}{2}n + 1$ terms for even n can be written as $\prod_r^n \eta_r(\sum_s \eta_s - 1)$. For odd n the sequence ends with the two terms $\prod_r \eta_r(\sum_{r<s} \eta_{rs} - \sum_r \eta_r)$ and $\prod_r \eta_r$ with a total of $\frac{1}{2}(n + 3)$ terms in all; however, in this case there is an *additional* commuting element $\sum_P \delta_P P \prod^n \beta_r$ analogous to that in the Dirac algebra, so that the number of commuting elements for odd n is increased to $\frac{1}{2}(n + 5)$. The general case of even n is perfectly regular, but, as mentioned before, the case of odd n introduces a new feature. Let us assume that we know the irreducible representations for $n = 2\nu$, and wish to construct those for $n = 2\nu + 1$. As for the Dirac algebra, we form the element

$$\omega_{2\nu} = \frac{i^\nu}{(\nu!)^2} \sum_P \delta_P P \prod_r^{2\nu} \beta_r = \frac{i^\nu}{\nu!} \epsilon^{r_1 \ldots r_{2\nu}} \beta_{r_1} \ldots \beta_{r_{2\nu}} \tag{39.23}$$

which can substitute for any β_r in the commutation rules (39.10), *but* it can be shown that $\omega_{2\nu}$ vanishes except for the representation of highest dimension for $\mathfrak{A}_{2\nu}(\beta)$. Thus, starting from this representation, $\beta_{2\nu+1}$ can be identified with the matrix $\pm\omega_{2\nu}$, and we have the phenomenon described by Kemmer (43) as *twinning*, since for any odd n there are *two* representations of $\mathfrak{A}_{2\nu+1}(\beta)$ of the same dimension as that of the senior representation for $\mathfrak{A}_{2\nu}(\beta)$. The scheme of representations has then the appearance:

Dimension n				Dimensions of Representations					Order	
0							1		1	(39.24)
1						1		(1, 1)	3	
2					1		3		10	
3				1		4		(3, 3)	35	
4			1		5		10		126	
5		1		6		15		(10, 10)	462	
6	1		7		21		35		1716	

in which the twin representations are bracketed together. As in the four-dimensional case, the representations can all be obtained quite simply by transcription of the now generalized tensor field equations to the matrix form.

The β-algebra in five dimensions deserves further brief mention here because it presents additional aspects of mathematical interest and has been suggested as a possible basis for limiting the number of arbitrary or disposable constants in the various meson theories. Though this connection is very formal in character, we see directly that, with respect to the five-dimensional (proper) orthogonal group, the irreducible representations of $\mathfrak{A}_5(\beta)$ of dimensions 6, 15, and 10 correspond to fields of scalar, vector, and skew-tensor (of special kind) character which, under restraint of the 5-axis, reduce to the usual four-dimensional scalar, mixed pseudoscalar-vector [10], and pseudovector [10] meson theories. The transcription here devolves on the respective equations (the † symbol is used in equations (39.25, 26) to emphasize again the five-dimensional character and index range involved)

$$\left.\begin{aligned}\partial^\mu \chi_\mu &= \kappa\phi\\ \partial_\mu \phi &= \kappa\chi_\mu\end{aligned}\right\} \quad \text{(cf. 26.12)} \qquad (39.25a)^\dagger$$

$$\left.\begin{aligned}\partial_\mu U_\nu - \partial_\nu U_\mu &= \kappa G_{\mu\nu}\\ \partial^\nu G_{\mu\nu} &= \kappa U_\mu\end{aligned}\right\} \quad \text{(cf. 26.21)} \qquad (32.25b)^\dagger$$

$$\left.\begin{aligned}\partial_\lambda G_{\mu\nu} + \partial_\mu G_{\nu\lambda} + \partial_\nu G_{\lambda\mu} &= \kappa U_{\lambda\mu\nu}\\ \partial^\nu U_{\lambda\mu\nu} &= \kappa G_{\lambda\mu}\end{aligned}\right\} \quad \text{(cf. 26.36)} \qquad (39.25c)^\dagger$$

which, under the stated restriction, reduce to the four-dimensional field equations described.

The numbers and types of elements are

Element	Number	Element	Number
1	1	$\eta_\kappa \beta_\lambda$	20
β_κ	5	$\eta_\kappa \beta_{\lambda\mu}$	60
$\beta_{\kappa\lambda}$	20	$\eta_\kappa \beta_{\lambda\mu\nu}$	60
$\beta_{\kappa\lambda\mu}$	30	$\eta_\kappa \beta_{\lambda\mu\nu\sigma}$	30
$\beta_{\kappa\lambda\mu\nu}$	30	$\eta_{\kappa\lambda} \beta_\mu$	30
$\beta_{\kappa\lambda\mu\nu\sigma}$	10	$\eta_{\kappa\lambda} \beta_{\mu\nu}$	60
η_κ	5	$\eta_{\kappa\lambda} \beta_{\mu\nu\sigma}$	30
$\eta_{\kappa\lambda}$	10	$\eta_{\kappa\lambda\mu} \beta_\nu$	20
$\eta_{\kappa\lambda\mu}$	10	$\eta_{\kappa\lambda\mu} \beta_{\nu\sigma}$	20
$\eta_{\kappa\lambda\mu\nu}$	5	$\eta_{\kappa\lambda\mu\nu} \beta_\sigma$	5
$\eta_{\kappa\lambda\mu\nu\sigma}$	1		

(39.26)†

and the determination of a linearly independent basis of the centrum and the irreducible representations proceeds on the lines discussed.[11] A point of interest is that the system of (39.25*c*) corresponds, strictly, to the combined twin representations denoted as (10, 10) which form an irreducible representation of degree 20 of the six-dimensional orthogonal group, *including* reflections. It is readily verified that the four independent scalar elements of the centrum of $\mathfrak{A}_5(\beta)$ constitute at the same time a commutative subalgebra, the centrum of the *extended* algebra related to the full orthogonal group in *six* dimensions. This algebra has 662 elements, to which correspond irreducible representations of dimensions 1, 6, 15, and 20, the first three being carried over from $\mathfrak{A}_5(\beta)$ because the *total* reflection operator $R = \Sigma \eta_\mu - 1$ *anticommutes* with all β_μ's *but* vanishes in the parent ten-dimensional representations.

[10] Recall that in the reduction to the four-dimensional group, a 5-skew-tensor $G^\dagger_{\mu\nu}$ reduces to the 4-skew-tensor $G_{\mu\nu}$ and 4-pseudovector $G_{\mu 5}$, a 5-vector $F^\dagger_\mu$ to the 4-vector F_μ and pseudoscalar F_5. For the related problem of possible nuclear source densities, see the preceding references and Bibliography.

[11] As remarked by Lubanski and Rosenfeld (42), the first of the scalar elements, denoted by M (cf. 39.22), is (apart from an additive constant) simply the *invariant* $\mathfrak{T}_{\kappa\lambda}\mathfrak{T}^{\kappa\lambda}$ of the orthogonal group in the next dimension above that of the algebra—*in so far as the connection* $\mathfrak{T}_{\mu\nu} = [\beta_\mu, \beta_\nu]_-$ *is assumed to hold.* Since this quantity is easily calculated, and its successive powers are linearly independent—up to the order of its characteristic equation—these give *essentially* the scalar elements of the sequence (39.22).

The twin representations in $\mathfrak{A}_5(\beta)$ differ one from the other only in reversal of sign of all the β_μ's, but are inequivalent because no similarity transformation can carry the one set over into the other; this is also evident from the properties of the pseudoscalar element Ω, which would reverse its sign under such transformation.

Before taking up the problem of the Peirce decomposition of the β-ring, it remains to verify the remark as to the scalar nature of the basis (39.21a) of the centrum of $\mathfrak{A}_4(\beta)$. From (39.19a) we have, with $\mathscr{B} = \beta_\kappa \beta^\kappa$,

$$\beta_\kappa \mathscr{B} = (5 - \mathscr{B})\beta_\kappa \qquad (39.27)$$
$$\beta^\kappa \beta_\lambda \beta_\kappa = \beta_\lambda$$

and, since two contracted indices in a scalar quantity can, by (39.10), be brought either adjacent to one another or immediately to the left and right of a β_κ, it follows that every scalar quantity of the β-ring is a polynomial $f(\mathscr{B})$ in $\mathscr{B}$. If $f(\mathscr{B})$ belongs to the centrum,

$$f(\mathscr{B})\beta_\kappa = \beta_\kappa f(5 - \mathscr{B}) = \beta_\kappa f(\mathscr{B}) \qquad (39.28a)$$

and

$$f(5 - \mathscr{B}) = f(\mathscr{B}) = \tfrac{1}{2}\{f(5 - \mathscr{B}) + f(\mathscr{B})\} \qquad (39.28b)$$

so that, in fact, every scalar quantity of the centrum is expressible as a polynomial in $\mathscr{B}(5 - \mathscr{B}) = \vartheta$.[12] By a similar argument we may show that every pseudoscalar quantity in the centrum is of the form $\omega f(\mathscr{B})$, and since ω *does not* commute with the β_κ in the present case (or any *even* n) no pseudoscalars will appear. Further, $\omega\mathscr{B}$ is a numerical multiple of ω so that when pseudoscalar quantities occur in the centrum at all (that is, for all *odd* n) they are simply numerical multiples of ω.

To obtain a linearly independent complete scalar basis for the centrum of $\mathfrak{A}_4(\beta)$ (we do not consider its straightforward generalization to arbitrary n), we consider the quantity

$$\mathscr{B}_4 = \sum_{\kappa\lambda\mu\nu} \beta_\kappa^2 \ldots \beta_\nu^2 \qquad (39.29)$$

where the summation is over all possible values of κ, λ, μ, ν such that no two indices are equal. Now β_κ^2 and β_λ^2 always commute and $\beta_\kappa^4 = \beta_\kappa^2$, so that

$$\mathscr{B}_4\mathscr{B} = \sum_{\kappa\lambda\mu\nu\sigma} \beta_\kappa^2 \ldots \beta_\sigma^2 + 4 \sum_{\kappa\lambda\mu\nu} \beta_\kappa^2 \ldots \beta_\nu^2 \qquad (39.30a)$$

or

$$\mathscr{B}_5 = \mathscr{B}_4(\mathscr{B} - 4) = \mathscr{B}(\mathscr{B} - 1) \ldots (\mathscr{B} - 4) \qquad (39.30b)^{13}$$

since $\mathscr{B}_1 = \mathscr{B}$. Clearly $\mathscr{B}_5 = 0$, since the *five* indices κ, λ, μ, ν, σ cannot all be different, and we note that $\mathscr{B}_1$, $\mathscr{B}_2$, $\mathscr{B}_3$, $\mathscr{B}_4$ are linearly indepenent. If then we put

[12] If we put $x = \frac{5}{2} - \mathscr{B}$ and $\phi(x) = f(\frac{5}{2} - x)$, the statement that such elements of the centrum are of the form $\frac{1}{2}f(\mathscr{B}) + \frac{1}{2}f(5 - \mathscr{B})$ becomes the statement that they are of the form $\frac{1}{2}\phi(x) + \frac{1}{2}\phi(-x)$, that is, a polynomial in $x^2 = \frac{25}{4} - \mathscr{B}(5 - \mathscr{B})$.

[13] Note that, since $(\mathscr{B} - \frac{5}{2})^2$ commutes with the β_μ's, it is a multiple of the unit matrix in any irreducible representation, and, correspondingly, $\mathscr{B}$ satisfies the equation $(\mathscr{B} - b_1)(\mathscr{B} - b_2) = 0$, with numbers $b_1 + b_2 = 5$. Together with the characteristic equation (39.30b) this permits only the two possibilities $b_1 = 1$, $b_2 = 4$, and $b_1 = 2$, $b_2 = 3$—apart from the trivial case. ω is represented by a nonzero matrix only in that representation in which $\mathscr{B}$ can have the eigenvalue 2 (in general $[\frac{1}{2}(n + 1)]$ where $[n]$ is the integral part of n), and this is just the senior presentation of $\mathfrak{A}_4(\beta)$.

$f(\mathscr{B}) = \mathscr{B}_5$, $f(\mathscr{B}) = 0 = f(5 - \mathscr{B})$ and so belongs to the centrum, and it follows from the characteristic equation (39.30*b*) for $\mathscr{B}$ that 1, ϑ, and ϑ^2 are linearly independent scalars forming a *complete* basis for the centrum. These are, of course, expressible in terms of the previous set I, M and N.

. . . .

(i) *The Decomposition of the Meson Equation**

The doubling of the dimensionality of the underlying function-space of electron theory amounts to taking the *direct* product of two spaces, the two-dimensional spin space and a similar two-dimensional energy-signature space—this separation thus corresponding to the synthesis of the Dirac operators γ_μ (§ 39*c*) in terms of two commuting sets of Pauli operators σ_j, σ_j', with respect to which the single-index 4-spinor Ψ^a can be replaced by a two-index symbol $\Psi_{r,r'}$ in which r, r' independently take on two values, say ± 1. The σ's operate on the first index and the σ''s on the second, the whole process being merely a particular statement of a simple but basic property of commuting systems of linear transformations.[14] In modified form, this type of resolution naturally applies to all theories related to half-integral spin and hence indefinite energy.

Again, a similar doubling occurs in all theories related to integral spin (now, of course, corresponding to the indefinite charge) so that for spin 1, for example, the analogous *basis* is the direct product of a three-dimensional spin space and a two-dimensional charge-signature or isotopic spin space. In this case the indices r, r' have different ranges, the spin-matrix transformations S_i operating on the first, and the isotopic-spin-matrix transformations τ_i on the second.

This comparison clearly brings out the far-reaching analogy in the roles of energy (charge) *character* with respect to half-integral (integral) spin, which evidently is basic to the close connection (if not full interchangeability) of these characters in relation to spin and statistics. Moreover, it is the essential physical correlate to the *non-occurrence* for integral spin of the so-called negative-energy difficulty, as well as being the physical basis of the Peirce decomposition to which we now return.

The usual matrix- or equivalent tensor-form of meson theory does not correspond to the preceding enumeration of components, simple because of the occurrence of redundant components for integral spin. As we shall see, these can be eliminated in a quite direct and obvious way, and the more formal algebraic process is merely a more informative, and incidentally elegant, realization of the same result—the slight increase in methodological complexity being justified by the insight gained. From the above discussion, the essential problem is clear: to decompose the β-ring in such a way that the ten- and five-dimensional representations corresponding to spins 1 and 0 give way to six- and two-dimensional representations, respectively. By the physical evidence this must be possible, so that in the first instance we must, by appropriate specification of subrings, delimit the number of elements such that irreducible representations of these dimensions can be realized; this is precisely what is accomplished by the Peirce decomposition, the principles involved being

* Sakata-Taketani (40*a*, *b*); Heitler (43); Schrödinger (43*a*, *b*).

[14] Cf. Van der Waerden (32), especially § 13.

contained in general terms in § 38*b* (equations 38.12–14, *et seq.*). Apart from the details of the reduction of the wave-equation (cf. *infra*), the essence of its application to the ring $\mathfrak{A}_4(\beta)$ is contained in the observation that β_4^2 (any β_μ^2)[15] is an idempotent of this matrix ring which can be identified with J (cf. 38.12)—whence, on restricting ourselves to the subring $\mathfrak{B} = J\mathfrak{A}J$, we deal only with matrices of the form $J\beta_\kappa J \subset \mathfrak{B}$,[16] and in a coordinate system adapted to this ring-decomposition, J is represented by the unit matrix and β_4 has an inverse: $\beta_4^{-1} = \beta_4$.

It will be more instructive if we consider first the ring $\mathfrak{A}_3(\beta)$, which is clearly related to the problem, and, since we shall be simultaneously considering the β-rings in three and four dimensions, we shall denote the former by $\mathfrak{A}_3(S)$. Thus, instead of (39.10) we now write

$$S_i S_j S_k + S_k S_j S_i = S_i \delta_{jk} + S_k \delta_{ij} \qquad (39.31)$$

and similarly $2S_k^2 - 1 = \xi_k$. This algebra has the 35 linearly independent elements

Element	Number	Element	Number	
1	1	ξ_{ij}	3	(39.32)
S_i	3	ξ_{ijk}	1	
S_{ij}	6	$\xi_i S_j$	6	
S_{ijk}	3	$\xi_i S_{jk}$	6	
ξ_i	3	$\xi_{ij} S_k$	3	

and the four irreducible representations (cf. 39.24) of dimensions 1, 4, (3, 3) correspond to the realization of the centrum by the scalar elements

$$I \qquad M = \sum_i \xi_i - \sum_{i<j} \xi_{ij} \qquad N = \xi_1 \xi_2 \xi_3 \qquad (39.33a)$$

and the pseudoscalar

$$\Omega = \sum_P P \delta_P S_1 S_2 S_3 = \epsilon^{ijk} S_i S_j S_k \qquad (39.33b)$$

This system, though associated *in abstracto* with spins 1 and 0, obviously does not satisfy the further commutation rules

$$S_i S_j - S_j S_i = i\epsilon_{ijk} S^k \qquad (39.34)$$

which we expect of any components of angular momentum. If we impose the conditions (39.34) in addition to (39.31), the system of independent elements becomes

Element	Number	
1	1	(39.35)
S_i	3	
S_{ij}	3	
ξ_i	3	

[15] The choice of β_4^2 is only important for physical applications.

[16] Note that $J\beta_4 J = \beta_4$, $J\beta_\lambda \beta_\mu = J\beta_\lambda \beta_\mu J$.

which correspond to the contracted centrum I, M and two irreducible representations of dimensions 1, 3. For the non-trivial representation we find, for example,

$$S_1 = \begin{Vmatrix} 0 & 0 & 0 \\ 0 & 0 & -i \\ 0 & i & 0 \end{Vmatrix} \qquad S_2 = \begin{Vmatrix} 0 & 0 & -i \\ 0 & 0 & 0 \\ i & 0 & 0 \end{Vmatrix} \qquad S_3 = \begin{Vmatrix} 0 & -i & 0 \\ i & 0 & 0 \\ 0 & 0 & 0 \end{Vmatrix} \tag{39.36}$$

that is, the nucleus of the orthogonal group in three dimensions.

Returning now to the Peirce decomposition we note, with the help of equations (39.19a, b), that the subring $\mathfrak{B} = J\mathfrak{A}J$ is simply $\mathfrak{A}_3(S)$, or rather, the automorphism: $\mathfrak{A}(S) \to J\mathfrak{A}(S)J$. This is not at all surprising since the operators S_k, defined by

$$\beta_i\beta_j - \beta_j\beta_i = i\epsilon_{ijk}S^k \tag{39.37}$$

satisfy the commutation rules (39.31) *and* (39.34),[17] and are just the appropriate spin operators, according to the work of Chapter III; their representative matrices in the two irreducible representations of $\mathfrak{A}_4(\beta)$ being fully reducible. From (39.37) we can obtain the relations (k, l, m = 1, 2, 3 cyclically):

$$\eta_k(\beta_k\beta_l + \beta_l\beta_k) = -\eta_l(\beta_k\beta_l + \beta_l\beta_k) = i\epsilon_{klm}S^m \tag{39.38a}$$

$$\left.\begin{aligned} \eta_k\eta_l S_m &= -S_m \\ S_m{}^2 &= \tfrac{1}{2}(1 - \eta_k\eta_l) \end{aligned}\right\} \tag{39.38b}$$

$$(\beta_k\beta_l + \beta_l\beta_k) = -\eta(S_kS_l + S_lS_k) \qquad (k \neq l) \tag{39.38c}$$

$$\left.\begin{aligned} \eta_k &= \eta(1 - 2S_k{}^2) \\ \beta_k{}^2 &= \tfrac{1}{2}(1 + \eta_k) = \tfrac{1}{2}(1 + \eta) - \eta S_k{}^2 \end{aligned}\right\} \tag{39.38d}$$

with

$$\eta = \eta_1\eta_2\eta_3$$

Then the independent elements of $\mathfrak{B}$ can be expressed in terms of the S_k and their multiple products (that is, in the sense of $JS_kJ = JS_k$) and the quantities $J\eta J$, $J\beta J = J\beta_4 J$, and $J\varepsilon J = -iJ\beta\eta J$, the triad η, β, ε satisfying *essentially* the Pauli relations

$$\begin{aligned} \beta^2 &= \eta^2 = \varepsilon^2 = 1 \\ \eta\varepsilon &= -\varepsilon\eta = i\beta \\ \beta\eta &= -\eta\beta = i\varepsilon \\ \varepsilon\beta &= -\beta\varepsilon = i\beta^2\eta \end{aligned} \tag{39.39a}$$

The qualification on the realization of the Pauli ring is removed in $\mathfrak{B}$, since $J = I$, so that we may in fact put

$$\begin{aligned} \varepsilon = \sigma_1 &= \begin{Vmatrix} 0 & 1 \\ 1 & 0 \end{Vmatrix} \\ \beta = \sigma_2 &= \begin{Vmatrix} 0 & -i \\ i & 0 \end{Vmatrix} \\ \eta = \sigma_3 &= \begin{Vmatrix} 1 & 0 \\ 0 & -1 \end{Vmatrix} \end{aligned} \tag{39.39b}$$

and this system of transformations evidently corresponds to the required isotopic-spin space.

[17] Note that both (39.10) and (39.31) are required to establish (39.34).

Thus we find a separation into two types of elements, the η, β, ε, and the S_k and their products; the latter system, with J, again constituting a subring, that is, the system of (39.34, 35). The unit matrix of the system of S_k-matrices is given by

$$E = \sum_i S_i^2 - 1 \tag{39.40a}$$

which satisfies

$$E^2 = E \tag{39.40b}$$
$$ES_k = S_k E = S_k$$

and is, of course, an idempotent of the $\mathfrak{A}$-ring. The elements η, β, ε commute with the S_k, so that these two sets belong to *distinct* systems of linear transformations and therefore operate on *different* indices of the vectors in this space. This corresponds to our dealing with and reducing these two systems separately, the resultant reduced matrices being direct products of reduced matrices belonging to these different systems.

According to this separation there will be, within the two irreducible representations of $\mathfrak{A}_4(\beta)$, a separation into parts, one of which will be the six-dimensional representation in question and the other the two-dimensional. In both instances these correspond to vectors of the type $\Psi_1 = J\Psi$, the vectors $\Psi_2 = (1 - J)\Psi$ in the alternate subspace defining the *redundant* components in terms of the *essential* components and their spatial derivatives. Hence if we consider the meson equation, for either spin, in the form

$$\mathscr{E}\Psi = \mathscr{H}\Psi \tag{39.41}$$

and decompose this in accordance with (38.12, 13), we obtain

$$\mathscr{E}J\Psi = \{J\mathscr{H}J + J\mathscr{H}(1 - J)\}\Psi \tag{39.42a}$$
$$\mathscr{E}(1 - J)\Psi = \{(1 - J)\mathscr{H}J + (1 - J)\mathscr{H}(1 - J)\}\Psi \tag{39.42b}$$

the second part of (39.42a) producing *some* Ψ_2 expressible in terms of Ψ_1. From the direct physical side, we have the wave equation and its adjoint

$$(\beta_\mu D^\mu + \kappa)\Psi = 0 \tag{39.43a}$$
$$\Psi^+(\beta_\mu \bar{D}^\mu - \kappa) = 0 \tag{39.43b}$$

where $\Psi^+ = i\Psi^* \eta_4$ and $D_\mu = \partial_\mu - \dfrac{ie}{\hbar c}\phi_\mu$, thus allowing for electromagnetic interaction, and we see that the operator $(1 - \beta_4^2)$ eliminates the time component on premultiplying (39.43a) and postmultiplying (39.43b) to give:

$$[\beta_k \beta_4^2 D^k + \kappa(1 - \beta_4^2)]\Psi = 0 \tag{39.44a}$$
$$\Psi^+[\beta_4^2 \beta_k \bar{D}^k - \kappa(1 - \beta_4^2)] = 0 \tag{39.44b}$$

If, in particular, we consider the ten-dimensional representation in which the eigenvalues of β_4^2 are 1 and 0—with multiplicities 6 and 4, respectively—then with β_4^2 diagonal these equations express the redundant components, belonging to eigenvalues 0, in terms of the essential components, belonging to eigenvalues 1, as stated above. Then from (39.42a) the decomposed six-component wave-equation is

$$J\mathscr{E}\Psi = \left\{J\mathscr{H}J - \frac{1}{\kappa}J\mathscr{H}D^k\beta_k J\right\}\Psi \tag{39.45}$$

where we have dropped the index labelling Ψ, since only operators from the subring $\mathfrak{B}$ occur, and all operations are now confined to the subspace P_1. Equation (39.42*b*) then becomes unimportant since it only gives the development in time of Ψ_2.

From the wave equation (39.13) we can readily obtain the Hamiltonian form (cf. Kemmer 39)

$$\mathscr{H} = mc^2\beta_4 + e\phi_4 + \frac{\hbar c}{i} D^k \left(\frac{\beta_k\beta_4 - \beta_4\beta_k}{i}\right) + \frac{ie}{\kappa}\{F^{\mu\nu}\beta_\mu\beta_4\beta_\nu + F^{\mu 4}\beta_\mu\} \quad (39.46)$$

where, as usual, $F^{\mu\nu}$ denotes the electromagnetic field strengths; and the *decomposed* Hamiltonian of (39.45) then takes the form

$$\mathfrak{H} = mc^2\beta_4 + e\phi_4 J - \frac{\hbar c}{\kappa} D^k D^l \beta_4\beta_k\beta_l \quad (39.47)$$

With our previous results and the definition of the magnetic field components, $D^kD^l - D^lD^k = -\frac{ie}{\hbar c}F^{kl}$, this may be fully expressed in terms of the charge and spin operators:

$$\mathfrak{H} = \hbar c\kappa\beta + e\phi_4 - \frac{\hbar c}{\kappa}\beta\left(\frac{1+\eta}{2}\right)\boldsymbol{D}^2 + \frac{\hbar c}{\kappa}\beta\eta(\boldsymbol{S}\cdot\boldsymbol{D})^2 - \frac{\beta(1+\eta)}{2mc}(\boldsymbol{S}\cdot\boldsymbol{H}) \quad (39.47b)$$

or

$$\mathfrak{H} = mc^2\beta + \frac{1}{m}\left\{\beta\left(\frac{1+\eta}{2}\right)\boldsymbol{p}^2 - \beta\eta(\boldsymbol{S}\cdot\boldsymbol{p})^2\right\} \quad (39.47c)$$

for the field-free case. These Hamiltonian forms represent, in effect, a new way of obtaining roots of the classical form $c(\boldsymbol{p}^2 + m^2c^2)^{1/2}$, and though the elimination of the redundant components is not a Lorentz invariant process, the theory is none the less Lorentz invariant—if not in appearance. The striking deviation from more familiar forms is the unsymmetrical footing of the spatial-temporal components and the non-Hermiticity of $\mathfrak{H}$.[18]

[18] As already remarked, the form (39.47) may also be obtained from the tensor field equations (39.11, 12) by eliminating directly the redundant components as in the 3-vector transcription of § 26*b*, viz. eliminating $G_{kl} = \text{curl}_{kl}\boldsymbol{U}$, $U_4 = -\frac{1}{\kappa}\text{div}\,\boldsymbol{G}_{4k}$:

$$\left(-\kappa + \frac{1}{\kappa}\text{curl curl} + \frac{1}{\kappa}\nabla^2\right)\boldsymbol{G}_{4k} = i\boldsymbol{U}_{k,4}$$
$$-(\kappa + \text{curl curl})\,\boldsymbol{U}_k = i\boldsymbol{G}_{4k,4}$$

Clearly, the isotopic-spin space corresponds to combining the *two* 3-vectors $\boldsymbol{U}_k$, $i\boldsymbol{G}_{4k}$ into *one* field quantity ψ in six components and *two* indices, that is, a 2×2 matrix. The first index takes the role of k, the second serves to distinguish between $\boldsymbol{U}_k$ and $i\boldsymbol{G}_{4k}$ which occupy, respectively, the position in the first row and column, and second row and column. Then with the particular representation of the spin matrices

$$S_1 = \begin{Vmatrix} 0 & 0 & 0 \\ 0 & 0 & -i \\ 0 & i & 0 \end{Vmatrix} \qquad S_2 = \begin{Vmatrix} 0 & 0 & -i \\ 0 & 0 & 0 \\ i & 0 & 0 \end{Vmatrix} \qquad S_3 = \begin{Vmatrix} 0 & -i & 0 \\ i & 0 & 0 \\ 0 & 0 & 0 \end{Vmatrix}$$

we see that the curl curl operation can be represented (apart from an obvious factor) as $(\boldsymbol{S}\cdot\boldsymbol{p})^2$, in the same sense as $\boldsymbol{p}^2$ represents ∇^2. The isotopic-spin space operators, now in the representation

$$\beta = \begin{Vmatrix} 0 & i \\ -i & 0 \end{Vmatrix} \qquad \eta = \begin{Vmatrix} -1 & 0 \\ 0 & 1 \end{Vmatrix} \qquad \beta\eta = \begin{Vmatrix} 0 & i \\ i & 0 \end{Vmatrix}$$

send $\boldsymbol{U}_k$ into $i\boldsymbol{G}_{4k}$ and conversely when operating only on the second or *charge* index. This leads immediately to (39.47), and a similar device, obtained by replacing G_{4k}, U_k by χ_4, ϕ respectively, then gives the same matrix Hamiltonian equations of the scalar field—except for the vanishing of the spin term.

It may be noted here that in this theory β plays the role of the density operator (which in the Dirac electron theory happens to be expressible as the unit matrix) so that we might be tempted to identify $\beta\mathfrak{H}$ as the Hamiltonian because of its Hermiticity. This would be quite incorrect, since, as is easily verified using the properties $\beta\eta = -\eta\beta$, $\{\beta(1+\eta)\}^2 = 0$, $(\boldsymbol{S}\cdot\boldsymbol{p})^4 = \boldsymbol{p}^2(\boldsymbol{S}\cdot\boldsymbol{p})$ and any convenient coordinate frame, say that one for which $p_2 = p_3 = 0$, we find that the classical quadratic Hamiltonian form is given by $\mathfrak{H}^2$, not $(\beta\mathfrak{H})^2$.

One further remark: the two- and six-component wave functions may be again considered as an eight-component quantity which synthesizes the theories for spins 0 and 1 and, since $\mathfrak{H}$ is (spatial) reflection-invariant, the eight-component ψ must be either symmetric or antisymmetric, so that ψ comprises two 3-vectors and two pseudoscalars or two pseudo-3-vectors and two scalars, respectively. *Per se* this observation is not very significant, but it is interesting in regard to the pseudoscalar-vector meson field theory of nuclear forces proposed by Møller-Rosenfeld (40), since we find an automatic reduction of coupling parameters and a certain amount of support for their proposal.

(*e*) *The Chandra Ring*

In § 34*c* we discussed the matrix conditions corresponding to the positive definiteness of the total charge *or* total energy, and as criteria on *permissible* α-algebras we may ask whether any, other than the Dirac and Duffin-Kemmer algebras, exist satisfying these conditions. Apart from the separate issue of the presence or absence of *subsidiary conditions*, we might well also infer that the class of elementary particles is limited to spins 0, 1, and $\frac{1}{2}$—*if* no such further algebras can be constructed. In continuation of this research, Harish-Chandra (47*c*) has constructed a *gegenbeispiel* which shows that the conditions of § 34*c* do not, *per se*, so limit the possible species of particles. We label this matrix ring, which is an example of a direct ring product, the Chandra ring.

This ring $\mathfrak{A}(\alpha)$ is *essentially* the Kronecker product of the Dirac and Duffin-Kemmer rings $\mathfrak{A}(\gamma)$, $\mathfrak{A}(\beta)$ as seen from the postulated commutation rules:

$$[\alpha_\kappa, \alpha_\lambda]_+ = 2\delta_{\kappa\lambda} - [\beta_\kappa, \beta_\lambda]_+ \tag{39.48a}$$

$$\alpha_\kappa \omega \beta_\lambda + \beta_\lambda \omega \alpha_\kappa = [\beta_\kappa, \beta_\lambda]_+ \tag{39.48b}$$

$$\beta_\kappa \beta_\lambda \beta_\mu + \beta_\mu \beta_\lambda \beta_\kappa = \beta_\kappa \delta_{\lambda\mu} + \beta_\mu \delta_{\kappa\lambda} \tag{39.48c}$$

$$\left.\begin{aligned} \alpha_\kappa &= \gamma_\kappa + \omega\beta_\kappa \\ \omega &= \gamma_1\gamma_2\gamma_3\gamma_4 \end{aligned}\right\} \tag{39.48d}$$

since equations (39.48) all reduce to the Dirac case, if we set all $\beta_\kappa = 0$.[19]

The stated structure of $\mathfrak{A}(\alpha)$ will be verified if we show that

$$\mathfrak{A}(\alpha) \subset \mathfrak{A}(\gamma) \times \mathfrak{A}(\beta) \tag{39.49a}$$

$$\mathfrak{A}(\gamma) \times \mathfrak{A}(\beta) \subset \mathfrak{A}(\alpha) \tag{39.49b}$$

whence

$$\mathfrak{A}(\alpha) = \mathfrak{A}(\gamma) \times \mathfrak{A}(\beta) \tag{39.49c}$$

[19] This definition of α_κ, instead of the more obvious one, is a matter of convenience.

To this end we note first that there exists an element R in $\mathfrak{A}(\beta)$ such that

$$[R, \beta_\kappa]_+ = 0 \quad , \quad R^2 = 1 \tag{39.50a}$$ [20]

and therefore an $R' \epsilon \mathfrak{A}(\beta)$ such that

$$-\beta_\kappa = R'\beta_\kappa R'^{-1} \tag{39.50b}$$

$$[\beta_\kappa, C]_- = 0 \qquad C = R'^{-2}$$

whence C belongs to the centrum. From examination of the regular representation it follows that there exists an element c in the centrum such that $c^2 = C$, and therefore $R = R'c$ satisfies (39.50a). Moreover, R is a scalar since it commutes with the nucleus of proper rotations $\mathfrak{T}_{\kappa\lambda} = [\beta_\kappa, \beta_\lambda]_-$ and the reflection operator $2\beta_4{}^2 - 1$. From (39.48d, 39.50a) we see that γ_κ and β_λ commute, and $\mathfrak{A}(\gamma)$ is just the Dirac algebra. Similarly, $R \epsilon \mathfrak{A}(\alpha)$ so that

$$R\gamma_\kappa = \tfrac{1}{2}(R\alpha_\kappa + \alpha_\kappa R) \,\epsilon\, \mathfrak{A}(\alpha) \tag{39.51a}$$

$$\beta_\kappa = \omega(\alpha_\kappa - \gamma_\kappa) \,\epsilon\, \mathfrak{A}(\alpha) \tag{39.51b}$$

whence (39.49b) follows because $\mathfrak{A}(\gamma) \subset \mathfrak{A}(\alpha)$ and $\mathfrak{A}(\beta) \subset \mathfrak{A}(\alpha)$, the converse (39.49$a$) then obtaining from (39.48d) and therefore (39.49c) follows.

Therefore, if we choose an *irreducible* representation of the β_κ in (39.48d) so that the corresponding $\mathfrak{A}(B) : \beta_\kappa \rightarrow B_\kappa$ is simple, then, since $\mathfrak{A}(\Gamma) : \gamma_\kappa \rightarrow \Gamma_\kappa$ is also simple, it follows from (39.49c) that $\mathfrak{A}(\alpha)$ or more precisely $\mathfrak{A}(A) : \alpha_\kappa \rightarrow A_\kappa$ is also simple, and the representation so obtained is irreducible.

The tensor form of equations (39.48) assures the invariance of $\mathfrak{A}(\alpha)$ under the Lorentz group, the nucleus of the proper rotations being given by

$$\mathfrak{T}_{\kappa\lambda} = \tfrac{1}{4}[\gamma_\kappa, \gamma_\lambda]_- + [\beta_\kappa, \beta_\lambda]_- \tag{39.52}$$

the reflection operator by $\gamma_4(2\beta_4{}^2 - 1)R$, and it readily follows that the minimum equation of α_4 is of the fourth degree (provided the underlying representation of the β_κ is not the trivial one). Moreover, if Λ_γ, Λ_β are non-singular Hermitian Λ matrices (cf. § 34b) in the γ- and β-spaces respectively: $\gamma_\kappa{}^* = \Lambda_\gamma \gamma_\kappa \Lambda_\gamma{}^{-1}$, $\beta_\kappa{}^* = \Lambda_\beta \beta_\kappa \Lambda_\beta{}^{-1}$, then $\Lambda_\alpha = \Lambda_\gamma \times \Lambda_\beta$ is the corresponding matrix in the α-space; and for the representation $\Lambda_\gamma = \gamma_4$, $\Lambda_\beta = 2\beta_4{}^2 - 1$, $\Lambda_\alpha \alpha_4{}^3$ is a non-singular matrix with eigenvalues 0 and 1, so that by the criteria of § 34c the total charge is positive definite, and plus bracket (Fermi-Dirac) quantization is possible.

Corresponding to the two non-trivial representations of the β_κ, the α_κ are either 40×40 or 20×20 matrices and, according to the work of § 17a (cf. especially equations 17.17, 18), the respective product spaces $\mathfrak{R}(\gamma) \times \mathfrak{R}(\beta)$ reduce (with respect to the proper Lorentz group) as:

$$\begin{aligned}&\{\mathfrak{D}(\tfrac{1}{2}, 0) \oplus \mathfrak{D}(0, \tfrac{1}{2})\} \times \{\mathfrak{D}(1, 0) \oplus \mathfrak{D}(\tfrac{1}{2}, \tfrac{1}{2}) \oplus \mathfrak{D}(0, 1)\}\\ &\quad = \mathfrak{D}(\tfrac{3}{2}, 0) \oplus 2\mathfrak{D}(1, \tfrac{1}{2}) \oplus 2\mathfrak{D}(\tfrac{1}{2}, 1) \oplus \mathfrak{D}(0, \tfrac{3}{2}) \oplus 2\mathfrak{D}(\tfrac{1}{2}, 0) \oplus 2\mathfrak{D}(0, \tfrac{1}{2})\end{aligned} \tag{39.53a}$$

and

$$\begin{aligned}&\{\mathfrak{D}(\tfrac{1}{2}, 0) \oplus \mathfrak{D}(0, \tfrac{1}{2})\} \times \{\mathfrak{D}(0, 0) \oplus \mathfrak{D}(\tfrac{1}{2}, \tfrac{1}{2})\}\\ &\quad = \mathfrak{D}(1, \tfrac{1}{2}) \oplus \mathfrak{D}(\tfrac{1}{2}, 1) \oplus 2\mathfrak{D}(\tfrac{1}{2}, 0) \oplus 2\mathfrak{D}(0, \tfrac{1}{2})\end{aligned} \tag{39.53b}$$

[20] The $-\beta_\kappa$ satisfy the same C.R. as the β_κ so that the automorphism $\beta_\kappa \rightarrow -\beta_\kappa$, which leaves the centrum invariant, is of the type called inner, $\mathfrak{A}(\beta)$ being semisimple.

thus implying that the corresponding wave-equations describe particles which are capable of *both* spins $\frac{3}{2}$ and $\frac{1}{2}$. The wave equation

$$(\alpha^\kappa \partial_\kappa + \kappa)\psi = 0 \tag{39.54a}$$

leads directly to

$$\left.\begin{aligned} (\alpha^\kappa \alpha^\lambda \partial_\kappa \partial_\lambda - \kappa^2)\psi &= 0 \\ (\delta^{\kappa\lambda} \partial_\kappa \partial_\lambda - \kappa^2)\psi &= 0 \end{aligned}\right\} \tag{39.54b}$$

or,

$$(\delta^{\kappa\lambda} - \alpha^\kappa \alpha^\lambda)\partial_\kappa \partial_\lambda \psi = 0 \tag{39.54c}$$

With (39.48*a*) this reads as the subsidiary condition

$$\beta^\kappa \beta^\lambda \partial_\kappa \partial_\lambda \psi = 0 \tag{39.54d}$$

or, multiplying by $\beta^\mu \partial_\mu$ and using (39.48*c*):

$$\left.\begin{aligned} \beta^\lambda \partial_\lambda \psi &= 0 \\ (\gamma^\kappa \partial_\kappa + \kappa)\psi &= 0 \end{aligned}\right\} \tag{39.54e}$$

This system may be shown to possess, in a special sense, a structure beyond ordinary spin in that we find effectively *two* spin quantities combined to a given resultant, and, although (as for the system (39.55*a*, 56*a*) of the next section) the non-relativistic approximation is the same as that of the Dirac electron, it must, for reasons already noted (cf. also § 40), be considered doubtful that it corresponds to anything in Nature.[21]

(*f*) *The Bhabha Ring*

This last example of a *finite* matrix ring again corresponds to what we have labelled (*compound*) spin ($\frac{3}{2}$, $\frac{1}{2}$), but differs from the preceding case in that it leads to wave equations of type II, classified under the coupling schemes of § 35*b* and therefore the 5-dimensional orthogonal group according to § 38*a*. The two possible coupling schemes are (cf. §§ 35*a*, *b*)

$$\begin{array}{ccc} (\frac{1}{2}, 1) & \rightleftharpoons & (1, \frac{1}{2}) \\ \upharpoonleft\downharpoonright & & \upharpoonleft\downharpoonright \\ (0, \frac{1}{2}) & \rightleftharpoons & (\frac{1}{2}, 0) \end{array} \tag{39.55a}$$

$$(0, \tfrac{3}{2}) \rightleftharpoons (\tfrac{1}{2}, 1) \rightleftharpoons (1, \tfrac{1}{2}) \rightleftharpoons (\tfrac{3}{2}, 0) \tag{39.55b}$$

corresponding to wave functions of dimensions 16 and 20 respectively; or, relative to the group transition (cf. § 38*a*, especially equations 38.8–38.11):

5-dimensional orthogonal $\to$ 4-dimensional Lorentz $\to$ proper Lorentz

$$D^{(5)}(\tfrac{3}{2}, \tfrac{1}{2}) \to \{\tfrac{3}{2}, \tfrac{1}{2}\} \oplus \{\tfrac{1}{2}, \tfrac{1}{2}\} \to \mathfrak{D}(1, \tfrac{1}{2}) \oplus \mathfrak{D}(\tfrac{1}{2}, 1) \oplus \mathfrak{D}(\tfrac{1}{2}, 0) \oplus \mathfrak{D}(0, \tfrac{1}{2}) \tag{39.56a}$$

$$D^{(5)}(\tfrac{3}{2}, \tfrac{3}{2}) \to \{\tfrac{3}{2}, \tfrac{3}{2}\} \oplus \{\tfrac{3}{2}, \tfrac{1}{2}\} \to \mathfrak{D}(\tfrac{3}{2}, 0) \oplus \mathfrak{D}(0, \tfrac{3}{2}) \oplus \mathfrak{D}(1, \tfrac{1}{2}) \oplus \mathfrak{D}(\tfrac{1}{2}, 1) \tag{39.56b}$$

Again these systems are primarily of interest as illustrations of algebraic principles and the *direct method* of obtaining explicit realizations of the irreducible repre-

[21] The field-theoretic aspect of this theory is none the less of interest because of its unusual *mixed* tensor-spinor form, which also brings out more clearly the character of the subsidiary conditions.

sentations. We shall see that Bhabha's algebra $\mathfrak{A}(\alpha)$, like Chandra's, is the direct product[22] of two algebras—now the Dirac algebra $\mathfrak{A}(\gamma)$ and another, say $\mathfrak{A}(\xi)$, of order 42 so that $\mathfrak{A}(\alpha)$ is of order $16 \times 42 = 672$ (cf. 38.11) instead of the previous $16 \times 126 = 2016$.

There is nothing very special about the C.R. of the present case, but they are so excessively tedious as to again imply that they cannot be a correct description of physical fact; we find ($\alpha_{\kappa\lambda\mu\nu} = \alpha_\kappa \alpha_\lambda \alpha_\mu \alpha_\nu$, etc.)

$$\begin{aligned}2(\alpha_{\kappa\lambda\mu\nu} + \alpha_{\kappa\nu\mu\lambda} + \alpha_{\lambda\mu\nu\kappa} + \alpha_{\nu\mu\lambda\kappa}) \\ = 3(\alpha_{\kappa\lambda} + \alpha_{\lambda\kappa})\delta_{\mu\nu} + 3(\alpha_{\kappa\nu} + \alpha_{\nu\kappa})\delta_{\lambda\mu} \\ + (\alpha_{\mu\nu} + \alpha_{\nu\mu})\delta_{\kappa\lambda} + (\alpha_{\lambda\mu} + \alpha_{\mu\lambda})\delta_{\kappa\nu} + (\alpha_{\lambda\nu} + \alpha_{\nu\lambda})\delta_{\kappa\mu} + (\alpha_{\kappa\mu} + \alpha_{\mu\kappa})\delta_{\lambda\nu} \\ - \tfrac{3}{2}(\delta_{\kappa\lambda}\delta_{\mu\nu} + \delta_{\kappa\mu}\delta_{\lambda\nu} + \delta_{\lambda\mu}\delta_{\kappa\nu})\end{aligned} \tag{39.57}$$

of which the characteristic equation $\alpha_\kappa^4 - \frac{5}{2}\alpha_\kappa^2 + \frac{9}{16} = 0$ shows that the maximum spin is $\frac{3}{2}$, the case $s = \frac{1}{2}$ being included. As for spins 0, 1, and $\frac{1}{2}$, there exist auxiliary quantities η_μ (containing only odd or even powers of the α_μ according as the spin is integral or half-integral) such that $[\eta_\mu, \alpha_\nu]_+ = 0$, in general, and

$$\left.\begin{aligned} [\eta_\mu, \eta_\nu]_+ &= 0 \quad \text{for half-integral spin} \\ [\eta_\mu, \eta_\nu]_- &= 0 \quad \text{for integral spin} \end{aligned}\right\} \tag{39.58}$$

We now introduce the further auxiliary quantities ξ_μ defined by

$$\xi_\mu = \eta_\mu \alpha_\mu = \alpha_\mu \eta_\mu \tag{38.59a}$$

whence, in virtue of $\eta_\mu^2 = 1$ and $[\eta_\mu, \alpha_\nu]_+ = 0$,

$$\xi_\mu \eta_\mu = \eta_\mu \xi_\mu = \alpha_\mu \tag{39.59b}$$

and

$$\left.\begin{aligned} [\xi_\mu, \eta_\nu]_- &= 0 \quad \text{for half-integral spin} \\ [\xi_\mu, \eta_\nu]_+ &= 0 \quad \text{for integral spin} \end{aligned}\right\} \tag{39.60}$$ [23]

It immediately follows that for half-integral spin the algebra $\mathfrak{A}(\alpha)$ is a direct product, since the above relations show that for this class any product of the α's separates into a product of two disjoint parts, one in the ξ's and the other in the η's. It is further evident that the η's (with unity) form a subalgebra abstractly identical with Dirac's $\mathfrak{A}(\gamma)$—whence, as stated, $\mathfrak{A}(\alpha) = \mathfrak{A}(\gamma) \times \mathfrak{A}(\xi)$.

In this example,

$$\eta_\mu = \tfrac{1}{3}(4\alpha_\mu^3 - 7\alpha_\mu) \tag{39.61a}$$

so that the ξ's satisfy the relations

$$\left.\begin{aligned} &\xi_\mu^2 = \alpha_\mu^2 = \xi_\mu + \tfrac{3}{4} \\ &\xi_{\mu\nu} + \xi_{\nu\mu} + 2\xi_{\mu\nu\mu} = -\tfrac{1}{2}\xi_\mu && (\mu \neq \nu) \\ &\xi_{\mu\nu\rho} - \xi_{\rho\nu\mu} = \xi_{\nu\rho\mu} - \xi_{\mu\rho\nu} = \xi_{\rho\mu\nu} - \xi_{\nu\mu\rho} && (\mu \neq \nu \neq \rho) \\ &\xi_\mu(\xi_{\nu\rho\sigma} - \xi_{\sigma\rho\nu}) = (\xi_{\nu\rho\sigma} - \xi_{\sigma\rho\nu})\xi_\mu && (\mu \neq \nu \neq \rho \neq \sigma) \end{aligned}\right\} \tag{39.61b}$$

[22] Within the limits of this theory the formation by a direct product of rings is characteristic of *all* wave equations associated with half-integral spin. We do not require the general case, for which see Madhava Rao, *et. al.* (46). Cf. Madhava Rao (42, 45, 47).

[23] As for the special cases already studied, the η_μ are interpretable as rotations in the μ, 5-plane through an angle π.

The independent elements are then given by

Element	Number
I	1
ξ_μ	4
$\xi_{\mu\nu}$	12
$\xi_{\mu\nu\rho}$	16
$\xi_{\mu\nu\rho\sigma}$	9

(39.62*a*)

and the *primitive* idempotents by

$$\left.\begin{aligned} e_1 &= \tfrac{1}{15}N + \tfrac{1}{2} \\ e_2 &= \tfrac{1}{8}M - \tfrac{1}{24}N + \tfrac{5}{16} \\ e_3 &= -\tfrac{1}{8}M - \tfrac{1}{40}N + \tfrac{3}{16} \end{aligned}\right\} \tag{39.62b}$$

$$\left.\begin{aligned} &M = \Sigma\xi_\mu - \Sigma\xi_{\mu\nu} \qquad N = 2\Sigma\xi_{\mu\nu\rho} - \Sigma\xi_{\mu\nu\rho\sigma} \quad \text{(unequal indices)} \\ &M^2 = \tfrac{9}{2} - M - N \qquad N^2 = \tfrac{225}{4} \qquad MN = -\tfrac{15}{2}M \end{aligned}\right\} \tag{39.62c}$$

corresponding to the three irreducible representations of dimensions 1, 4, and 5; $\mathfrak{A}(\xi)$ as the direct sum of left ideals $\mathfrak{A}(\xi)\{e_1, e_2, e_3\}$ being semisimple.[24]

The trivial representation of the ξ_μ assigns $-\frac{1}{2}$ to each of them and simply leads to the Dirac matrices for $\mathfrak{A}(\alpha)$. The representations of dimensions 4 and 5 are uniquely determined by the commutation rules, starting with any ξ, say ξ_4, diagonal with elements $-\frac{1}{2}$ and $\frac{3}{2}$.[25] Needless to say, there are other possible choices (not to mention methods), and we find, for example (cf. Madhava Rao (46)),

$$\xi_1 \to \Xi_1 = \begin{Vmatrix} \frac{\sqrt{5}+1}{4} & \frac{1}{2} & 0 & \frac{\sqrt{5}+1}{4} \\ \frac{1}{2} & -\frac{\sqrt{5}-1}{4} & 0 & \frac{\sqrt{5}-1}{4} \\ 0 & 0 & -\frac{1}{2} & 0 \\ \frac{\sqrt{5}+1}{4} & \frac{\sqrt{5}-1}{4} & 0 & 0 \end{Vmatrix} \qquad \xi_2 \to \Xi_2 = \begin{Vmatrix} -\frac{\sqrt{5}-1}{4} & 0 & \frac{1}{2} & \frac{\sqrt{5}-1}{4} \\ 0 & -\frac{1}{2} & 0 & 0 \\ \frac{1}{2} & 0 & \frac{\sqrt{5}+1}{4} & \frac{\sqrt{5}+1}{4} \\ \frac{\sqrt{5}-1}{4} & 0 & \frac{\sqrt{5}+1}{4} & 0 \end{Vmatrix} \tag{39.63}$$

$$\xi_3 \to \Xi_3 = \begin{Vmatrix} -\frac{1}{2} & 0 & 0 & 0 \\ 0 & \frac{\sqrt{5}+1}{4} & \frac{1}{2} & \frac{\sqrt{5}+1}{4} \\ 0 & \frac{1}{2} & -\frac{\sqrt{5}-1}{4} & \frac{\sqrt{5}-1}{4} \\ 0 & \frac{\sqrt{5}+1}{4} & \frac{\sqrt{5}-1}{4} & 0 \end{Vmatrix} \qquad \xi_4 \to \Xi_4 = \begin{Vmatrix} -\frac{1}{2} & 0 & 0 & 0 \\ 0 & -\frac{1}{2} & 0 & 0 \\ 0 & 0 & -\frac{1}{2} & 0 \\ 0 & 0 & 0 & \frac{3}{2} \end{Vmatrix}$$

[24] These left ideals are, respectively, of ranks 16, 25, and 1, from which it follows that the corresponding idempotents generate the representations of dimensions 4, 5, and 1.

[25] In the first case the corresponding multiplicities are 3, 1, and in the second 3, 2.

(g) The Spurs

In physical applications using the particle-like aspect (cf. for example the references of footnote 2) the knowledge of the non-zero spurs of the basis elements of the various algebras $\mathfrak{A}(\alpha)$ is essential. These we list here for convenient reference, noting that for the Dirac algebra (§ 39*c*) all spurs are zero except that spur $I = 4$, so that the knowledge of the non-zero spurs for the Duffin-Kemmer (§ 39*d*) and Madhava Rao (§ 39*f*) algebras immediately gives those of the Chandra algebra (§ 39*e*) and the Bhabha algebra (§ 39*f*).

Duffin-Kemmer (§ 39*d*) (39.64)

Representation	1-row	5-row	10-row
spur I	1	5	10
spur η_μ	-1	-1	2
spur $\eta_{\mu\nu}$	1	1	-2
spur $\eta_{\mu\nu\rho}$	-1	3	-2
spur $\eta_{\mu\nu\rho\sigma}$	1	-3	2

Madhava Rao (§ 39*f*) (39.65)

Representation	1-row	4-row	5-row
spur I	1	4	5
spur ξ_μ	$-\frac{1}{2}$	0	$\frac{3}{2}$
spur $\xi_{\mu\nu}$	$\frac{1}{4}$	0	$-\frac{3}{4}$
spur $\xi_{\mu\nu\rho}$	$-\frac{1}{8}$	$\frac{1}{2}$	$-\frac{5}{8}$
spur $\xi_{\mu\nu\rho\sigma}$	$\frac{1}{16}$	$-\frac{1}{4}$	$\frac{5}{16}$

(h) Example of Infinite Rings

As a last algebraic example, though purely of formal character, we wish to mention that we can, of course, construct theories (of either type I or II) on *infinite* α-algebras. One simple example will serve as illustration, viz. one proposed by Bhabha and characterized by the C.R. (corresponding to spin 1)

$$\Sigma\,(\mathrm{Perm})\,\alpha_\kappa\alpha_\lambda\alpha_\mu = 2(\alpha_\mu\delta_{\kappa\lambda} + \alpha_\kappa\delta_{\lambda\mu} + \alpha_\lambda\delta_{\mu\kappa}) \qquad (39.66)$$

which are more general than the Duffin-Kemmer C.R., but are also satisfied by the latter matrices—as is to be expected. However, in contrast to the Duffin-Kemmer algebra, the system (39.66) *does not* lead to either of the relations

$$\mathfrak{T}_{\kappa\lambda} \sim [\alpha_\kappa, \alpha_\lambda]_-$$

$$\big[\alpha_\kappa, [\alpha_\lambda, \alpha_\mu]_-\big]_- = \alpha_\mu\delta_{\kappa\lambda} - \alpha_\lambda\delta_{\kappa\mu}$$

nor does (39.66) permit us to move any particular α_κ two or more places to left or right in a product of α's, since every α_κ appears in (39.66) at *both* extreme left and right in more than one term. This abstract algebra is infinite; only two of its non-trivial irreducible representations satisfy the Duffin-Kemmer C.R., and it cannot be the direct sum of algebras of the latter type.

The field equations of this class will, of course, all be in pure tensor form, of which one example is the system:

$$\partial_\lambda G^{\kappa\lambda} = -\kappa U^\kappa \tag{39.67}$$

$$\partial_\kappa G_{\lambda\mu} + \partial_\lambda G_{\mu\kappa} + \partial_\mu G_{\kappa\lambda} = \kappa U_{\kappa\lambda\mu}$$

$$\partial_\kappa U_\lambda - \partial_\lambda U_\kappa + \partial^\mu U_{\kappa\lambda\mu} = \kappa G_{\kappa\lambda}$$

in which U_κ, $G_{\kappa\lambda}$, and $U_{\kappa\lambda\mu}$ represent, respectively, a 4-vector, skew-tensor of second rank, and skew-tensor of third rank—all of which satisfy the second-order wave equation. The particle-like aspect then devolves on 14×14 matrices, but the system must be considered as *compound* in any case, and further, *non-elementary* with respect to the criterion of absence of subsidiary conditions—of which there are evidently several.[26]

§ 40. Wave Equations Irreducible under the Inhomogeneous Lorentz Group

(*a*) *The Development of the Invariant-Theoretic Viewpoint*

The Lorentz invariant systems heretofore considered have usually been classified principally in relation to the proper subgroup which as we know (cf. § 1) is but one of the disjoint pieces of the entire *homogeneous* group $\mathscr{L}$; and this emphasis, also in the course of historical development, is perhaps natural in view of the connection with spin and the characterization of different fields. Yet, as remarked in §§ 33*a*, *b*, this restriction of viewpoint is at the same time also rather surprising, since it is clear that the infinitesimal operations of the Lorentz group—in their most primitive physical context [1]—include the four translations of origin in space- and time-like sense; and it is only the complete set of *ten* generators and their invariant manifolds[2] which comprise the essential content of the *dynamical* aspect of quantum theory.

The restriction to the homogeneous group means, physically, restriction to a fixed space-time point, and in the quantal interpretation the states of the corresponding particles are invariant (in the literal sense) under translations and, in particular, independent of time. Thus, even though the (irreducible) representations of the homogeneous group are, in general, also (reducible) representations of the inhomogeneous group, it is likely (with few exceptions) [3] that the corresponding states have no immediate or simple physical significance. The recognition of this basic fact (particularly emphasized by Wigner, *loc. cit.*) does not at all vitiate the other theories which we have considered. These have been essential steps in the sequence of development of the theory, and are also necessary *if* we cannot be absolutely certain that

[26] Note that the total energy is not positive definite.

[1] Recall also the derivation of the momentum-energy conservation theorem.

[2] Equivalently, the infinite (unitary) irreducible representations.

[3] N.B. Among the vast variety of field equations which we have considered, we have also met *all* the systems (up to equivalence) irreducible under the inhomogeneous Lorentz group—except for two special cases of continuous spin (cf. §§ 40*b*, *c*). We do not enumerate the relevant sections here, since the equivalent *prototype* equations which appear in § 40*c* serve this purpose.

only particles *defined as elementary* with respect to irreducibility under the inhomogeneous group are realized in Nature—however likely or desirable this might appear.

Purely on grounds of *logical simplicity* we should prefer to delimit sharply the possible species of elementary particles allowed in present quantum theory, rather than lament the closure of the confusing array of spin s, $s-1$, . . . , charge 0, $\pm e$, $\pm 2e$, . . . , and composite rest-mass m_1, m_2, . . . particles. These are not allowed by the above general theory and should they appear in Nature we would have a certain sign of further need for radical revision of our basic concepts along *necessarily* non-Hamiltonian principles;[4] nor does the recent introduction of the concept of *non-localizable* systems modify this conclusion.

Again, from the physical *and* mathematical side, there is the plain fact that all is not well—even in the simple theory of fields *without* interactions. As foreshadowed in the discussion of § 32, the beautiful theorem on the relativistic connection between spin and statistics (i.e. Pauli's theorem) can no longer be maintained within the narrow confines of the proper subgroup—if only because the infinite representations (§§ 17*b*, iv, v) allow *all possible* correlations as a result of there being corresponding manifolds which allow positive definite energy-(charge-)density to be associated with half-integral (integral) spin fields; a complete reversal of Pauli's result.[5] As if matters could be any worse, there are also representations which simultaneously permit definite energy- and charge-density for either class of quanta.[5] For the details we must refer to the indicated original papers,[6] but we may here mention the result that these undesirable newcomers are excluded by the general theory.

(*b*) *Classes of Invariant Manifolds*

Apart from the two exceptions (§§ 40*b*(iii), 40*c*), the manifolds invariant (strictly *covariant*) under the *inhomogeneous* Lorentz group (which we here denote by $\mathfrak{L}$) were essentially obtained by Majorana (32) and independently by Dirac (36); the complete classification, which is only outlined here, follows from the general analysis by Wigner (39, 47, 48). The essential concepts and constructs are already contained in our previous work, in particular § 25 and §§ 33*a*, *b*, and we need merely recapitulate that the basic classification proceeds on the determination of the unitary operator corresponding to each transformation in $\mathfrak{L}$ and hence the irreducible representations $U(\mathfrak{L})$ of $\mathfrak{L}$; these devolve principally on the *little groups*, i.e. the various subgroups of the *homogeneous* Lorentz transformations which leave a prototype momentum vector invariant (cf. § 25). As is well known both abstractly and from elementary quantum theory, the momenta (but not the coordinates) are defined by the Lorentz group $\mathfrak{L}$ as infinitesimal translations,[6] so that it is both logical and simple to operate now in the *momentum*-representation.

[4] It has been shown by J. von Neumann (Princeton Lectures 1936) that, as long as the Hamiltonian exists and the hierarchy of so-called observables includes all bounded operators, the wave-mechanical framework is essentially inescapable. For a brief sketch of the proof see, for example, the author's book *Perturbation Methods, etc.* (Blackie & Son, Ltd., and Hafner, New York), Chapter XII.

[5] Cf. Gelfand (48*a*, *b*, *c*), Harish-Chandra (47*b*). The latter possibility, implying *inter alia* a description of particles of only one sign, can no longer be viewed as particularly relevant in view of the now well-known phenomenon of pair-creation and -annihilation.

[6] See also the author's forthcoming volume, *Invariantive Aspects of the Elementary Particle Problem*, Fulbright Lectures.

In brief, the (system) states may be expressed as functions $\psi(p, \xi)$ of the momentum vector p, the momenta varying over the manifold $p_\kappa p^\kappa = P$ ($P = m^2c^2 > 0$ or $P = 0$ in the cases of physical significance), and of auxiliary spin variable(s) ξ, which may assume a finite or infinite number of values.[7] The unitary operators U, which represent the finite Lorentz transformations, follow by exponentiation: every inhomogeneous transformation $x^\mu \to \mathfrak{L}x^\mu$ ($\mathfrak{L} \sim \Lambda + a$, cf. equations 1.9) has its *image* in the manifold of ψ's:

$$U(\mathfrak{L})\psi(p, \xi) = e^{-ip_\mu a^\mu} Q(p, \Lambda)\psi(\Lambda^{-1}p, \xi) \tag{40.1}$$

where $Q(p, \Lambda)$ is a unitary operator which may depend on p but affects only the variable(s) ξ; and, as is necessary to quantization (with definite metric), there is defined a positive definite, Lorentz invariant, inner product (ϕ, ψ) obtained by an integration over the manifold $p^\kappa p_\kappa = P$ and summation or integration over the variable(s) ξ.[8] The unitary operators $Q(p_0, \Lambda)$, where p_0 is the chosen fixed momentum vector ($\Lambda p_0 = p_0$), form an irreducible representation of the little group in any particular case. The invariants P and W of § 25 (see especially equations 25.2–7) now clearly play the fundamental role in the classification.

Three general classes of fields are found:

(i) (P_s). *Quanta of discrete spin s and non-zero rest-mass.*

$P > 0$, and in the rest system the momentum vector has but one non-zero component $p_4 = \pm mc$, so that (cf. 25.5, 7) $W = (\hbar mc)^2 s(s + 1)$. The operator $P^{-1}W$ represents the square of the spin angular momentum and in an irreducible representation has the usual values, there being $2s + 1$ independent states. The representation $U(\mathfrak{L})$ is single- or double-valued according as s is integral or half-integral; the first few cases ($s = 0, \frac{1}{2}, 1$) correspond to the Klein-Gordon, Dirac, and Duffin-Kemmer[9] equations.

(ii) (O_s). *Quanta of discrete spin s and zero rest-mass.*

These representations may, in the sense noted in § 25 (cf. especially footnote 7, p. 95), be considered as limiting cases of the representations (P_s) for $m \to 0$.[10] For a given momentum vector, there exist but two independent states ($s \neq 0$) corresponding to the two different states of polarization. Right and left circularly polarized states are discriminated by the operator equations $\omega_\kappa = sp_\kappa$ and $\omega_\kappa = -sp_\kappa$, respectively, so that the representation (O_s) is characterized by $P = 0$, $\omega_\kappa\omega_\lambda = s^2 p_\kappa p_\lambda$. Again the first few cases ($s = 0, \frac{1}{2}, 1$) correspond to the scalar (Klein-Gordon or Duffin-Kemmer), neutrino (Dirac), and photon (Maxwell) wave-equations.

(iii) ($O(\Xi)$ and $O'(\Xi)$). *Quanta of continuous spin and zero rest-mass.*

Here, $P = 0$ and $W = \Xi^2$, where Ξ is a real positive number. The significance of the term *continuous* spin is that there exist infinitely many different states of

[7] For the two continuous spin cases (§ 40c), the variable ξ includes more than simply spin.

[8] The possibility of defining this inner product is clearly a basic element of the usual c-number interpretation of quantum theory, and is by the same token a possibly more fundamental reason for the necessity of the existence of a Hermitian Λ matrix (§ 34*b*). This Λ is, of course, not to be confused with the above Λ which represents the homogeneous component contained in any element of $\mathfrak{L}$.

[9] The first field-theoretic form of this system was obtained by Proca (36, 38).

[10] P and W are both equal to zero and hence do not suffice to characterize these representations.

polarization which may be described by a continuous variable. $O(\Xi)$ and $O'(\Xi)$ correspond to integral and half-integral spin character in the basic sense that the representations are single- and double-valued, respectively. There are several unusual features intrinsic to this class of quanta: there appears a length Ξ^{-1} (in appropriate units) in the characterization of the manifolds, so that, if these quanta were to exist and transmit interactions, the energy would be expected to involve this length;[11] in a *Hohlraum* there would exist infinitely many states in a given energy range which would correspond to infinite heat capacity. Most surprising is the result that it is not possible to localize the wave functions inside a finite domain, but a simple *positive definite* form of scalar product exists in coordinate space.

(c) *Further Considerations on Individual Classes*

The explicit realization of each of these irreducible representations requires only a choice (among equivalent sets) of a convenient set of wave functions, and verification that, in the manifold so-defined, a Lorentz invariant inner product can be constructed. This we shall briefly sketch for each class, and it will simplify the notation and printing if we set $\hbar = c = 1$ in this section.

In the class (P_s) the case $s = 0$ is particularly simple, since ξ assumes but one value and may therefore be omitted. Correspondingly, $Q(p, \Lambda) = 1$, and the little group is realized by the trivial one-dimensional representation ($S_{\kappa\lambda} = 0$, $\omega_\kappa = 0$). The wave equation reduces to $p^\kappa p_\kappa = m^2$, and the inner product is determined by the norm (ψ, ψ),

$$(\psi, \psi) = \mathcal{N}(\psi) = \int |\psi(p)|^2 \, d\Omega \qquad d\Omega = d^3p/|p_4| \tag{40.2}$$

the integration extending over both sheets of the hyperboloid $p^\kappa p_\kappa = P = m^2$.[12] The wave function in configuration space follows by the familiar Fourier theorem, and as is well known does not lead to definite density; the integral over the density in configuration space coincides with (40.2) only if $\psi(p) = 0$ for $p_4 < 0$.

For particles of higher spin, $s = \frac{1}{2}N$ $(N = 1, 2, \ldots)$, a convenient form is the 4-spinor formulation of Dirac's generalized equations first given by Kramers, Belinfante, and Lubanski (cf. § 36*b*, equations 36.4; § 38*a*, equations 38.1 *et seq.*). However, following Wigner and Bargmann (47), we take this system in the form:

$$\gamma^\kappa_{(r)}\gamma^\lambda_{(r)} + \gamma^\lambda_{(r)}\gamma^\kappa_{(r)} = 2g^{\kappa\lambda} \qquad \text{(cf. 36.4)} \tag{40.3}$$

$$(\gamma^\kappa_{(r)} p_\kappa - m)\psi(p, \xi_1, \xi_2, \ldots, \xi_N) = 0 \qquad \text{(cf. 38.1)} \tag{40.4}$$

where $r = 1, 2, \ldots, N$ and ψ is a symmetric function of the N $(= 2s)$ *four-valued* spin variables ξ_r. This form differs from (38.1) only in the absence of the factor N in the mass term, and, apart from convenience, leads to the second-order wave-equation without separate postulate. Here again $\mathcal{S}^{\kappa\lambda} = \frac{1}{2}i \sum_r \gamma^\kappa_{(r)}\gamma^\lambda_{(r)}$, $(\kappa \neq \lambda)$.

[11] The covariance of these systems *does not* support a non-vanishing rest-mass.

[12] This expression, as a typical form, is Lorentz invariant because $d\Omega$ is an invariant volume element in momentum space. For a most illuminating discussion on the general case, from both the geometrical standpoint and that of the relevance of states of positive *and* negative energy in the determination of $\mathcal{N}(\psi)$, see Wigner (48).

The norm
$$\mathscr{N}(\psi) = \int \left| \sum_{\xi} \psi^* \gamma^4_{(1)} \gamma^4_{(2)} \ldots \gamma^4_{(N)} \psi \right| d\Omega \tag{40.5a}$$

is easily shown to be invariant as required. For example, with respect to the rotations, $\mathscr{J}_{\kappa\lambda} = \mathscr{M}_{\kappa\lambda} + \mathscr{S}_{\kappa\lambda}$, we see that it is only necessary to verify the so-called spin part, since $\mathscr{N}(\psi)$ is, in fact, invariant with respect to the component parts. For the latter this requires
$$\mathscr{N}\big((1 + i\epsilon\mathscr{S}_{\kappa\lambda})\psi\big) = \mathscr{N}(\psi) \tag{40.5b}$$

to terms of order ϵ^2, and this is seen to be so, since the components $\mathscr{S}_{kl}$ are Hermitian and commute with $\prod_r \gamma^4_{(r)}$, while the components $\mathscr{S}_{k4}$ are skew-Hermitian but anticommute with $\prod_r \gamma^4_{(r)}$. The invariance against reflections is also easily seen. From the properties of the γ's (40.5a) can be written as
$$\mathscr{N}(\psi) = \int \left| \frac{m}{p_4} \right|^N \sum_{\xi} |\psi|^2 d\Omega \tag{40.5c}$$

It is worth noting that the absolute signs above, as in the definition of $d\Omega$, are unnecessary in the case of odd N, which corresponds to the possibility of defining a simple *positive definite* scalar product in *configuration space* for this case only.

If we choose the invariant momentum vector as $p_0 = (0, 0, 0, m)$, this corresponds to the rest system and, as before (§ 25), the little group is simply the subgroup of rotation in 3-dimensional space; it is also convenient for the determination of the irreducible representations to take the γ^4 as *diagonal* with elements diag $\| 1, 1, -1, -1 \|$.[13] Equations (40.4) then show that only the components of ψ corresponding to the first two rows of $\gamma_{(r)}$ can be different from zero, so that, of the 4^N components, only the 2^N of this type can remain and even these are not all independent. Since ψ is symmetric in the ξ's, it will have as equal components all those in which the same number, say τ, of the N indices correspond to the first row of the γ's, the $N - \tau$ other indices to the second row; hence, there are but $N + 1 = 2s + 1$ independent components. The usual duplication of components arises because the same considerations apply on the last two rows of the γ's for $p_4 = -m$.

The transformation properties of these independent elements under the little (rotation) group follow simply by considering the effect of the $\mathscr{S}_{\kappa\lambda}$ on ψ, since the space-like $\mathscr{M}_{\kappa\lambda}$ give a zero resultant when operating on a ψ with pure time-like $p = p_0$. For this purpose we may assume, for all r, that $\frac{1}{2}i\gamma_1\gamma_2$ is represented by diag $\| \frac{1}{2}, -\frac{1}{2}, \frac{1}{2}, -\frac{1}{2} \|$, since it commutes with γ^4; and if the sum of these is applied to the component of ψ in which τ of the ξ's correspond to the first row, $N - \tau$ to the second—then this component is multiplied by $\tau - s$. Since τ ranges from 0 to $2s$, the $\mathscr{J}_{12}\psi = \mathscr{S}_{12}\psi$ range from $-s\psi$ to $s\psi$, corresponding to $D(s)$.

. . . .

For the class $[O_s]$ the case $s = 0$ differs from the preceding only in the obvious absence of the mass term and modification of the domain of integration to the light-cone.

[13] N.B. The realization of the representation in all but class (iii) could be omitted, since necessarily we obtain nothing more than a more formal and precise *systematization* of previous results. Still this is not trivial.

The discussion for higher spins needs but little qualification, the wave-equations follow from (40.3, 4), the infinitesimal operators are the same, but the scalar product must be defined by (40.5*c*), omitting the mass factor (its invariance follows directly even now, since the wave function is continuous in m except for $p_1 = p_2 = p_3 = 0$). The essentially new feature is that now the wave equation, as well as the infinitesimal operators, is invariant under any one of the operators $\gamma_{5(r)} = i\gamma^1_{(r)}\gamma^2_{(r)}\gamma^3_{(r)}\gamma^4_{(r)}$—whence the previous manifold is restricted by the conditions

$$\gamma_{5(r)}\psi = \psi \tag{40.6a}$$

$$\gamma_{5(r)}\psi = -\psi \tag{40.6b}$$

These manifolds correspond physically to right and left polarization, being invariant under proper Lorentz transformations and interchanged by reflections. As anticipated in § 25, a convenient invariant momentum vector is $p_0 = (0, 0, 1, 1)$ with which the wave equations can be expressed as

$$\gamma^3_{(r)}\gamma^4_{(r)}\psi = \psi \qquad (r = 1, 2, \ldots, N) \tag{40.7}$$

With $\gamma^3_{(r)}\gamma^4_{(r)} = \text{diag}\,\|\,1,\ 1,\ -1,\ -1\,\|$ this equation states that ψ is non-zero only if all ξ's have values corresponding to the first two rows of the γ's. Since the γ_5's commute with the $\gamma^3\gamma^4$'s, they may be assumed to be represented by $\text{diag}\,\|\,1,\ -1,\ 1,\ -1\,\|$, so that in the manifold defined by (40.6*a*) and (40.7) all components of $\psi(p_0, \xi)$ vanish unless they correspond to the stated values of the ξ's. Thus, the manifold (40.6*a*, 7) is one-dimensional for any given momentum 4-vector, and similarly for the manifold (40.6*b*, 7), as follows by a previous argument.

The infinitesimal operators of the little group are $\mathscr{I}_{12}$, $\mathscr{I}_{13} - \mathscr{I}_{14}$, $\mathscr{I}_{23} - \mathscr{I}_{24}$ and again we need only consider the effect of the $\mathscr{S}_{\kappa\lambda}$. It is readily seen from (40.7) that the components of $\mathscr{S}_{\kappa\lambda}$ corresponding to the last two generators (e.g. $\frac{1}{2}i\sum_r(\gamma^1_{(r)}\gamma^3_{(r)} - \gamma^1_{(r)}\gamma^4_{(r)})$ vanish on application to $\psi(p = p_0)$, while $\mathscr{S}_{12}$ gives $\frac{1}{2}N$ if applied to the ψ of (40.6*a*, 7) and $-\frac{1}{2}N$ if applied to the ψ of (40.6*b*, 7). Although these manifolds cannot be discriminated by the invariants P and W which are both zero, they can be characterized by $P = 0$, and the set $\omega_\kappa = sp_\kappa$, $\omega_\kappa = -sp_\kappa$ or the combined invariant $\omega_\kappa\omega_\lambda = s^2 p_\kappa p_\lambda$ if reflections are taken into account.

. . . .

For the class $[O(\Xi)]$ the auxiliary variable is a space-like 4-vector ξ_μ of length λ —orthogonal to p_μ.[14] The *scalar* function $\psi(p, \xi)$ is determined by the equations

$$p^\kappa p_\kappa \psi = 0 \tag{40.8a}$$

$$p^\kappa \xi_\kappa \psi = 0 \tag{40.8b}$$

$$\xi^\kappa \xi_\kappa \psi = -\lambda^2 \psi \tag{40.8c}$$[15]

$$p_\kappa \frac{\partial \psi}{\partial \xi_\kappa} = -i\Xi\psi \tag{40.8d}$$

[14] N.B. Quite generally, for two orthogonal 4-vectors: if one is time-like the other is space-like, and if one is null, the other is *either* proportional to it or space-like. The proof devolves on Schwarz's inequality and is essentially obvious.

We also require the simple theorem which results from this, viz. if four 4-vectors are mutually orthogonal *and* linearly independent, one of them is time-like, three are space-like. It is important to note the distinction that four mutually orthogonal 4-vectors are not necessarily independent, since a null vector is self-orthogonal. Cf. Wigner, *loc. cit.* (39).

or, transforming to the coordinate-representation,

$$\Box \Phi = 0 \tag{40.9a}$$

$$\xi^\kappa \frac{\partial}{\partial x^\kappa} \Phi = 0 \tag{40.9b}$$

$$(\xi^\kappa \xi_\kappa + \lambda^2)\Phi = 0 \tag{40.9c}$$[15]

$$\left(\frac{\partial^2}{\partial x^\kappa \partial \xi_\kappa} - \Xi\right)\Phi = 0 \tag{40.9d}$$

According to (40.8d)

$$\psi(p, \xi + \rho p) = e^{-i\rho\Xi}\, \psi(p, \xi) \tag{40.10}$$

and correspondingly the infinitesimal displacement operators are the p_κ, the rotations being the $\mathscr{M}_{\kappa\lambda}$ plus the rotation operators in ξ-space

$$\mathscr{S}_{\kappa\lambda} = i\left(\xi_\kappa \frac{\partial}{\partial \xi^\lambda} - \xi_\lambda \frac{\partial}{\partial \xi^\kappa}\right) = i(\xi_\kappa g_{\lambda\mu} - \xi_\lambda g_{\kappa\mu}) \frac{\partial}{\partial \xi_\mu} \tag{40.11}$$

In order to determine the invariant inner product, we need to introduce, for every vector p on the light cone, two real space-like vectors $u^{(1)}(p)$, $u^{(2)}(p)$ of unit length and mutually orthogonal as well as orthogonal to p:

$$u^\mu_{(r)}\, u_{(s)\mu} = -\delta_{rs} \qquad u^\mu_{(r)}\, p_\mu = 0 \qquad (r, s = 1, 2) \tag{40.12}$$

Then, ξ is the linear combination with real coefficients (cf. footnote 14)

$$\xi = \alpha p + \beta_1 u^{(1)} + \beta_2 u^{(2)} \tag{40.13}$$

which implies (note the case $\xi^\mu \xi_\mu = -1$) that $\beta_1{}^2 + \beta_2{}^2 = 1$, whence $\beta_1 + i\beta_2 = e^{i\tau}$; or, $\psi(p, \xi)$ is a function of p, α, τ,

$$\psi(p, \xi) = \phi(p, \alpha, \tau) \tag{40.14}$$

Evidently the choice of the $u^{(r)}(p)$ is not unique, but it follows from (40.10) that $| \phi(p, \alpha, \tau) | = | \phi(p, 0, \tau |$ and the norm defined by

$$\mathscr{N}(\psi) = \int | \phi(p, 0, \tau) |^2 \, d\Omega \, d\tau \tag{40.15}$$

may be shown to be independent of this choice. The Lorentz invariance of this norm follows on noting that if a homogeneous Lorentz transformation maps p on $\Lambda^{-1}p$ and ξ on $\Lambda^{-1}\xi$, it is permissible to choose the $u^{(r)}(p)$ in the new system to be the transforms of the original ones, whence α, β_1, β_2 (equation 40.13) and τ remain unchanged.

With p_0 again given by (0, 0, 1, 1) the infinitesimal operators are also again

[15] By suitable adjustment of units we can replace λ^2 in (40.8c, 9c) by 1, which is a convenient variability range for ξ. It is seen that the consistency and invariance requirements on this set prevent the introduction of a mass term.

$\mathscr{J}_{12}$, $\mathscr{J}_{13} - \mathscr{J}_{14}$, $\mathscr{J}_{23} - \mathscr{J}_{24}$, and for $p = p_0$ only the spin parts of the latter two give non-zero resultants; viz.

$$\mathscr{S}_{13} - \mathscr{S}_{14} = -i\xi_1\left(\frac{\partial}{\partial\xi_3} + \frac{\partial}{\partial\xi_4}\right) + i(\xi_3 - \xi_4)\frac{\partial}{\partial\xi_1} \tag{40.16a}$$

$$\mathscr{S}_{23} - \mathscr{S}_{24} = -i\xi_2\left(\frac{\partial}{\partial\xi_3} + \frac{\partial}{\partial\xi_4}\right) + i(\xi_3 - \xi_4)\frac{\partial}{\partial\xi_2} \tag{40.16b}$$

$$\left.\begin{aligned}(\mathscr{S}_{13} - \mathscr{S}_{14})\,\psi(p_0, \xi) &= \Xi\,\xi_1\psi\\ (\mathscr{S}_{23} - \mathscr{S}_{24})\,\psi(p_0, \xi) &= \Xi\,\xi_2\psi\end{aligned}\right\} \tag{40.17}$$

Thus, ψ is not invariant under the *rotations* in ξ-space, and the corresponding squared *momentum* has the value $(\xi_1^2 + \xi_2^2)\Xi^2 = \Xi^2 = W$.

The discussion of the class $[O'(\Xi)]$ is virtually identical to the preceding, except that we introduce a discrete four-valued spin variable ζ—in addition to ξ. The wave equations then become

$$\gamma^\kappa p_\kappa \psi = 0 \tag{40.18a}$$

$$p^\kappa \xi_\kappa \psi = 0 \tag{40.18b}$$

$$\xi^\kappa \xi_\kappa \psi = -\lambda^2\psi \tag{40.18c}$$

$$p_\kappa \frac{\partial}{\partial\xi_\kappa}\psi = -i\Xi\psi \tag{40.18d}$$

and the norm

$$\mathscr{N}(\psi) = \int p_4^{-2}\Sigma\,|\,\phi(p, 0, \tau)\,|^2\, d^3p\, d\tau \qquad \text{(cf. 40.5c, 15)} \tag{40.19}$$

however, $W\psi = \Xi^2\psi$ and $P\psi = 0$, as before.

BIBLIOGRAPHY

ALBERT, A. A.
"Modern Higher Algebra", University of Chicago Press (1937).

ARTIN, E. (and NESBITT, C. J.; THRALL, R. M.)
Univ. Mich. Publ. in Math., No. 1 (1946), "Rings with Minimum Condition".

BARGMANN, V.
Sitz. Ber. der Preuss. Akad., No. 24, 346 (1932), "Bemerkungen zur allegemein-relativistischen Fassung der Quantentheorie".
Helv. Phys. Acta, **vii**, 57 (1933), "Uber den Zusammenhang zwischen Semivektoren und Spinoren und die Reduktion der Diracgleichungen fur Semivektoren".
Annals of Math., **48**, No. 3, 568 (1947), "Irreducible Unitary Representations of the Lorentz Group".
—— (and WIGNER, E. P.).
Proc. Nat. Acad. Sci., **34**, No. 5, 211 (1948), "Group Theoretical Discussion of Relativistic Wave Equations".

BELINFANTE, F. J.
Physica, **vi**, 849 (1939*a*), "Undor Calculus and Charge Conjugation".
ibid., 870 (1939*b*), "The Undor Equation of the Meson Field".
ibid., 887 (1939*c*), "On the Spin Angular Momentum of Mesons".
—— (and PAULI, W.).
ibid., **vii**, 177 (1940*a*), "On the Statistical Behaviour of Known and Unknown Particles".
ibid., 449 (1940*b*), "On the Current and the Density of the Electric Charge, the Energy, the Linear Momentum and the Angular Momentum of Arbitrary Fields".
ibid., **vii**, 765 (1940*c*), "On the Quantum Theory of Wave Fields".
ibid., **x**, 720 (1943), "On the Commutativity of the Dirac Electron Wave Function with the Electromagnetic Field".
ibid., **xii**, 1 (1946*a*), "On the Longitudinal and Transversal Delta Function, with some Applications".
ibid., 17 (1946*b*), "On the Vanishing of div $\mathscr{E} - 4\pi\rho$ in Quantum Electrodynamics".

BETHE, H. A.
Phys. Rev., **57**, 390 (1940*a*), "The Meson Theory of Nuclear Forces".
—— (and NORDHEIM, L. W.).
bid., 998 (1940*b*), "On the Theory of Meson Decay".

BHABHA, H. J.
Proc. Roy. Soc. A, **166**, 501 (1938), "On the Theory of Heavy Electrons and Nuclear Forces".
Nature, **143**, 276 (1939*a*), "The Fundamental Length Introduced by the Theory of the Mesotron (Meson)".
Proc. Roy. Soc. A, **172**, 384 (1939*b*), "Classical Theory of Mesons".

BHABHA, H. J.
—— (and CORBEN, H. C.).
ibid., A, **178**, 273 (1941*a*), " General Classical Theory of Spinning Particles in a Maxwell Field ".
ibid., A, **178**, 314 (1941*b*), " General Classical Theory of Spinning Particles in a Meson Field ".
—— (and MADHAVA RAO, B. S.).
Proc. Indian Acad. Sci., **139** (1941*c*), " The Scattering of Charged Mesons ".
Rev. Mod. Phys., **17**, 200 (1945*a*), " Relativistic Wave Equations for the Elementary Particles ".
Proc. Indian Acad. Sci., **21**, No. 6, 241 (1945*b*), " Relativistic Wave Equations for the Proton ".
—— (and HARISH-CHANDRA).
Proc. Roy. Soc. A, **185**, 250 (1946), " On the Fields and Equations of Motion of Point Particles ".
Report Int. Conf. (Camb. 1946), 22 (1947), " Relativistic Wave Equations for the Elementary Particles ".
Rev. Mod. Phys., **21**, No. 3, 451 (1949), " On the Postulational Basis of the Theory of Elementary Particles ".

BOPP, F.
Ann. der Phys., No. 5, **38**, 345 (1940), " Eine Lineare Theorie des Elektrons ".
ibid., **42**, 573 (1943), " Lineare Theorie des Elektrons, II ".

BORN, M.
—— (and FUCHS, K.).
Proc. Roy. Soc. Edin., **60**, 100 (1940*a*), " Reciprocity. Pt. II: Scalar Wave Functions ".
—— (and FUCHS, K.).
ibid., 141 (1940*b*), " Reciprocity. Pt. III: Reciprocal Wave Functions ".
—— (and PENG, H. W.).
ibid., 40 (1944*a*), " Quantum Theory of Fields. I. Pure Fields ".
—— (and PENG, H. W.).
ibid., 92 (1944*b*), " Quantum Theory of Fields. II. Statistics of Pure Fields ".
—— (and PENG, H. W.).
ibid., 127 (1944*c*), " Quantum Theory of Fields. III. Electromagnetic Field and Electron in Interaction ".
Report Int. Conf. (Camb. 1946), 14 (1947), " Relativistic Quantum Mechanics and the Principle of Reciprocity ".
Nature, **163**, 207 (1949*a*), " Elementary Particles and the Principle of Reciprocity ".
—— (and CHENG, K. C.; GREEN, H. S.).
ibid., **164**, 281 (1949*b*), " Reciprocity Theory of Electrodynamics ".
—— (and RODRIGUEZ, A. E.).
ibid., **163**, 320 (1949*c*), " Meson Masses and the Principle of Reciprocity ".
—— (and GREEN, H. S.).
Proc. Roy. Soc. Edin., **62**, 470 (1949*d*), " Quantum Theory of Rest Masses ".

BRAUER, R. (and WEYL, H.).
Amer. Jour. of Math., **57**, 425 (1935), " Spinors in n-Dimensions ".

DE BROGLIE, L.

Comptes Rendus, **208**, 1697 (1939*a*), " Sur un cas de reducibilité en Mecanique ondulatoire des particules de spin 1 ".

ibid., **209**, 265 (1939*b*), " Sur la Théorie des Particules de Spin Quelconque ".

" Une nouvelle Théorie de la Lumière ", Hermann, Paris (1940).

" Théorie Générale des Particules A Spin ", Gauthier-Villars, Paris (1943).

BRUNO, B.

Ark. f. Mat. Ast. Fys., **35**, No. 1 (1947), " Transformation Properties and Spin in the Meson Theory of Eriksson ".

CARTAN, E.

Bull. Soc. Math. de France, **41**, 53 (1913), " Les Groupes Projectifs qui ne laissent Invariante Aucune Multiplicité Plane ".

Actualités Sci. et Ind., Hermann, Paris, No. 643 (1938*a*), " La Théorie des Spineurs I ".

ibid., No. 701 (1938*b*), " La Théorie des Spineurs II ".

CASIMIR, H.

Neder. Akad. van Wetensch., Proc., **34**, 844 (1931).

Helv. Phys. Acta, **6**, 287 (1933), " Uber die Intensität der Streustrahlung gebundener Elektronen ".

—— (and VAN DER WAERDEN, B. L.).

Math. Ann. **111**, 1 (1935).

CHANG, T. S.

Proc. Roy. Soc. A, **182**, 302 (1944), " The Impulse-Energy Tensor of Material Particles ".

Proc. Camb. Phil. Soc., **42**, 132 (1946), " A Note on the Hamiltonian Equations of Motion ".

ibid., **43**, 196 (1947), " A Note on the Hamiltonian Theory of Quantization II ".

ibid., **44**, 76 (1948), " Field Theories with Higher Derivatives ".

CHENG, K. C. (and BORN, M.; GREEN, H. S.).

Nature, **164**, 281 (1949*b*). " Reciprocity Theory of Electrodynamics ".

—— (and RODRIGUEZ, A. E.).

ibid., **163**, 367 (1949), " Particles with Spin and the Principle of Reciprocity ".

CORBEN, H. C. (and MASSEY, H. S. W.).

Proc. Camb. Phil. Soc., **35**, 84 (1939), " The Emission and Absorption of Heavy Electrons ".

—— (and SCHWINGER, J.).

Phys. Rev., **58**, 953 (1940), " The Electromagnetic Properties of Mesotrons ".

—— (and BHABHA, H. J.).

Proc. Roy. Soc. A, **178**, 273 (1941), " General Classical Theory of Spinning Particles in a Maxwell Field ".

Phys. Rev., **69**, 225 (1946*a*). " A Classical Theory of Electromagnetism and Gravitation ".

ibid., **70**, 947 (1946*b*), " Special Relativistic Field Theories in Five Dimensions ".

COURANT, R. (and HILBERT, D).

" Methoden der Mathematischen Physik I, II ", Interscience Publishers, Inc., New York (1937).

DARWIN, C. G.

Proc. Roy. Soc. A, **118**, 654 (1928), " The Wave Equation of the Electron ".

DAVYDOV, A. S.
Akad. Nauk. Zhurnal Eksp. Teoret. Fys., **17**, 427 (1947), " On the Theory of Particles of Spin $\frac{3}{2}$."

DIRAC, P. A. M.
Proc. Roy. Soc. A, **136**, 453 (1932), " Relativistic Quantum Mechanics ".
—— (and FOCK, V. A.; PODOLSKY, B.).
Physik. Zeits. Sowjetunion, **2**, 468 (1932), " On Quantum Electrodynamics ".
Proc. Roy. Soc. A, **155**, 447 (1936), " Relativistic Wave Equations ".
Ann. d. l'Inst. H. Poincaré, **9**, 13 (1939*a*), " La Théorie de l'Electron et du Champ Electromagnetique ".
—— (and PEIERLS, R.; PRYCE, M. H. L.).
Proc. Camb. Phil. Soc., **38**, 193 (1941), " On Lorentz Invariance in the Quantum Theory ".
Proc. Roy. Soc. A, **180**, 1 (1942), " The Physical Interpretation of Quantum Mechanics ".
Comm. Dublin Inst. Adv. Studies. A, No. 1 (1943), " Quantum Electrodynamics ".
Proc. Roy. Soc. A, **183**, 284 (1945), " Unitary Representations of the Lorentz Group ".
Comm. Dublin Inst. Adv. Studies. A, No. 3 (1946). " Developments in Quantum Electrodynamics ".
" Quantum Mechanics ", Oxford Univ. Press, 3rd edition (1947).
Phys. Rev., **73**, 1092 (1948), " Quantum Theory of Localizable Dynamical Systems ".

DE DONDER, T.
" Théorie Invariantive du Calcul des Variations ", Gauthier-Villars, Paris (1935).

DUFFIN, R. J.
Phys. Rev., **54**, 1114 (1938), " On the Characteristic Matrices of Covariant Systems ".

EDDINGTON, A. E. S.
" Mathematical Theory of Relativity ", Cambridge Univ. Press (1937).

EINSTEIN, A.
Sitz. Ber. der Preuss. Akad., 522 (1932), " Semivektoren und Spinoren ".
Proc. Neder. Akad. van Wetensch., **36**, 497 (1933*a*), " Die Diracgleichungen fuer Semivektoren ".
ibid., 615 (1933*b*), " Spaltung der Natuerlichsten Feldgleichungen fuer Semivektoren in Spinor Gleichungen vom Diracschen Typus ".
Ann. Math., No. 1, **35**, 104 (1934), " Darstellung der Semivektoren als gewöhnliche Vektoren von Besonderen Differentiations Charakter ".

EISENBUD, L. (and WIGNER, E. P.).
Proc. Nat. Acad. Sci., **27**, 281 (1941), " Invariant Forms of Interaction between Nuclear Particles ".

ERIKSSON, H. A. S.
Ark. f. Mat. Ast. Fys., No. 10, **29** (1942), " Vector-pseudovector Meson Theory ".
ibid., No. 14, **29** (1943), " Spinor Representations of Rotations and Dirac's Equations in Five-Dimensional Space ".
ibid., No. 6, **33** (1946), " On a Generalization of Dirac's Equations ".
ibid., No. 21, **34** (1947), " On the Particle Aspect of the General Mixed Meson Theory ".

FERMI, E.
Zeits. f. Phys., **88**, 161 (1934), " Versuch einer Theorie der Beta-Strahlen ".

FEYNMAN, R. P.
Phys. Rev., **76**, 749 (1949*a*), " The Theory of Positrons ".
ibid., 769 (1949*b*), " Space-Time Approach to Quantum Electrodynamics ".

FIERZ, M.
Helv. Phys. Acta., **12**, 3 (1938), " Uber die Relativistische Theorie kräftefreier Teilchen mit beliebigem Spin ".
—— (and PAULI, W.).
ibid., 297 (1939*a*), " Uber Relativistische Feldgleichungen von Teilchen mit beliebigem Spin im Elektromagnetischen Feld ".
—— (and PAULI, W.).
Proc. Roy. Soc. A, **173**, 211 (1939*b*), " On the Relativistic Wave Equations for Particles of Arbitrary Spin in an Electromagnetic Field ".
Helv. Phys. Acta., 257 (1940), " Uber den Drehimpuls von Teilchen mit Ruhemasse Null und beliebigem Spin ".

FOCK, V. A.
Physik. Zeits. Sowjetunion, **2**, 468 (1932), " On Quantum Electrodynamics ".

FRÖHLICH, H. (and HEITLER, W.; KEMMER, N.).
Proc. Roy. Soc. A, **166**, 154 (1938), " On the Nuclear Forces and the Magnetic Moments of the Neutron and the Proton ".

FUCHS, K.
Proc. Roy. Soc. Edin. A, **59**, 109 (1939), " On the Invariance of Quantized Field Equations ".
—— (and BORN, M.).
ibid., **60**, 100 (1940*a*), " Reciprocity. Pt. II: Scalar Wave Functions ".
—— (and BORN, M.).
ibid., 141 (1940*b*), " Reciprocity. Pt. III: Reciprocal Wave Functions ".
ibid., 147 (1940*c*), " Reciprocity. Pt. IV: Spinor Wave Functions ".
ibid., **61**, 26 (1941), " Reciprocity. Pt. V: Reciprocal Spinor Functions ".

FURRY, W. H.
Phys. Rev., **54**, 56 (1938), " Note on the Theory of the Neutral Particle ".

GARDING, L.
Proc. Camb. Phil. Soc., **41**, 49 (1945), " Relativistic Wave Equations for Zero Rest-Mass ".

GEHENIAU, J.
Comptes Rendus, **207**, 1173 (1938).
Bull. Acad. Roy. de Belg. Cl. d. Sci., No. 1, **18** (1938), " Mécanique Ondulatoire de l'Electron et du Photon ".

GELFAND, I. M.
—— (and NEUMARK, M.).
Jour. of Phys. U.S.S.R., **10**, 93 (1946), " Unitary Representations of the Lorentz Group ".
—— (and YAGLON, A. M.).
Akad. Nauk. S.S.S.R. Zhurnal Eksp. Teoret. Fys., **18**, 703 (1948*a*), " General Lorentz Invariant Equations and Infinite-Dimensional Representations of the Lorentz Group ".
—— (and YAGLON, A. M.).
ibid., 1096 (1948*b*), " Pauli's Theorem for General Lorentz Invariant Equations ".

GELFAND, I. M.

—— (and YAGLON, A. M.).

ibid., 1105 (1948*c*), " Charge Conjugation for General Lorentz Invariant Equations ".

GREEN, H. S.

Proc. Roy. Soc. A, **197**, 73 (1949), " On the Self-energies and Cross-sections of Orthodox Quantum Mechanics ".

Proc. Camb. Phil. Soc., **45**, 263 (1949), " The Relativistic Quantum Mechanics of the Elementary Particles ".

—— (and BORN, M.).

Proc. Roy. Soc. Edin. A, **62**, 470 (1949), " Quantum Theory of Rest Masses ".

—— (and BORN M.; CHENG, K. C.).

Nature., **164**, 281 (1949), " Reciprocity Theory of Electrodynamics ".

HARISH-CHANDRA.

Proc. Roy. Soc. A, **183**, 142 (1944), " On the Removal of the Infinite Self-Energies of Point Particles ".

Proc. Ind. Acad. Sci., **21**, 135 (1945*a*), " On the Scattering of Scalar Mesons ".

ibid., **22**, 30 (1945*b*), " Algebra of the Dirac Matrices ".

Proc. Roy. Soc. A, **185**, 269 (1946*a*), " On the Equations of Motion of Point Particles ".

ibid., **186**, 502 (1946*b*), " The Correspondence between the Particle and the Wave Aspects of the Meson and the Photon ".

Proc. Ind. Acad. Sci., **23**, 152 (1946*c*), " A Note on the σ-Symbols ".

Proc. Camb. Phil. Soc., **43**, 414 (1946*d*), " On the Algebra of the Meson Matrices ".

Phys. Rev., **71**, 793 (1947*a*), " On Relativistic Wave Equations ".

Proc. Roy. Soc. A, **189**, 372 (1947*b*), " Infinite Irreducible Representations of the Lorentz Group ".

ibid., **192**, 195 (1947*c*), " Relativistic Equations for Elementary Particles ".

HEISENBERG, W.

—— (and PAULI, W.).

Zeits. f. Physik., **56**, 1 (1929), " Zur Quantendynamik der Wellenfelder ".

—— (and PAULI, W.).

ibid., **59**, 168 (1930), " Zur Quantendynamik der Wellenfelder, II ".

ibid., **113**, 61 (1939). " Zur Theorie der Explosionsartigen Schauer in der kosmischen Strahlung ".

" The Present Situation in the Theory of Elementary Particles ", Cambridge Univ. Press (1948).

HEITLER, W.

Proc. Roy. Soc. A, **166**, 154 (1938), " On the Nuclear Forces and the Magnetic Moments of the Neutron and Proton ".

ibid., **176**, 368 (1940), " Inner Excited States of the Proton and Neutron ".

Proc. Roy. Irish Acad., **49**, 1 (1943), " On the Particle Equation of the Meson ".

—— (and PENG, H. W.).

ibid., 101 (1943), " On the Production of Mesons by Proton-Proton Collisions ".

ibid., **51**, 33 (1946), " A Theorem in the Charge-Symmetrical Meson Theory ".

—— (and HU, N.).

ibid., 123 (1947), " Proton Isobars in the Theory of Radiation Damping ".

Hilbert, D.

Gött. Nach. Math. Phys., 395 (1915).

Hoyle, F.

Proc. Camb. Phil. Soc., **33**, 277 (1937), " The Generalized Fermi Interaction ".

Proc. Roy. Soc. A, **166**, 249 (1938), " Beta Transitions in a Coulomb Field ".

Hund, F.

Zeits. f. Physik., **118**, 426 (1940), " Uber eine Symmetrieeigenschaft der Wellentheorie der Materie ".

Infeld, L. (and van der Waerden, B. L.).

Sitz. Ber. der Preuss. Akad., No. 9, 380 (1933), " Die Wellengleichung des Elektrons in der allgemeinen Relativitätstheorie ".

Joos, G.

" Theoretical Physics ", Blackie & Son, Glasgow, 2nd edition (1951).

Jordan, P. (and Pauli, W.).

Zeits. f. Physik., **47**, 151 (1928), " Zur Quantenelektrodynamik Ladungsfreier Felder ".

Kemmer, N.

Helv. Phys. Acta., **10**, 47 (1937), " Zur Theorie der Neutron-Proton Wechselwirkung ".

Proc. Roy. Soc. A, **166**, 127 (1938*a*), " Quantum Theory of Einstein-Bose Particles and Nuclear Interaction ".

Proc. Camb. Phil. Soc., **34**, 354 (1938*b*), " The Charge Dependence of Nuclear Forces ".

Proc. Roy. Soc. A, **173**, 91 (1939), " The Particle Aspect of Meson Theory ".

Proc. Camb. Phil. Soc., **39**, 189 (1943), " The Algebra of Meson Matrices ".

Kramers, H. A.

Proc. Neder. Akad. van Wetensch., **40**, 814 (1937), " The Use of Charge-Conjugated Wave Functions in the Hole Theory of the Electron ".

—— (and Belinfante, F. J.; Lubanski, J. K.).

Physica, **8**, 597 (1941), " Uber Freie Teilchen mit Nichtverschwindenen Masse und beliebigen Spinquantenzahl ".

Laporte, O. (and Uhlenbeck, G.).

Phys. Rev., **37**, 1380 (1931), " Application of Spinor Analysis to the Maxwell and Dirac Equations ".

Le Couteur, K. J.

Proc. Camb. Phil. Soc., **44**, 63 (1948*a*), " Particles of Half-Odd Integral Spin ".

ibid., 229 (1948*b*), " The Theory of Neutral Particles ".

Proc. Roy. Soc. A, **196**, 251 (1949), " The Indefinite Metric in Relativistic Quantum Mechanics ".

Lubanski, J. K.

Physica, **8**, 44 (1941*a*), " Sur le Spin des Particules Elementaires ".

—— (and Kramers, H. A.; Belinfante, F. J.).

ibid., 597 (1941*b*), " Uber Freie Teilchen mit Nichtverschwindenen Masse und beliebigen Spinquantenzahl ".

—— (and Rosenfeld, L.).

ibid., **9**, 117 (1942*a*), " Sur la Representation des Champs Mesiques dans l'Espace à Cinq Dimensions ".

ibid., 310, 325 (1942*b*, *c*), " Sur le Théorie des Particules Elementaires de Spin Quelconque I, II ".

MADHAVA RAO, B. S.

—— (and BHABHA, H. J.).

Proc. Ind. Acad. Sci., **13**, 9 (1941), " The Scattering of Charged Mesons ".

Jour. Mysore Univ., **3**, 59 (1942*a*), " Commutation Rules Related to Particles of Spin Half and One ".

Proc. Ind. Acad. Sci., **15**, 139 (1942*b*), " Commutation Rules for Matrices Related to Particles of Higher Spin ".

Jour. Mysore Univ., **6**, 57 (1945), " Commutation Rules for Matrices Related to Particles of Higher Spin—III ".

Proc. Roy. Soc. A, **187**, 385 (1946), " Algebra Related to Elementary Particles of Spin $\frac{3}{2}$ ".

Proc. Ind. Acad. Sci., **26**, 221 (1947), " Generalized Algebra of Elementary Particles ".

MAJORANA, E.

Nuovo Cim., **9**, 335 (1932), " Teoria Relativistica Di Particelle con Momento Intrinseco Arbitrario ".

ibid., **14**, 171 (1937), " Teoria Simmetrica Dell' Elettrone e Del Positrone ".

MØLLER, C. (and ROSENFELD, L.).

D. Kgl. Danske Vid. Sels. Mat. Fys. Medd., **17**, 8 (1940), " On the Field Theory of Nuclear Forces ".

ibid., **18**, 6 (1941), " On the Theory of Mesons ".

MURNAGHAN, F. D.

" The Theory of Group Representations ", John Hopkins Univ. Press (1938).

NILSSON, S. B.

Ark. f. Mat. Ast. Fys., **36**, No. 5 (1948), " The e^4 Approximation of the Self Energy of the Electron Treated by Analytic Continuation ".

PAIS, A.

Physica, **9**, 407 (1942), " On the Electric Charge-Current Density of a Nuclear System ".

PAULI, W.

Enzykl. der Math. Wissensch., **V**, § 19 (1922), " Relativitätstheorie ".

—— (and JORDAN, P).

Zeits. f. Physik, **47**, 151 (1928), " Zur Quantenelektrodynamik Ladungfreier Felder ".

—— (and HEISENBERG, W.).

ibid., **56**, 1 (1929), " Zur Quantendynamik der Wellenfelder ".

—— (and HEISENBERG, W.).

ibid., **59**, 168 (1930), " Zur Quantendynamik der Wellenfelder II ".

Handb. d. Physik, **24**, Pt. I (1933), " Wellenmechanik ".

—— (and WEISSKOPF, V).

Helv. Phys. Acta, **7**, 709 (1934), " Uber die Quantisierung der Skalaren Relativistischen Wellengleichung ".

Zeeman Verhandelingen, 31-Nijhoff (1935), " Beiträge Zur Mathematischen Theorie der Diracschen Matrizen ".

—— (and FIERZ, M.).

Proc. Roy. Soc. A, **173**, 211 (1939*a*), " On the Relativistic Wave Equations for Particles of Arbitrary Spin in an Electromagnetic Field ".

—— (and FIERZ, M.).

Helv. Phys. Acta, **12**, 297 (1939*b*), " Uber Relativistischen Feldgleichungen von Teilchen mit beliebigem Spin im Elektromagnetischen Feld ".

PAULI, W.
—— (and BELINFANTE, F. J.).
Physica, **7**, 177 (1940), " On the Statistical Behaviour of Known and Unknown Elementary Particles ".
Phys. Rev., **58**, 716 (1940), " The Connection Between Spin and Statistics ".
Rev. Mod. Phys., **13**, 203 (1941), " Relativistic Field Theories ".
ibid., **15**, 175 (1943).

PENG, H. W.
Proc. Roy. Irish Acad., **51**, 113 (1947), " On the Representation of the Wave Function of a Quantized Field by Means of a Generating Function ".

PETIAU, G.
Acad. Roy. Belgique, **16** (1936), " Contribution à la Théorie des Equations d'Ondes Corpusculaires ".
Jour. Phys. Rad., 134 (1941), " Sur les Equations d'Ondes du Second Ordre pour les Corpuscules de Spin $h/4\pi$ Dans les Champs Extérieurs ".
ibid., **7**, 124, 181 (1946), " Sur les Equations D'Ondes des Corpuscules de Spin Quelconque I, II ".

PODOLSKY, B. (and DIRAC, P. A. M.; FOCK, V. A.).
Physik Zeits. Sowjetunion, **2**, 468 (1932), " On Quantum Electrodynamics ".

PROCA, A.
Jour. Phys. Rad., **7**, 347 (1936), " Sur la Théorie Ondulatoire des Electrons Positifs et Négatifs ".
ibid, **6**, 61 (1938), " Théorie non Relativiste des Particules A Spin Entier ".

RACAH, G.
Nuovo Cim., **14**, 322 (1937), " Sulla Simmetria Tra Particelle e Antiparticelle ".

ROSENFELD, L.
Zeits. f. Physik, **63**, 574 (1930*a*), " Invariance of the Canonical Exchange Relations ".
Ann. der Physik, **5**, 113 (1930*b*), " Quantization of Wave Field ".
Acad. Roy. Belgique, **18**, No. 6 (1940), " Sur le Tenseur d'Impulsion-Energie ".

ROZENTHAL, S.
D. Kgl. Danske Vid. Sels. Mat. Fys. Medd., **18**, 7 (1941), " On the Theory of Beta Decay ".

RUSE, H. S.
Proc. Roy. Soc. Edin., **57**, 97 (1937), " On the Geometry of Dirac's Equations and their Expression in Tensor Form ".

SAKATA, S.
—— (and YUKAWA, H.).
Proc. Phys. Math. Soc. Japan, **19**, 91 (1937*a*), " Note on Dirac's Generalized Wave Equations ".
—— (and YUKAWA, H.).
ibid., 1084 (1937*b*), " On the Interaction of Elementary Particles II ".
—— (and YUKAWA, H.; TAKETANI, M.).
ibid., **20**, 319 (1938*a*), " On the Interaction of Elementary Particles III ".
—— (and YUKAWA, H.; TAKETANI, M.).
ibid., 720 (1938*b*), " On the Interaction of Elementary Particles IV ".

SAKATA, S.
—— (and TAKETANI, M.).
ibid., **22**, 757 (1940*a*), " On the Wave Equation of the Meson ".
—— (and TAKETANI, M.).
Sci. Papers I.P.C.R. Tokyo, **38**, 1 (1940*b*), " Note on Casimir's Method of Spin Summation in the Case of the Meson ".
—— (and INOUE, T.).
Prog. Theoret. Phys. Japan, **1**, 143 (1946), " On the Correlation between Mesons and Yukawa Particles ".

SCHOUTEN, J. A.
Kon. Neder. Akad. van Wetensch, **52**, 178, 217 (1949*a*, *b*), " On the Geometry of Spin Spaces I, II ".

SCHRÖDINGER, E.
Proc. Roy. Irish Acad., **48**, 135 (1943*a*), " Pentads, Tetrads, and Triads of Meson Matrices ".
ibid., **49**, 29 (1943*b*), " Systematics of Meson Matrices ".

SCHWINGER, J.
Phys. Rev., **74**, 1439 (1948), " Quantum Electrodynamics. I. A Covariant Formulation ".
ibid., **75**, 651 (1949*a*). " Quantum Electrodynamics. II. Vacuum Polarization and Self-Energy ".
ibid., **76**, 790 (1949*b*), " Quantum Electrodynamics. III. The Electromagnetic Properties of the Electron—Radiative Corrections to Scattering ".

TAUB, A. H.
Phys. Rev., **56**, 799 (1939*a*), " Spinor Equations for the Meson and Their Solution When No Field is Present ".
Ann. of Math., **40**, 937 (1939*b*), " Tensor Equations Equivalent to the Dirac Equations ".
Phys. Rev., **57**, 807 (1940), " Solutions of Equations for Particles of Spin Zero or One When No Field Is Present ".

TOMONAGA, S.
Prog. Theoret. Phys. Japan, **1**, 27 (1946), " On a Relativistically Invariant Formulation of the Quantum Theory of Wave Fields ".

VEBLEN, O.
Proc. Nat. Acad. Sci., **19**, 979 (1933), " Spinors in Projective Relativity ".
Princeton Institute for Advanced Study—Lectures (1935–36), " Geometry of Complex Domains ".

VOLTERRA, V.
" Theory of Functionals ", Blackie & Son, Glasgow (1931).

VAN DER WAERDEN, B. L.
Gött. Nachr., 100 (1929), " Spinoranalyse ".
" Moderne Algebra I ", Springer, Berlin (1930).
" Moderne Algebra II ", Springer, Berlin (1931).
" Die Gruppentheoretische Methode in der Quantenmechanik ", Springer, Berlin (1932).

WEISS, P.

Proc. Roy. Soc. A, **156**, 192 (1936), " On the Quantization of a Theory Arising From a Variation Principle for Multiple Integrals—With Applications to Born's Electrodynamics ".

ibid., **169**, 102 (1938*a*), " On the Hamiltonian–Jacobi Theory and Quantization of a Dynamical Continuum ".

ibid., 119, (1938*b*), " On the Hamiltonian–Jacobi Theory and Quantization of Generalized Electrodynamics ".

WENTZEL, G.

" Quantum Theory of Fields ", Deuticke, Wien (1943)—Interscience, New York (1949).

DE WET, J. S.

Phys. Rev., **58**, 236 (1940), " On Spinor Equations for Particles with Arbitrary Spin and Rest Mass Zero ".

Proc. Camb. Phil. Soc., **43**, 511 (1947), " Symmetric Energy-Momentum Tensors in Relativistic Field Theories ".

ibid., **44**, 546 (1948*a*), " On the Quantization of Field Theories Derived from Higher Order Lagrangians ".

Proc. Roy. Soc. A, **195**, 365 (1948*b*), " On the Relativistic Invariance of Quantized Field Theories ".

WEITZENBÖCK, R.

" Invarianten-Theorie ", Noordhoff, Groningen (1923).

WEYL, H.

" Raum-Zeit-Materie ", Springer, Berlin (1923).

Math. Zeits., **23**, 271 (1925*a*); **24**, 328, 377, 789 (1925*b*), " Theorie der Darstellung Kontinuierlicher Halb-einfacher Gruppen durch Lineare Transformationen ".

" The Theory of Groups and Quantum Mechanics ", Dutton, New York (1931).

—— (and BRAUER, R.).

Amer. Jour. of Math., **57**, 425 (1935), " Spinors in n-Dimensions ".

" The Classical Groups ", Princeton Univ. Press (1939).

WHITTAKER, E. T.

Proc. Roy. Soc. A, **158**, 38 (1937), " On the Relation of the Tensor Calculus to the Spinor Calculus ".

WIGNER, E. P.

Annals of Math., **40**, 149 (1939), " On Unitary Representations of the Inhomogeneous Lorentz Group ".

—— (and EISENBUD L.).

Proc. Nat. Acad. Sci., **27**, 281 (1941), " Invariant Forms of Interaction between Nuclear Particles ".

Zeits. f. Physik, **124**, 665 (1947), " Relativistische Wellengleichungen ".

—— (and BARGMANN, V.).

Proc. Nat. Acad. Sci., **34**, 211 (1948), " Group Theoretical Discussion of Relativistic Wave Equations ".

WILD, E.

Proc. Roy. Soc. A, **191**, 253 (1947), " On First Order Wave Equations for Elementary Particles Without Subsidiary Conditions ".

Wilson, A. H.
Proc. Camb. Phil. Soc., **36**, 363 (1940*a*), " The Calculation of Processes Involving Mesons by Matrix Methods ".
—— (and Booth, F.).
ibid., 446 (1940*b*), " The Scattering of Neutral Mesons ".
—— (and Booth, F.).
Proc. Roy. Soc. A, **175**, 483 (1940*c*), " Radiative Processes Involving Fast Mesons ".
Proc. Camb. Phil. Soc., **37**, 302 (1941), " The Quantum Theory of Radiation Damping ".

Yukawa, H.
Proc. Phys. Math. Soc. Japan, **17**, 48 (1935), " On the Interaction of Elementary Particles I ".
ibid., **19**, 91 (1937*a*), " Note on Dirac's Generalized Wave Equations ".
ibid., 712 (1937*b*), " Short Note on the Possible Interpretation of the Penetrating Component of the Cosmic Ray ".
ibid., 1084 (1937*c*), " On the Interaction of Elementary Particles II ".
ibid., **20**, 319 (1938*a*), " On the Interaction of Elementary Particles III ".
ibid., 720 (1938*b*), " On the Interaction of Elementary Particles ".

INDEX